T0203011

Lecture Notes in Computer Science 11614

More information about this series at http://www.springer.com/series/7412

Lucio Tommaso De Paolis ·
Patrick Bourdot (Eds.)

Augmented Reality, Virtual Reality, and Computer Graphics

6th International Conference, AVR 2019
Santa Maria al Bagno, Italy, June 24–27, 2019
Proceedings, Part II

 Springer

Editors
Lucio Tommaso De Paolis 🆔
University of Salento
Lecce, Italy

Patrick Bourdot 🆔
University of Paris-Sud
Orsay, France

ISSN 0302-9743   ISSN 1611-3349   (electronic)
Lecture Notes in Computer Science
ISBN 978-3-030-25998-3   ISBN 978-3-030-25999-0   (eBook)
https://doi.org/10.1007/978-3-030-25999-0

LNCS Sublibrary: SL6 – Image Processing, Computer Vision, Pattern Recognition, and Graphics

This Springer imprint is published by the registered company Springer Nature Switzerland AG
The registered company address is: Gewerbestrasse 11, 6330 Cham, Switzerland

Preface

Virtual Reality (VR) technology permits the creation of realistic-looking worlds where the user inputs are used to modify in real time the digital environment. Interactivity contributes to the feeling of immersion in the virtual world, of being part of the action that the user experiences. It is not only possible to see and manipulate a virtual object, but also to feel and touch them using specific devices.

Mixed Reality (MR) and Augmented Reality (AR) technologies permit the real-time fusion of computer-generated digital contents with the real world and allow for the creation of fascinating new types of user interfaces. AR enhances the users' perception and improves their interaction in the real environment. The virtual objects help users to perform real-world tasks better by displaying information that they cannot directly detect with their own senses. Unlike the VR technology that completely immerses users inside a synthetic environment where they cannot see the real world around them, AR technology allows them to see 3D virtual objects superimposed upon the real environment. AR and MR supplement reality rather than completely replacing it and the user is under the impression that the virtual and real objects coexist in the same space.

Human-Computer Interaction technology (HCI) is a research area concerned with the design, implementation, and evaluation of interactive systems that make more simple and intuitive the interaction between user and computer.

This book contains the contributions to the 6th International Conference on Augmented Reality, Virtual Reality and Computer Graphics (SALENTO AVR 2019) that has held in Santa Maria al Bagno (Lecce, Italy) during June 24–27, 2019. Organized by the Augmented and Virtual Reality Laboratory at the University of Salento, SALENTO AVR 2019 intended to bring together the community of researchers and scientists in order to discuss key issues, approaches, ideas, open problems, innovative applications, and trends in virtual and augmented reality, 3D visualization, and computer graphics in the areas of medicine, cultural heritage, arts, education, entertainment, military, and industrial applications. We cordially invite you to visit the SALENTO AVR website (www.salentoavr.it) where you can find all relevant information about this event.

We are very grateful to the Program Committee and Local Organizing Committee members for their support and for the time spent to review and discuss the submitted papers and for doing so in a timely and professional manner.

We would like to sincerely thank the keynote speakers who willingly accepted our invitation and shared their expertise through illuminating talks, helping us to fully meet the conference objectives. In this edition of SALENTO AVR we were honored to have the following invited speakers:

- Luigi Gallo – ICAR-CNR, Italy
- Danijel Skočaj – University of Ljubljana, Slovenia
- Pasquale Arpaia – University of Naples Federico II, Italy

We extend our thanks to the University of Salento and the Banca Popolare Pugliese for the enthusiastic acceptance to sponsor the conference and to provide support in the organization of the event.

SALENTO AVR attracted high-quality paper submissions from many countries. We would like to thank the authors of all accepted papers for submitting and presenting their works at the conference and all the conference attendees for making SALENTO AVR an excellent forum on virtual and augmented reality, facilitating the exchange of ideas, fostering new collaborations, and shaping the future of this exciting research field.

We hope the readers will find in these pages interesting material and fruitful ideas for their future work.

June 2019   Lucio Tommaso De Paolis
                                                      Patrick Bourdot

Organization

Conference Chair

Lucio Tommaso De Paolis   University of Salento, Italy

Conference Co-chairs

Patrick Bourdot   CNRS/LIMSI, University of Paris-Sud, France
Marco Sacco   ITIA-CNR, Italy
Paolo Proietti   MIMOS, Italy

Scientific Program Committee

Andrea Abate   University of Salerno, Italy
Giuseppe Anastasi   University of Pisa, Italy
Selim Balcisoy   Sabancı University, Turkey
Vitoantonio Bevilacqua   Polytechnic of Bari, Italy
Monica Bordegoni   Politecnico di Milano, Italy
Davide Borra   NoReal.it, Italy
Andrea Bottino   Politecnico di Torino, Italy
Pierre Boulanger   University of Alberta, Canada
Andres Bustillo   University of Burgos, Spain
Massimo Cafaro   University of Salento, Italy
Bruno Carpentieri   University of Salerno, Italy
Sergio Casciaro   IFC-CNR, Italy
Marcello Carrozzino   Scuola Superiore Sant'Anna, Italy
Mario Ciampi   ICAR/CNR, Italy
Pietro Cipresso   IRCCS Istituto Auxologico Italiano, Italy
Arnis Cirulis   Vidzeme University of Applied Sciences, Latvia
Mario Covarrubias   Politecnico di Milano, Italy
Rita Cucchiara   University of Modena, Italy
Yuri Dekhtyar   Riga Technical University, Latvia
Matteo Dellepiane   National Research Council (CNR), Italy
Giorgio De Nunzio   University of Salento, Italy
Francisco José   University of Seville, Spain
   Domínguez Mayo
Aldo Franco Dragoni   Università Politecnica delle Marche, Italy
Italo Epicoco   University of Salento, Italy
Ben Falchuk   Perspecta Labs Inc., USA
Vincenzo Ferrari   EndoCAS Center, Italy
Francesco Ferrise   Politecnico di Milano, Italy
Dimitrios Fotiadis   University of Ioannina, Greece

Emanuele Frontoni	Università Politecnica delle Marche, Italy
Francesco Gabellone	IBAM ITLab, CNR, Italy
Damianos Gavalas	University of the Aegean, Greece
Osvaldo Gervasi	University of Perugia, Italy
Luigi Gallo	ICAR/CNR, Italy
Viktors Gopejenko	ISMA University, Latvia
Mirko Grimaldi	University of Salento, Italy
Heiko Herrmann	Tallinn University of Technology, Estonia
Sara Invitto	University of Salento, Italy
Fabrizio Lamberti	Politecnico di Torino, Italy
Leo Joskowicz	Hebrew University of Jerusalem, Israel
Tomas Krilavičius	Vytautas Magnus University, Lithuania
Salvatore Livatino	University of Hertfordshire, UK
Silvia Mabel Castro	Universidad Nacional del Sur, Argentina
Luca Mainetti	University of Salento, Italy
Eva Savina Malinverni	Università Politecnica delle Marche, Italy
Matija Marolt	University of Ljubljana, Slovenia
Daniel R. Mestre	Aix-Marseille University/CNRS, France
Sven Nomm	Tallinn University of Technology, Estonia
Fabrizio Nunnari	German Research Center for Artificial Intelligence (DFKI), Germany
Roberto Paiano	University of Salento, Italy
Andrea Pandurino	University of Salento, Italy
Giorgos Papadourakis	Technological Educational Institute (TEI) of Crete, Greece
Gianfranco Parlangeli	University of Salento, Italy
Gianluca Paravati	Politecnico di Torino, Italy
Nikolaos Pellas	University of the Aegean, Greece
Eduard Petlenkov	Tallinn University of Technology, Estonia
Roberto Pierdicca	Università Politecnica delle Marche, Italy
Sofia Pescarin	CNR ITABC, Italy
Arcadio Reyes Lecuona	Universidad de Malaga, Spain
James Ritchie	Heriot-Watt University, UK
Giuseppe Riva	Università Cattolica del Sacro Cuore, Italy
Andrea Sanna	Politecnico di Torino, Italy
Jaume Segura Garcia	Universitat de València, Spain
Paolo Sernani	Università Politecnica delle Marche, Italy
Robert Stone	University of Birmingham, UK
João Manuel R. S. Tavares	Universidade do Porto, Portugal
Daniel Thalmann	Nanyang Technological University, Singapore
Nadia Magnenat-Thalmann	University of Geneva, Switzerland
Franco Tecchia	Scuola Superiore Sant'Anna, Italy
Carlos M. Travieso-González	Universidad de Las Palmas de Gran Canaria, Spain
Manolis Tsiknaki	Technological Educational Institute of Crete (TEI), Greece

Antonio Emmanuele Uva	Polytechnic of Bari, Italy
Volker Paelke	Bremen University of Applied Sciences, Germany
Aleksei Tepljakov	Tallinn University of Technology, Estonia
Kristina Vassiljeva	Tallinn University of Technology, Estonia
Krzysztof Walczak	Poznań University of Economics and Business, Poland
Anthony Whitehead	Carleton University, Canada

Local Organizing Committee

Giovanna Ilenia Paladini	University of Salento, Italy
Silke Miss	Virtech, Italy
Valerio De Luca	University of Salento, Italy
Cristina Barba	University of Salento, Italy
Giovanni D'Errico	University of Salento, Italy

Keynote Speakers

Interactive Virtual Environments:
From the Laboratory to the Field

Luigi Gallo

ICAR-CNR, Italy

Virtual reality technology has the potential to change the way information is retrieved, processed and shared, making it possible to merge several layers of knowledge in a coherent space that can be comprehensively queried and explored. In realizing this potential, the human–computer interface plays a crucial role. Humans learn and perceive by following an interactive process, but the interaction occurs in different places and contexts, between people, and with people. Accordingly, the interactive system components have to be tailored to the application and its users, targeting robustness, multimodality, and familiarity. In this presentation, I aim to explore the promise and challenges of interactive virtual technologies by discussing generally the design and the evaluation results of selected real-world applications, developed by multi-disciplinary research groups, in very different domains, ranging from medicine and finance to cultural heritage.

Luigi Gallo is a research scientist at the National Research Council of Italy (CNR) – Institute for High-Performance Computing and Networking (ICAR). He graduated in 2006 in Computer Engineering, and received a PhD degree in Information Technology Engineering in 2010. Between 2012 and 2018, he worked as Adjunct Professor of Informatics at the University of Naples Federico II. He holds leadership roles within nationally funded research projects dealing with the development of ICT solutions for medical, financial, and cultural heritage applications. His fields of interest include computer vision, natural user interfaces, and the human interface aspects of virtual and augmented reality. He has authored and co-authored more than 90 publications in international journals, conference proceedings, and book chapters, and serves on the Organizing Committee of several international conferences and workshops.

Computer Vision as an Enabling Technology for Interactive Systems

Danijel Skočaj

University of Ljubljana, Slovenia

Computer vision research has made tremendous progress in recent years. Solutions for object detection and classification, semantic segmentation, visual tracking, and other computer vision tasks have become readily available. Most importantly, these solutions have become more general and robust, allowing for a significantly broader use of computer vision. Applications are less limited to constrained environments and are getting ready to be used in the wild. Efficient, robust, and accurate interpretation of scenes is also a key requirement of advanced interactive systems as well as augmented and mixed reality applications. Efficient understanding of scenes in 2D as well as in 3D and detection of key semantic elements in images allow for a wider use of such systems and more intelligent interaction with a user. In this talk, I will present some recent achievements in developing core computer vision algorithms as well as the application of these algorithms for interaction between a human, real environment and a virtual world.

Danijel Skočaj is an associate professor at the University of Ljubljana, Faculty of Computer and Information Science. He is the head of the Visual Cognitive Systems Laboratory. He obtained a PhD degree in computer and information science from the University of Ljubljana in 2003. His main research interests lie in the fields of computer vision, machine learning, and cognitive robotics; he is involved in basic and applied research in visually enabled cognitive systems, with emphasis on visual learning, recognition, and segmentation. He is also interested in the ethical aspect of artificial intelligence. He has led or collaborated in a number of research projects, such as EU projects (CogX, CoSy, CogVis), national projects (DIVID, GOSTOP, VILLarD), and several industry-funded projects. He has served as the president of the IEEE Slovenia Computer Society, and the president of the Slovenian Pattern Recognition Society.

Wearable Brain–Computer Interface
for Augmented Reality-Based Inspection
in Industry 4.0

Pasquale Arpaia

University of Naples Federico II, Italy

In the past two decades, augmented reality (AR) has gained great interest in the technical–scientific community and much effort has been made to overcome its limitations in daily use. The main industrial operations in which AR is applied are training, inspections, diagnostics, assembly–disassembly, and repair. These operations usually require the user's hands to be free from the AR device controller. Despite hand-held devices, such as tablets, smart glasses can guarantee hand-free operations with their high wearability. The combination of AR with a brain–computer interface (BCI) can provide the solution: BCI is capable of interpreting human intentions by measuring user neuronal activity. In this talk, the most interesting results of this technological research effort, as well as its further most recent developments, are reviewed. In particular, after a short survey on research at the University of Naples Federico II in cooperation with CERN, the presentation focuses mainly on state-of-the-art research on a wearable monitoring system. AR glasses are integrated with a trainingless non-invasive single-channel BCI, for inspection in the framework of Industry 4.0. A case study at CERN for robotic inspection in hazardous sites is also reported.

Pasquale Arpaia obtained a master's degree and PhD degree in Electrical Engineering at the University of Naples Federico II (Italy), where he is full professor of Instrumentation and Measurements. He is also Team Manager at the European Organization for Nuclear Research (CERN). He is Associate Editor of the Institute of Physics *Journal of Instrumentation*, Elsevier journal *Computer Standards & Interfaces*, and MDPI journal Instruments. He is Editor at Momentum Press of the book series *Emerging Technologies in Measurements, Instrumentation, and Sensors*. In the past few years, he was scientifically responsible for more than 30 awarded research projects in cooperation with industry, with related patents and international licences, and funded four academic spin-off companies. He acted as scientific evaluator in several international research call panels. He has served as organizing and scientific committee member in several IEEE and IMEKO conferences. He has been plenary speaker in several scientific conferences. His main research interests include digital instrumentation and measurement techniques for particle accelerators, evolutionary diagnostics, distributed measurement systems, ADC modelling and testing.

Contents – Part II

Augmented Reality

Application of Physical Interactive Mixed Reality System Based
on MLAT in the Field of Stage Performance 3
  Yanxiang Zhang, Pengfei Ma, Ali Raja Gulfraz, Li Kong,
  and Xuelian Sun

Application Advantages and Prospects of Web-Based AR Technology
in Publishing 13
  YanXiang Zhang and Yaping Lu

Generation of Action Recognition Training Data Through Rotoscoping
and Augmentation of Synthetic Animations 23
  Nicola Covre, Fabrizio Nunnari, Alberto Fornaser,
  and Mariolino De Cecco

Debugging Quadrocopter Trajectories in Mixed Reality 43
  Burkhard Hoppenstedt, Thomas Witte, Jona Ruof, Klaus Kammerer,
  Matthias Tichy, Manfred Reichert, and Rüdiger Pryss

Engaging Citizens with Urban Planning Using City Blocks,
a Mixed Reality Design and Visualisation Platform 51
  Lee Kent, Chris Snider, and Ben Hicks

Convolutional Neural Networks for Image Recognition in Mixed Reality
Using Voice Command Labeling 63
  Burkhard Hoppenstedt, Klaus Kammerer, Manfred Reichert,
  Myra Spiliopoulou, and Rüdiger Pryss

A Framework for Data-Driven Augmented Reality 71
  Georgia Albuquerque, Dörte Sonntag, Oliver Bodensiek,
  Manuel Behlen, Nils Wendorff, and Marcus Magnor

Usability of Direct Manipulation Interaction Methods for Augmented
Reality Environments Using Smartphones and Smartglasses 84
  Alexander Ohlei, Daniel Wessel, and Michael Herczeg

An Augmented Reality Tool to Detect Design Discrepancies:
A Comparison Test with Traditional Methods 99
  Loris Barbieri and Emanuele Marino

A New Loose-Coupling Method for Vision-Inertial Systems
Based on Retro-Correction and Inconsistency Treatment   111
    Marwene Kechiche, Ioan-Alexandru Ivan, Patrick Baert,
    Rolnd Fortunier, and Rosario Toscano

Ultra Wideband Tracking Potential for Augmented Reality Environments. . . .   126
    Arnis Cirulis

HoloHome: An Augmented Reality Framework to Manage
the Smart Home   137
    Atieh Mahroo, Luca Greci, and Marco Sacco

Training Assistant for Automotive Engineering Through
Augmented Reality   146
    Fernando R. Pusda, Francisco F. Valencia, Víctor H. Andaluz,
    and Víctor D. Zambrano

Microsoft HoloLens Evaluation Under Monochromatic
RGB Light Conditions.   161
    Marián Hudák, Štefan Korečko, and Branislav Sobota

Towards the Development of a Quasi-Orthoscopic Hybrid Video/Optical
See-Through HMD for Manual Tasks   170
    Fabrizio Cutolo, Nadia Cattari, Umberto Fontana,
    and Vincenzo Ferrari

An Empirical Evaluation of the Performance of Real-Time Illumination
Approaches: Realistic Scenes in Augmented Reality   179
    A'aeshah Alhakamy and Mihran Tuceryan

Cultural Heritage

Combining Image Targets and SLAM for AR-Based Cultural
Heritage Fruition.   199
    Paolo Sernani, Renato Angeloni, Aldo Franco Dragoni,
    Ramona Quattrini, and Paolo Clini

Optimization of 3D Object Placement in Augmented Reality Settings
in Museum Contexts   208
    Alexander Ohlei, Lennart Bundt, David Bouck-Standen,
    and Michael Herczeg

Transmedia Digital Storytelling for Cultural Heritage Visiting
Enhanced Experience.   221
    Angelo Corallo, Marco Esposito, Manuela Marra,
    and Claudio Pascarelli

Assessment of Virtual Guides' Credibility in Virtual
Museum Environments 230
*Stella Sylaiou, Vlasios Kasapakis, Elena Dzardanova,
and Damianos Gavalas*

Immersive Virtual System for the Operation of Tourist Circuits 239
*Aldrin G. Acosta, Víctor H. Andaluz, Hugo Oswaldo Moreno,
Mauricio Tamayo, Giovanny Cuzco, Mayra L. Villarroel,
and Jaime A. Santana*

Virtual Museums as a New Type of Cyber-Physical-Social System 256
Louis Nisiotis, Lyuba Alboul, and Martin Beer

Virtual Portals for a Smart Fruition of Historical
and Archaeological Contexts 264
*Doriana Cisternino, Carola Gatto, Giovanni D'Errico, Valerio De Luca,
Maria Cristina Barba, Giovanna Ilenia Paladini,
and Lucio Tommaso De Paolis*

Education

Intercultural Communication Research Based on CVR: An Empirical
Study of Chinese Users of CVR About Japanese Shrine Culture 277
Ying Li, YanXiang Zhang, and MeiTing Chin

Augmented Reality to Engage Preschool Children in Foreign
Language Learning 286
Elif Topsakal and Oguzhan Topsakal

A Study on Female Students' Attitude Towards the Use of Augmented
Reality to Learn Atoms and Molecules Reactions in Palestinian Schools 295
Ahmed Ewais, Olga De Troyer, Mumen Abu Arra, and Mohammed Romi

Intelligent System for the Learning of Sign Language Based on Artificial
Neural Networks.. 310
*D. Rivas, Marcelo Alvarez V., J. Guanoluisa, M. Zapata, E. Garcés,
M. Balseca, J. Perez, and R. Granizo*

Measuring and Assessing Augmented Reality Potential for Educational
Purposes: SmartMarca Project................................ 319
*Emanuele Frontoni, Marina Paolanti, Mariapaola Puggioni,
Roberto Pierdicca, and Michele Sasso*

Augmented Reality in Laboratory's Instruments, Teaching
and Interaction Learning 335
*Víctor H. Andaluz, Jorge Mora-Aguilar, Darwin S. Sarzosa,
Jaime A. Santana, Aldrin Acosta, and Cesar A. Naranjo*

Touchless Navigation in a Multimedia Application: The Effects
Perceived in an Educational Context   348
    Lucio Tommaso De Paolis, Valerio De Luca,
    and Giovanna Ilenia Paladini

Industry

Analysis of Fuel Cells Utilizing Mixed Reality and IoT Achievements   371
    Burkhard Hoppenstedt, Michael Schmid, Klaus Kammerer,
    Joachim Scholta, Manfred Reichert, and Rüdiger Pryss

Virtual Environment for Training Oil & Gas Industry Workers   379
    Carlos A. Garcia, Jose E. Naranjo, Fabian Gallardo-Cardenas,
    and Marcelo V. Garcia

Virtual Training for Industrial Process: Pumping System   393
    Edison P. Yugcha, Jonathan I. Ubilluz, and Víctor H. Andaluz

Virtual Training on Pumping Stations for Drinking Water Supply Systems. . .   410
    Juan E. Romo, Gissela R. Tipantasi, Víctor H. Andaluz,
    and Jorge S. Sanchez

Virtual Training System for an Industrial Pasteurization Process   430
    Alex P. Porras, Carlos R. Solis, Víctor H. Andaluz, Jorge S. Sánchez,
    and Cesar A. Naranjo

Virtual Environment for Teaching and Learning Robotics Applied
to Industrial Processes   442
    Víctor H. Andaluz, José A. Pérez, Christian P. Carvajal,
    and Jessica S. Ortiz

StreamFlowVR: A Tool for Learning Methodologies and Measurement
Instruments for River Flow Through Virtual Reality   456
    Nicola Capece, Ugo Erra, and Domenica Mirauda

Author Index   473

Contents – Part I

Virtual Reality

Design of a SCORM Courseware Player Based on Web AR and Web VR. . .   3
   YanXiang Zhang, WeiWei Zhang, and YiRun Shen

Animated Agents' Facial Emotions: Does the Agent Design
Make a Difference?.   10
   Nicoletta Adamo, Hazar N. Dib, and Nicholas J. Villani

Identifying Emotions Provoked by Unboxing in Virtual Reality   26
   Stefan H. Tanderup, Markku Reunanen, and Martin Kraus

Collaborative Web-Based Merged Volumetric and Mesh
Rendering Framework   36
   Ciril Bohak, Jan Aleksandrov, and Matija Marolt

Exploring the Benefits of the Virtual Reality Technologies for Assembly
Retrieval Applications   43
   Katia Lupinetti, Brigida Bonino, Franca Giannini, and Marina Monti

Metaphors for Software Visualization Systems Based on Virtual Reality   60
   Vladimir Averbukh, Natalya Averbukh, Pavel Vasev, Ilya Gvozdarev,
   Georgy Levchuk, Leonid Melkozerov, and Igor Mikhaylov

Design and Architecture of an Affordable Optical Routing - Multi-user VR
System with Lenticular Lenses   71
   Juan Sebastian Munoz-Arango, Dirk Reiners, and Carolina Cruz-Neira

Training Virtual Environment for Teaching Simulation and Control
of Pneumatic Systems   91
   Carlos A. Garcia, Jose E. Naranjo, Edison Alvarez-M.,
   and Marcelo V. Garcia

Real Time Simulation and Visualization of Particle Systems on GPU   105
   Bruno Ježek, Jiří Borecký, and Antonín Slabý

A Proof of Concept Integrated Multi-systems Approach for Large Scale
Tactile Feedback in VR   120
   Daniel Brice, Thomas McRoberts, and Karen Rafferty

Virtual Simulator for the Taking and Evaluation of Psychometric Tests
to Obtain a Driver's License.   138
    Jorge S. Sánchez, Jessica S. Ortiz, Oscar A. Mayorga,
    Carlos R. Sánchez, Gabrilea M. Andaluz, Edison L. Bonilla,
    and Víctor H. Andaluz

Development of the Multimedia Virtual Reality-Based Application
for Physics Study Using the Leap Motion Controller   150
    Yevgeniya Daineko, Madina Ipalakova, and Dana Tsoy

Virtual Reality and Logic Programming as Assistance
in Architectural Design   158
    Dominik Strugała and Krzysztof Walczak

Effects of Immersive Virtual Reality on the Heart Rate
of Athlete's Warm-Up   175
    José Varela-Aldás, Guillermo Palacios-Navarro,
    Iván García-Magariño, and Esteban M. Fuentes

The Transdisciplinary Nature of Virtual Space   186
    Josef Wideström

Automatic Generation of Point Cloud Synthetic Dataset for Historical
Building Representation   203
    Roberto Pierdicca, Marco Mameli, Eva Savina Malinverni,
    Marina Paolanti, and Emanuele Frontoni

Semantic Contextual Personalization of Virtual Stores   220
    Krzysztof Walczak, Jakub Flotyński, and Dominik Strugała

Towards Assessment of Behavioral Patterns in a Virtual
Reality Environment   237
    Ahmet Kose, Aleksei Tepljakov, Mihkel Abel, and Eduard Petlenkov

Design and Implementation of a Reactive Framework for the Development
of 3D Real-Time Applications   254
    Francesco Scarlato, Giovanni Palmitesta, Franco Tecchia,
    and Marcello Carrozzino

A Real-Time Video Stream Stabilization System Using Inertial Sensor   274
    Alessandro Longobardi, Franco Tecchia, Marcello Carrozzino,
    and Massimo Bergamasco

Using Proxy Haptic for a Pointing Task in the Virtual World:
A Usability Study   292
    Mina Abdi Oskouie and Pierre Boulanger

Medicine

Surgeries That Would Benefit from Augmented Reality and Their Unified
User Interface.   303
    Oguzhan Topsakal and M. Mazhar Çelikoyar

Assessment of an Immersive Virtual Supermarket to Train Post-stroke
Patients: A Pilot Study on Healthy People   313
    Marta Mondellini, Simone Pizzagalli, Luca Greci, and Marco Sacco

Upper Limb Rehabilitation with Virtual Environments.   330
    Gustavo Caiza, Cinthya Calapaqui, Fabricio Regalado, Lenin F. Saltos,
    Carlos A. Garcia, and Marcelo V. Garcia

Proof of Concept: VR Rehabilitation Game for People
with Shoulder Disorders.   344
    Rosanna Maria Viglialoro, Giuseppe Turini, Sara Condino,
    Vincenzo Ferrari, and Marco Gesi

GY MEDIC v2: Quantification of Facial Asymmetry in Patients
with Automated Bell's Palsy by AI.   351
    Gissela M. Guanoluisa, Jimmy A. Pilatasig, Leonardo A. Flores,
    and Víctor H. Andaluz

Machine Learning for Acquired Brain Damage Treatment   362
    Yaritza P. Erazo, Christian P. Chasi, María A. Latta,
    and Víctor H. Andaluz

Software Framework for VR-Enabled Transcatheter Valve
Implantation in Unity   376
    Giuseppe Turini, Sara Condino, Umberto Fontana, Roberta Piazza,
    John E. Howard, Simona Celi, Vincenzo Positano, Mauro Ferrari,
    and Vincenzo Ferrari

Virtual Reality Travel Training Simulator for People
with Intellectual Disabilities   385
    David Checa, Lydia Ramon, and Andres Bustillo

BRAVO: A Gaming Environment for the Treatment of ADHD.   394
    Maria Cristina Barba, Attilio Covino, Valerio De Luca,
    Lucio Tommaso De Paolis, Giovanni D'Errico, Pierpaolo Di Bitonto,
    Simona Di Gestore, Serena Magliaro, Fabrizio Nunnari,
    Giovanna Ilenia Paladini, Ada Potenza, and Annamaria Schena

Author Index   409

Augmented Reality

Application of Physical Interactive Mixed Reality System Based on MLAT in the Field of Stage Performance

Yanxiang Zhang[(✉)], Pengfei Ma, Ali Raja Gulfraz, Li Kong,
and Xuelian Sun

Department of Communication of Science and Technology,
University of Science and Technology of China, Hefei, Anhui, China
petrel@ustc.edu.cn

Abstract. At present, in some real-time occasion like as stage performances, mixed reality effects are often generated by a mechanical means, or only between physical objects and virtual objects. This will lead to many uncontrollable shortcomings. in order to overcome the above disadvantage, the authors developed a mixed reality system for physical interaction. The experiment unfolds between the experiencer and the controllable Four-axis unmanned aerial vehicle with some performance props. The first step of the study is to build the stereoscopic scene through the camera calibration, the target coordinates in image are obtained by infrared sensors deployed on the experimenter and the aircraft respectively by Camshift algorithm and it can be transformed into the spatial coordinates by projection transformation. And the second step is based on different preset working modes and the coordinate relationship expressions between them that have been configured, with the action of the experimenter, Ant Colony Algorithm is used to move the Four-axis UAV to a certain position, thereby realizing this kind of precisely controllable interaction between entities. The final experiment in this paper has proven that our work provided a great mixed reality effect for occasions such as stage performances.

Keywords: Infrared sensor · Camera calibration · Camshift · Route plan · Four-axis UAV · OpenCV

1 Introduction

In recent years, with the continuous development of technologies such as digitization, visualization and mixed reality, people have put forward higher requirements for the experience of real-time performance on the stage. In the traditional stage performance field that combines mixed reality technology, the audience's experience and satisfaction are mainly from the interaction between actors and virtual props [1]. But so far the technology of superimposing virtual elements on the exact position of a real scene by a computer is still immature and requires manual assistance to achieve. In practical applications, taking the stage performances of ancient Chinese mythology as an example, these mythological backgrounds require some special effects, which often need the interaction of real actors and props to produce. In the current means of producing these special effects, one of the means is to let the actors fly by using

L. T. De Paolis and P. Bourdot (Eds.): AVR 2019, LNCS 11614, pp. 3–12, 2019.
https://doi.org/10.1007/978-3-030-25999-0_1

mechanical techniques, and another current mainstream method uses actors to interact with virtual images in the background of the screen [2], the above means often have the characteristics of low dynamic response resolution, insufficient degree of freedom and process-uncontrollability. In view of the improvement of the above shortcomings, authors developed an entity mixed reality interactive system to apply to occasions such as stage performances.

In this system, experiment was unfolded between the experiencer and the controllable Four-axis unmanned aerial vehicle, which some physical items can be overlaid or hung underneath. The system mainly aims to accomplish such a function: Under different mode settings, the drone with the physical props can move according to a certain path as the entity moves. Of course, the entity here is a hand but is not limited to this part. In order to achieve this function, the system is mainly divided into the following sections:

I. The infrared binocular camera is used to detect the infrared light source deployed on the experimenter's hand and the drone. The detection algorithm uses the Camshift algorithm to obtain the 2D coordinates of the two moving targets on one frame of the video stream in real time.

II. Calibrating the binocular camera to obtain the stereo space where the moving target is located, and converting the 2D coordinates (x, y) into real-time 3D coordinates (x, y, z).

III. Through the movement of the experimenter's hand, the real-time $P_{hand} = (x_{hand}, y_{hand}, z_{hand})$ coordinates are obtained. According to the preset flight path relationship $P_{UAV} = f(P_{hand})$ between the two targets, we can combined with the flight control system command and Ant Colony Algorithm and adjusted the drone to $P_{UAV} = (x_{UAV}, y_{UAV}, z_{UAV})$ in real time.

Here, we will give the overall design framework of the system and elaborate on the theoretical details and engineering implementation details. Figure 1 describes it.

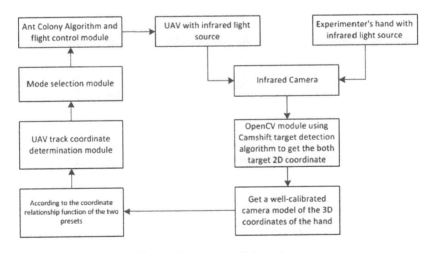

Fig. 1. System overall flow chart.

2 System Design

2.1 Camshift Moving Target Detection Algorithm

In our system, the 2D coordinate tracking of two infrared moving target points (hand and drone) uses the Camshift algorithm. As we all know, Camshift uses the target's color histogram model to convert the image into a color probability distribution map. It can initialize the size and position of a search window, and adaptively adjust the position and size of the search window based on the results obtained in the previous frame, thereby locate the central location of the target in the current image [3]. The system realizes the tracking of 2D coordinates mainly adopts the following three processes.

I. (1) To avoid image sensitivity to light, we convert the image from RGB space to HSV space.
   (2) By calculating the histogram of the H component, the probability or number of pixels representing the occurrence of different H component values is found in the histogram, which can get the color probability look-up table.
   (3) A color probability distribution map is obtained by replacing the value of each pixel in the image with the probability pair in which the color appears. Actually this is a back projection process and color probability distribution map is a grayscale image.
II. The second process of the Camshift algorithm uses meanshift as the kernel. The meanshift algorithm is a non-parametric method for density function gradient estimation. It detects the target by iteratively finding and optimizing the extreme value of the probability distribution. We use the following flow chart to represent this process.
III. Extending the meanshift algorithm to a continuous image sequence is achieved by the camshift algorithm [4]. It performs a meanshift operation on all frames of the images, and takes the result of the previous frame. That is, the size and center of the search window becomes the initial value of the search window of the next frame of the meanshift algorithm. With this iteration, we can track 2D coordinates of the experimenter's hand and drone on the image. Its main following process is based on the integration of I and II.

In the engineering implementation of the above algorithm, we developed a MFC-based PC software using the Camshift function in OpenCV as a kernel to display the 2D coordinates of the two moving target in real time.

2.2 3D Reconstruction

In order to obtain the 3D coordinates of the hand and drone in the world coordinate system, we carried out a 3D reconstruction experiment based on camera calibration. Before understanding 3D coordinates, we need to understand the four coordinate systems and the three-dimensional geometric relationship between them. First, we introduce 3D reconstruction based on binocular camera.

I. Regarding the calibration of the camera, we should first understand the four coordinate systems.

(1) Pixel coordinate system. The Cartesian coordinate system u-v is defined on the image, and the coordinates (u, v) of each pixel are the number of columns and the number of rows of the pixel in the array. Therefore, (u, v) is the coordinate of image coordinate system in pixels, which is also the (x, y) value obtained by the Camshift algorithm in our previous system module.

(2) Retinal coordinate system. Since the image coordinate system only indicates the number of columns and rows of pixels in the digital image [5], and the physical position of the pixel in the image is not represented by physical units, it is necessary to establish an retinal coordinate system x-y expressed in physical units (for example, centimeters). we use (x, y) to represent the coordinates of the retinal coordinate system measured in physical units. In the x-y coordinate system, the origin O_1 is defined at the intersection of the camera's optical axis and the image plane, and becomes the principal point of the image [6]. This point is generally located at the center of the image, but there may be some deviation due to camera production. O_1 becomes (u_0, v_0) in the coordinate system, and the physical size of each pixel in the x-axis and y-axis directions is dx, dy. The relationship between the two coordinate systems is as follows:

$$\begin{bmatrix} u \\ v \\ 1 \end{bmatrix} = \begin{bmatrix} 1/dx & s' & u_0 \\ 0 & 1/dy & v_0 \\ 0 & 0 & 1 \end{bmatrix} \begin{bmatrix} x \\ y \\ 1 \end{bmatrix}$$

where s' represents the skew factor because the camera retinal coordinate system axes are not orthogonal to each other.

(3) World coordinate system. The relationship between the camera coordinate system and the world coordinate system can be described using the rotation matrix R and the translation vector t. Thus, the homogeneous coordinates of the point P in the space in the world coordinate system and the camera coordinate system are $(X_w, Y_w, Z_w, 1)^T$ and $(X_c, Y_c, Z_c, 1)^T$, respectively, and the following relationship exists:

$$\begin{bmatrix} X_c \\ Y_c \\ Z_c \\ 1 \end{bmatrix} = \begin{bmatrix} R & t \\ 0^T & 1 \end{bmatrix} \begin{bmatrix} X_w \\ Y_w \\ Z_w \\ 1 \end{bmatrix} = M_1 \begin{bmatrix} X_w \\ Y_w \\ Z_w \\ 1 \end{bmatrix}$$

where R is a 3x3 orthogonal unit matrix, t is a 3-dimensional translation vector, and $0 = (0, 0, 0)^T$, M_1 is the contact matrix between two coordinate systems. In our opinion, there are twelve unknown parameters in the M_1 contact matrix to be calibrated. Our method is to use the twelve points randomly selected by space, and find the coordinates of these twelve points according to the world coordinates and camera coordinates, then it forms twelve equations, so the M_1 matrix can be uniquely determined.

(4) Camera linear model. Perspective projection is the most commonly used imaging model and can be approximated by a pinhole imaging model [7]. It is characterized in that all light from the scene passes through a projection center, which corresponds to the center of the lens. A line passing through the center of the projection and perpendicular to the plane of the image is called the projection axis or the optical axis. As shown in Fig. 2, x_1, y_1 and z_1 are fixed-angle coordinate systems fixed on the camera. Following the right-hand rule, the X_c axis and the Y_c axis are parallel to the coordinate axes x_1 and y_1 of the image plane, and the distance OO_1 between the planes of the X_c and Y_c and the image plane is the camera focal length f. In the actual camera, the image plane is located at the distance f from the center of the projection, and the projected image is inverted. To avoid image inversion, it is assumed that there is a virtual imaging x', y', z' plane in front of the projection center. The projection position (x, y) of $P(X_c, Y_c, Z_c)$ on the image plane can be obtained by calculating the intersection of the line of sight of point $P(X_c, Y_c, Z_c)$ and the virtual imaging plane.

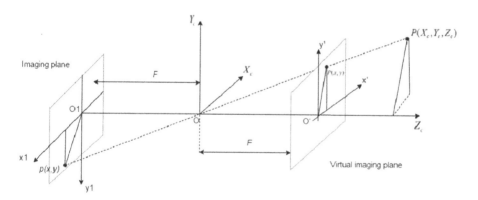

Fig. 2. Camera model.

The relationship between the camera coordinate system and the retinal coordinate system is:

$$x = \frac{fX_c}{Z_c}, y = \frac{fY_c}{Z_c}$$

where (x, y) is the coordinate of point P in the retinal coordinate system, and $P(X_c, Y_c, Z_c)$ is the coordinate of the space point P in the camera coordinate system, which is represented by the subordinate coordinate matrix:

$$Z_c \begin{bmatrix} x \\ y \\ 1 \end{bmatrix} = \begin{bmatrix} f & 0 & 0 & 0 \\ 0 & f & 0 & 0 \\ 0 & 0 & 1 & 0 \end{bmatrix} \begin{bmatrix} X_c \\ Y_c \\ Z_c \\ 1 \end{bmatrix}$$

By combining the above equations, we can get the relationship between the image coordinate system and the world coordinate system:

$$Z_c \begin{bmatrix} u \\ v \\ 1 \end{bmatrix} = \begin{bmatrix} 1/dx & s' & u_0 \\ 0 & 1/dy & v_0 \\ 0 & 0 & 1 \end{bmatrix} \begin{bmatrix} f & 0 & 0 & 0 \\ 0 & f & 0 & 0 \\ 0 & 0 & 1 & 0 \end{bmatrix} \begin{bmatrix} R & t \\ 0^T & 1 \end{bmatrix} \begin{bmatrix} X_w \\ Y_w \\ Z_w \\ 1 \end{bmatrix}$$

$$= \begin{bmatrix} \alpha_u & s & u_0 \\ 0 & \alpha_v & v_0 \\ 0 & 0 & 1 \end{bmatrix} [R \quad t] \begin{bmatrix} X_w \\ Y_w \\ Z_w \\ 1 \end{bmatrix} = K[R \quad t]\tilde{X} = P\tilde{X}$$

where $\alpha_u = \frac{f}{dx}$, $\alpha_v = \frac{f}{dy}$, $s = s'f$. $[R \quad t]$ is completely determined by the orientation of the camera relative to the world coordinate system, so it is called the camera external parameter matrix, which consists of the rotation matrix and the translation vector. K is only related to the internal structure of the camera, so it is called the camera internal parameter matrix. (u_0, v_0) is the coordinates of the main point, α_u, α_v are the scale factors on the u and v axes of the image, respectively, s is the parameter describing the degree of tilt of the two image coordinate axes [8]. P is the 3 × 4 matrix called the projection matrix, that is, conversion matrix of world coordinate system relative to image coordinate system. It can be seen that if the internal and external parameters of the camera are known, the projection matrix P can be obtained. For any spatial point, if the three-dimensional world coordinates (X_w, Y_w, Z_w) are already known, the position (u, v) at the image coordinate point can be obtained. However, if we know the image coordinates (u, v) at a certain point in the space even if the projection matrix is known, its spatial coordinates are not uniquely determined. In our system, it is mainly determined by using a binocular camera to form stereoscopic vision and depth information to get the position of any point in the world coordinate.

II. 3D reconstruction based on the binocular camera. As we know, when people's eyes are observing objects, the brain will naturally produce near and deep consciousness of the object. The effect of generating this consciousness is called stereo vision. By using a binocular camera to observe the same target from different angles, two images of the target can be acquired at the same time. And the three-dimensional information is restored by the relative parallax of the target in the imaging, thereby realizing the stereoscopic positioning effect.

As shown in Fig. 3, for any point P in space, two cameras C_1 and C_2 are used to observe point P at the same time. O_1 and O_2 are the optical centers of the two cameras respectively. P_1, P_2 are the imaging pixels of P in the imaging plane of two cameras. It

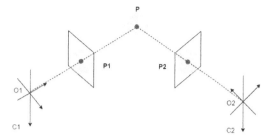

Fig. 3. Binocular vision imaging principle.

can be known that the straight line O_1P_1 and the straight line O_2P_2 intersect at the point P, so the point P is unique and the spatial position information is determined.

In this model, the three-dimensional coordinate calculation of the spatial point P can be solved by the least squares method according to the projection transformation matrix.

Assuming that in the world coordinate system, the image coordinates of the spatial point $P(x, y, z)$ on the imaging planes of the two cameras are $P_1(u_1, v_1)$ and $P_2(u_2, v_2)$. According to the camera pinhole imaging model, we can get:

$$Z_{c1}\begin{bmatrix} u_1 \\ v_1 \\ 1 \end{bmatrix} = M_1 \begin{bmatrix} x \\ y \\ z \\ 1 \end{bmatrix}, Z_{c2}\begin{bmatrix} u_2 \\ v_2 \\ 1 \end{bmatrix} = M_2 \begin{bmatrix} x \\ y \\ z \\ 1 \end{bmatrix},$$

$$M_1 = \begin{bmatrix} m_{111} & m_{112} & m_{113} & m_{114} \\ m_{121} & m_{122} & m_{123} & m_{124} \\ m_{131} & m_{132} & m_{133} & m_{134} \end{bmatrix}, M_2 = \begin{bmatrix} m_{211} & m_{212} & m_{213} & m_{214} \\ m_{221} & m_{222} & m_{223} & m_{224} \\ m_{231} & m_{232} & m_{233} & m_{234} \end{bmatrix}$$

where Z_{c1} and Z_{c2} are the Z coordinates of the P points in the left and right camera coordinate systems, and M_1 and M_2 are the projection matrices of the left and right cameras. The premise of these two formulas is that we must obtain the pixel coordinates (u_1, v_1) and (u_2, v_2) on the left and right images of P point in advance. We combine the two equations above and then eliminate Z_{c1} and Z_{c2}. Then we will get:

$$AP = b$$

where:

$$A = \begin{bmatrix} m_{131} - m_{111} & m_{132} - m_{112} & m_{133} - m_{113} \\ m_{131} - m_{121} & m_{132} - m_{122} & m_{133} - m_{123} \\ m_{231} - m_{211} & m_{232} - m_{212} & m_{233} - m_{213} \\ m_{231} - m_{221} & m_{232} - m_{222} & m_{233} - m_{223} \end{bmatrix}, P = \begin{bmatrix} x & y & z \end{bmatrix}^T, b = \begin{bmatrix} m_{114} - u_1 m_{134} \\ m_{124} - v_1 m_{134} \\ m_{214} - u_2 m_{234} \\ m_{224} - v_2 m_{234} \end{bmatrix}$$

According to the least squares method, the three-dimensional coordinates of the spatial point P under the world coordinate system can be obtained as:

$$P = (A^T A)^{-1} A^T b$$

Therefore, we firstly use Camshift to find the pixel coordinates (u_{hand}, v_{hand}) of the experimenter's hand on the image through the above series of algorithms, then 3D reconstruction is performed by binocular camera vision to obtain the 3D coordinates of the hand in the world coordinate system.

2.3   Ant Colony Algorithm for Path Planning

In view of the fact that the world coordinates of the space field points are already available, the next problem to be solved is to use the coordinates of the hand so that the drone can move with it. We extracted it as a path planning problem for drones. System uses the Ant Colony Algorithm to adjust the two moving targets for precise motion through the preset interaction path between the hand and the drone.

The algorithm in this paper is to set several points on the preset path, which are randomly generated by the movement of the hand, and the number is not fixed, so this is a typical TSP problem. In our system, the central controls the drone to fly in accordance with the path of the random point coordinates generated by the hand. When the UAV position is off the route or the target point is changed, the system will regenerate the path based on the changed UAV dynamic and static information using the Ant Colony algorithm. When the flight path of the drone changes, the system will respond quickly. According to the newly generated track of the drone system, the drone will be controlled to fly. System will convert the control command signal into a PWM signal to reach the drone and realize the purpose of attitude control. The implementation of the entire aircraft control system is shown in Fig. 4.

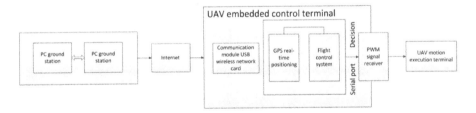

Fig. 4. UAV path planning system based on ant colony algorithm.

3   Field Experiment

The field experiment of the system is mainly divided into three parts. The first parts is the calibration of the binocular camera, then the second part are construction and deployment of the UAV path planning system and sensor components, which implement the two main modes: including that drone props flying around space and interact with actors' actions, and the third part is the physical interaction between the experimenter and the drone props. Here are some of our system experiments pictures.

First, the deployment of the UAV terminal control system and sensor system are designed by the authors. This work is the basis for the debugging of experimental hardware systems.

The Fig. 5 shows photos of an experiencer interacting with a drone with a balloon prop. The system detects the coordinates of the change of the hands of the experiencer and adjusts the drone to its corresponding coordinates in real time. The effect is that the experiencer can push the balloon to fly and realize the physical interaction between both.

Fig. 5. Experiments in which researchers interact with physical items.

After joint debugging, it is confirmed that the system can achieve dynamic interaction between entities within a certain range, but the time response still has certain limitations. The authors are prepared to solve this problem with a faster algorithm in the next work.

4 Result

This paper develops a mixed reality system for scene interaction between actors and solid props in the stage performance field. The implementation of the experiment combines multiple techniques of multi-point positioning, computer vision, drone control, and path planning. Through the previous technical details and field experiments, authors finally realized the scene of the drone props flying around the stage space and interacting with the actors according to the set mode. These implementations have greatly overcome some of the limitations of special effects in traditional stage performances and enhance the immersion of the audience. It has great potential for future business deployment.

References

1. Chatman, S.B.: Story and Discourse: Narrative Structure in Fiction and Film. Cornell University Press, Ithaca (1980)
2. Zhang, Y., Ma, P.F., Zhu, Z.Q.: Integrating performer into a real-time augmented reality performance spatially by using a multi-sensory prop. In: Proceedings of the 23rd ACM Symposium on Virtual Reality Software and Technology, p. 66. ACM (2017)

3. Baker, S., Scharstein, D., Lewis, J.P., Roth, S., Black, M.J., Szeliski, R.: A database and evaluation methodology for optical flow. Int. J. Comput. Vis. **92**(1), 1–31 (2011)
4. Allen, J.G., Xu, R.Y.D., Jin, J.S.: Object tracking using camshift algorithm and multiple quantized feature spaces. In: Proceedings of the Pan-Sydney area workshop on Visual information processing, pp. 3–7. Australian Computer Society, Inc. (2004)
5. Yimin, L., Naiguang, L., Xiaoping, L., Peng, S.: A novel approach to sub-pixel corner detection of the grid in camera calibration. In: 2010 International Conference on Computer Application and System Modeling (ICCASM), vol. 5, pp. V5–18. IEEE (2010)
6. Li, J., Wan, H., Zhang, H., Tian, M.: Current applications of molecular imaging and luminescence-based techniques in traditional Chinese medicine. J. Ethnopharmacol. **137**(1), 16–26 (2011)
7. Wen, Z., Cai, Z.: Global path planning approach based on ant colony optimization algorithm. J. Central South Univ. Technol. **13**(6), 707–712 (2006)
8. Bouguet, J.-Y.: Pyramidal implementation of the affine lucas kanade feature tracker description of the algorithm. Intel Corporation **5**(1–10), 4 (2001)

Application Advantages and Prospects of Web-Based AR Technology in Publishing

YanXiang Zhang$^{(\boxtimes)}$ and Yaping Lu

Department of Communication of Science and Technology,
University of Science and Technology of China, Hefei, Anhui, China
petrel@ustc.edu.cn, Luyp007@mail.ustc.edu.cn

Abstract. Web-based Augmented Reality technology (WebAR) is a new hot spot in the development of AR technology. Its lightweight, easy-to-propagate and cross-platform communication features will break through the limitations of the existing Application-based Augmented Reality (APP AR) mode. The paper first analyzes the development status and technical advantages of WebAR. Next, the publishing mode of WebAR application in publishing is compared with the current popular APP AR publishing mode to expound the characteristics of WebAR publishing mode. Then, combined with WebAR's technical advantages and characteristics of publishing and distribution mode, this paper analyzes the advantages its application in publishing brings to publishers and readers. Finally, the application prospect of WebAR in the publishing of three-dimensional textbooks and sci-tech periodicals is briefly discussed.

Keywords: WebAR · APP AR · Augmented reality · Publish ·
Three-dimensional textbook · Enhanced publishing

1 Introduction

In recent years, augmented reality publications have become more active in the bookselling landscape. Currently, AR books are basically implemented based on mobile application. However, this technology mode has many problems in the publication and dissemination of AR content. On the one hand, there are high technical thresholds for the development and release of application, and the cost is relatively high. On the other hand, different AR books often require different applications, and the data capacity of such applications tends to be large, which reduces the willingness of readers to download or retain to a certain extent. Application download for experiencing AR has become a problem for readers and an obstacle to the issuance of AR publications. These terminal costs have become the limiting factor in the development of AR publications.

Recently, web-based augmented reality technology (hereinafter referred to as WebAR), which eliminates many intermediate links such as application, breaks through the existing disadvantages of the existing AR publishing mode. It will break the limitations and shortcomings of the current application of augmented reality technology in publishing, and bring new development space for augmented reality publishing.

L. T. De Paolis and P. Bourdot (Eds.): AVR 2019, LNCS 11614, pp. 13–22, 2019.
https://doi.org/10.1007/978-3-030-25999-0_2

2 WebAR Development Status and Technical Advantages

2.1 WebAR Technology Model

The specific process of WebAR implementation is shown in Fig. 1. The browser uses WebRTC-technology (Web Real-Time Communication) to obtain the camera video stream to realize the identification of the AR, thereby realizing the virtual and real fusion, so that the AR experience can be realized without developing a hardware-based third-party application program. WebRTC-technology has also been applied to explore the WebVR framework [1].

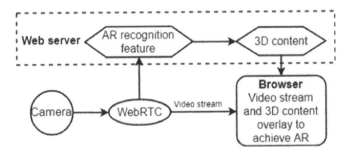

Fig. 1. Schematic diagram of the WebAR implementation process

Figure 1 Show the implementation process of WebAR.

With its technical advantages, WebVR has been initially applied in the field of commercial promotion. Earlier, Universal Pictures and OmniVirt have launched the AR advertising experience application Pitch Perfect 3, which uses facial recognition technology to place the AR object on the user's face. It needs not to download any programs or applications as long as the device itself and the browser version support it. For the publishing industry, WebAR can change the dilemma of current augmented reality publishing.

2.2 WebAR Development Status

WebAR is the latest exploration of augmented reality technology to realize the development of terminals. The web browser that has been ubiquitous is used as a terminal equipped with augmented reality technology. Users can realize the human-computer interaction of AR content through the browser by directly clicking the webpage link. Of course, WebAR technology is different from AR-browsers. The AR- browsers are not new as specialized, general-purpose augmented reality applications [2].

In the past two years, this field has developed rapidly. By the end of 2017 officially support Apple's Safari browser support for WebRTC so as to realize the WebAR. In early 2018, Google announced a new product named Article, which allows users to integrate augmented reality into their mobile phones or computer web pages using the Chrome browser. Mozilla FireFox's Android and PC versions support WebRTC earlier. By September 2018, Mozilla also launched a web browser Firefox Reality designed specifically for VR and AR experiences. Domestic technology companies have also achieved initial development. In October 2017, the U4 2.0 kernel of UC browser announced support for WebAR, providing it with a running ecosystem. At the beginning of 2018, WeChat version 6.6.6 also began to support WebRTC to support WebAR. Around April 2018, the well-known augmented reality technology company in China regarded the EasyAR team of Sight Puls AR to officially open the WebAR content publishing platform. To meet the diverse needs of the market and developers for AR applications, enabling more people to use augmented reality technology via WeChat or browser. With the support of many technical teams, WebAR technology is booming.

3   Comparison of the Influence of Web and APP AR on Publishing Mode

At present, augmented reality publications in the book market are mostly realized through the native augmented reality application developed by users' mobile phones. From the perspective of book selection, editing, processing and publishing, its publishing model is shown in Fig. 2: In the early stage of publication, the publishing house planned the topic of augmented reality books. In the editing and processing stage, the publisher first needs to provide the book content to the technology company for the production of virtual content and the development of native APP for different hardware devices and operating systems(For example, for IOS system and Android system, you need to develop different versions of APP respectively). Native language is used to development app is different from scripting language or compiler language. It is difficult to have a platform on the market to provide a framework for cross-platform implementation. At the same time, the publisher also edits the paper version of the book, carrying the marker of the augmented reality for readers to scan. In the book issuance stage, the publishing house should not only use the traditional distribution channels to promote the paper version of the augmented reality book, but also promote the exclusive augmented reality application of the book in the major mobile application markets; And readers will need to download these AR applications separately for different publications (especially from different publishers).

For example, CITIC Press Group has introduced the "Science Run Out" series of AR books, in which each AR book needs to download the different applications for different themes. There are "idinosaurAR" for the dinosaurs theme, "iSolarSystem AR" for the solar system theme, and so on. Users read the entire series of books will be a burden, and users will only download the app when they purchase the book. In the Apple Store, the single app downloads in this series are less than 100. Reader will not download the APP actively.

Figure 2 Show the publishing process of APP AR publications.

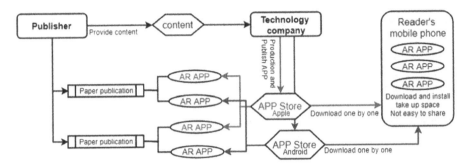

Fig. 2. Schematic diagram of the APP AR publication publishing model

If WebAR technology is applied to publications to achieve augmented reality, and the technical advantages of WebAR are utilized, the development of native application in the traditional publishing process will be omitted. WebAR book publishing mode is shown in Fig. 3: The publishing house is responsible for augmented reality publishing book selection plan. Editing the processing stage, the technology company provides WebAR's online technology platform, and publishers can publish and manage augmented reality content directly on the platform. Therefore, when the augmented reality publication is released, the publishers only need to print the QR codes of the website and the AR identifiers on the paper publication to realize the AR interactive reading. It is also more convenient for publishers to promote this. Link relevant web pages to social media, and users can directly experience it.

Figure 3 Show the publishing process of WebAR publications.

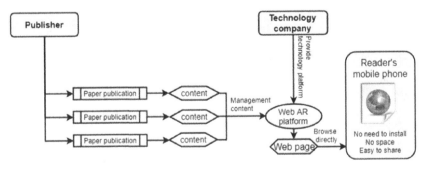

Fig. 3. Schematic diagram of the WebAR publication publishing model

4   Application Advantages of WebAR in Publishing

The efficiency of AR applications involves avoiding wasted time and effort. [3] WebAR technology enables users to read augmented reality content and interact with virtual content through a browser, which is highly efficient. Applied in publications, publishers do not need to develop specialized AR applications, so they will have some of the application advantages described below for publishers and readers.

4.1   Publisher

(1) Reduce development costs so that easy to make

For publishers, the reduction in development costs and production difficulty is a significant advantage of WebAR applications in publishing. First of all, WebAR is a part of Web front-end application development from a technical point of view. The current web-side development ecosystem in the field of computer technology has matured. For developers, WebAR development does not require learning a new language or purchasing native application development tools and publishing licenses. It is only based on web script development, which brings great convenience to development. Furthermore, WebAR launches browser-based rather than applications that rely on user hardware devices and operating systems, with a natural cross-platform advantage. Publishers use it in publications, and do not need to develop different augmented reality software clients for hardware using different operating systems on the market, which reduces development costs to some extent.

It is worth mentioning that with the increasing attention of WebAR technology, there are some network technology companies in the field of AR technology that develop specialized WebAR online production platforms. The EasyAR online development platform of Sight Plus AR is a relatively early technology platform in China. This type of development platform provides developers with development framework and technical support to reduce development difficulty. With the maturity of such platforms, publishers no longer need to invest in high-cost AR technology development, but can focus more on the creation of digital content, which will help to structurally reduce the cost of AR publishing in the long run.

(2) Reduce the cost of distribution promotion so that easy to spread

A major feature of WebAR is cross-platform operation and multi-platform application. In augmented reality publications, the promotion cost of publishers' augmented reality programs will be greatly reduced. At the same time, WebAR itself can spread like a virus. And for WebAR, sharing is communication. For WebAR, sharing is communication, so it is beneficial for augmented reality publications to promote themselves in social media.

In the early days, even if the publication augmented reality content was not completed, the publisher could directly post advertisements on web pages and social networks. By promoting the experiential demo link of WebAR publications, and cooperating with creative copywriting to attract audience attention and forwarding. This will spark a discussion boom on social networks.

(3) Improve the reuse rate of publishing resources so that easy to deploy

The development of WebAR is based on the web, so it has the characteristics of easy deployment, easy updating and easy maintenance of web applications. With the technical advantages of WebAR, the publisher can manage all AR virtual content on a unified server or cloud platform, which is beneficial to improve the publishing resource reuse rate of the publishing house and facilitate the deployment of AR content.

Virtual content such as AR models developed by publishers for publishing augmented reality books is a valuable publishing resource. The publisher develops augmented reality publications. If the augmented reality content developed for a book is integrated into an application or read hardware, the difficulty of augmented reality content resource management will limit the reuse of the resource. WebAR can support a unified online resource management platform, and publishers can re-develop and utilize the WebAR resources of a published book. In this case, if the book is republished, there will be higher quality virtual content; If the resources are redeployed to the same series of publications or other series of publications, the production cost will be reduced, which will help to enhance the serialization and industrialization of books and enhance the competitiveness of the book market.

4.2   Reader

(1) Click to experience, reduce reading costs

The WebAR effect experience does not require the user to download any new applications, but can be implemented using any native software such as a browser or WeChat that can mobilize the camera to scan the AR markers. This convenient mode of click-to-app experience is used in augmented reality publications to give readers a good reading experience and reduce the reader's reading burden.

Currently, augmented reality books on the market are limited by the high development cost of the application, and the augmented reality content richness and fidelity are different. The book pricing is high but the augmented reality reading experience is unstable. Reader experience augmented reality requires downloading the app, which undoubtedly increases the reader's reading cost. Once this reading value is less than expected by readers, it is easy for future readers to choose to reject such products. WebAR is the exact opposite. Readers only need to "scan it" to experience it, without downloading any products. This unburdened reading method can, to some extent, enhance the reader's acceptance of augmented reality books, and is more conducive to the sharing of augmented reality content among readers.

(2) Lightweight operation to increase reader loyalty

Another advantage that browser-based WebAR is more important to readers is their lightweight operation. In this lightweight mode, the reader does not need to download a large number of applications to read different augmented reality readings. Readers will not choose to uninstall the app because it won't take up memory on the phone and won't be forgotten over time. It is more conducive to increasing readers' loyalty to publications and even publishers.

Many publishers have now developed their own proprietary augmented reality applications, and even the same publisher has different augmented reality applications. These different applications, as well as a large number of 3D models and video audio in these applications, run huge amounts of memory and information space. This is a great burden for readers who consume augmented reality products. The APP designs currently use the embedded module function, which automatically triggers the download of a large amount of resources when the user clicks on the function module. These files are stored in mobile phones and are difficult to find by users who are not familiar with digital knowledge. Compared with APP, the WebAR runs on a browser, and various resources such as models and audio loaded when reading AR books are cached by the browser for next access. And the browser itself has the function of clearing the cache, so it's easy for users to delete these caches. In the lightweight operation mode of WebAR, readers can read the augmented reality products of different publishers like browsing the webpage, and even select the preferred publishers and their published products, which will produce reading stickiness and maintain a sense of loyalty.

5   The Application Prospect of WebAR in Publishing

Augmented reality technology has broad application prospects in scientific visualization and educational visualization since its development due to its intuitive, vivid and interactive features. However, owing to the difficulty and cost of augmented reality terminal development, augmented reality publications are mostly concentrated in the field of children's publishing, and there are few publications in other fields. If WebAR technology is applied to publishing, its low-cost, lightweight, cross-platform natural technology advantages will not only make AR shine in the field of children's publishing, but it can also create more possibilities in the development of three-dimensional textbooks and the enhanced publishing of scientific journals currently explored by the publishing industry.

5.1   WebAR and Three-Dimensional Textbook Publishing

Using cloud storage and cloud computing, the three-dimensional textbooks of the Internet and various terminals can better meet the learning needs of students in the new era.

Like many innovative technologies, augmented reality technology is based not only on the use of technology, but also on how AR is designed, implemented, and integrated into formal and informal learning environments [4]. The application of AR technology in the publication of textbooks, its powerful visual features can better meet the needs of teaching and improve teaching efficiency. However, with the current APP AR enhanced publishing model and a small number of independent ecosystem hardware publishing models, high development costs limit the possibility of popularization; the use of high reading costs is not conducive to multidisciplinary development; Low resource reuse rates are harmful to the optimization of publishing resources. In the case of unbalanced and multi-version textbooks in domestic education development, WebAR applications not only reduce the development cost, but also support the new media teaching resources that have been popularized, and optimize the original three-dimensional textbook resources such as MOOC and electronic textbooks in the form of WebAR links. Provide greater richness and interactivity.

5.2 WebAR and Sci-Tech Journals Enhanced Publishing

Research projects around researchers in the process of scientific research often generate "dark data" that is large and difficult to be discovered by users [5]. Enhanced Publication of sci-tech periodicals is to better promote the sharing and dissemination of research results, structuring publications, linking a large amount of relevant data and literature, and helping researchers discover a large number of grey literature. Thus, the investigator-focused "small data", locally generated "invisible data" and "incidental data" have been discovered [6]. The enhanced publication of sci-tech periodicals has matured abroad, and China National Knowledge Infrastructure is also exploring the enhancement of publishing of scientific journals. Enhanced Publishing adds more data and multimedia links to published articles that are more in line with the purpose of science popularization and dissemination. WebAR provides new ideas for the enhanced publishing of sci-tech journal publishers.

WebAR's link operation mode is fully capable of interfacing with the structured markers of scientific journal articles. When reading the paper, the reader jumps to the AR interactive page by clicking on the marker. So the reader can understand some of the data behind the paper in a visual and intuitive way. This also facilitates the publication of data papers in the scientific field. Chavan and Penev have suggested that mainstreaming data papers will be a step toward raising the level of data publishing to academic publishing, which is expected to significantly increase the efficiency of biodiversity science [7]. And due to the characteristics of AR itself, it is very suitable for making experimental multimedia 3D models that need to be supplemented by scientific, technical and medical aspects.

6 Discussion and Conclusion

Changes in society, culture, and technology are changing the way content is produced, distributed, displayed and accessed. So publishing practices need to be reconsidered and updated [8]. WebAR is in the market start-up period in 2018, and its natural technological advantages in publications will help open up new augmented reality publishing markets. However, WebAR technology is currently in the research and development stage, and there are technical defects such as limited rendering capabilities of augmented reality effects, and insufficient compatibility between augmented reality content and browsers. But at the same time, we must realize that the mature technology ecosystem of the Web front-end is the best soil for WebAR technology development. Some of WebAR's current technical defects will be overcome in the near future.

Of course, WebAR wants to develop in the publishing industry, and technology maturity is the foundation. More important is to improve the copyright protection of WebAR publications and to establish a complete WebAR publishing industry chain.

As a public operating environment, the code of the webpage is open, and the downloading and uploading of webpage files is relatively free. When publishers use WebAR to develop augmented reality publications, there may be copyright protection issues for augmented reality content. Some users download a copyrighted augmented reality model, audio, and other files from a publisher through a browser. When users use or disseminate such models and documents, they will infringe the copyright of the publisher to a certain extent, and will also affect the interests of the publisher. For this type of copyright protection, it is necessary to improve the content security through technical means when developing WebAR. And it is also necessary to think about how to strengthen the protection measures for such electronic publications from the perspective of laws and regulations.

The establishment of the WebAR publishing industry chain, the professional technical service platform and the content service platform are indispensable. The specialized technical service platform refers to a professional augmented reality development platform such as EasyAR, which provides development tools and technical support for publishers and other objects with AR development needs. The professional content service platform mainly refers to the AR three-dimensional material providing platform. The three-party cooperation of publishers, technical service platforms and content service platforms can establish a complete WebAR publishing ecosystem.

The application development of WebAR technology has just started, and the typical cases are limited. Therefore, the intention of this paper focuses on theoretical deduction and conceiving speculation. However, the application of augmented reality technology in the publishing industry has not yet established a mature publishing industry chain. The rise of WebAR will bring new development opportunities for the ecological improvement of the augmented reality publishing industry.

Acknowledgements. The research is supported by the Ministry of Education (China) Humanities and Social Sciences Research Foundation, number: 19A10358002. It is also supported by Research Foundation of Department of Communication of Science and Technology University of Science and Technology of China.

References

1. Gunkel, S., Prins, M., Stokking, H., Niamut, O.: WebVR meets WebRTC: towards 360-degree social VR experiences, pp. 457–458. IEEE (2017)
2. Grubert, J., Langlotz, T., Grasset, R.: Augmented reality browser survey. Institute for Computer Graphics and Vision, University of Technology Graz, Technical report (1101) (2011)
3. Parhizkar, B., Al-Modwahi, A.A.M., Lashkari, A.H., Bartaripou, M.M., Babae, H.R.: A survey on web-based AR applications. arXiv preprint arXiv:1111.2993 (2011)
4. Wu, H.K., Lee, S.W.Y., Chang, H.Y., Liang, J.C.: Current status, opportunities and challenges of augmented reality in education. Comput. Educ. **62**, 41–49 (2013)
5. Heidorn, P.B.: Shedding light on the dark data in the long tail of science. Libr. Trends **57**(2), 280–299 (2008)

6. Onsrud, H., Campbell, J.: Big opportunities in access to "Small Science" data. Data Sci. J. **6**, OD58–OD66 (2007)
7. Chavan, V., Penev, L.: The data paper: a mechanism to incentivize data publishing in biodiversity science. BMC Bioinform. **12**(15), S2 (2011)
8. Simeone, L., Iaconesi, S., Ruberti, F.: Fakepress: A Next-Step Publishing House. ESA Research Network Sociology of Culture Midterm Conference: Culture and the Making of Worlds. Step Publishing House, 15 October 2010

Generation of Action Recognition Training Data Through Rotoscoping and Augmentation of Synthetic Animations

Nicola Covre[1]($\boxtimes$)(iD), Fabrizio Nunnari[2](iD), Alberto Fornaser[1](iD),
and Mariolino De Cecco[1](iD)

[1] Department of Industrial Engineering, University of Trento, Trento, Italy
{nicola.covre,alberto.fornaser,mariolino.dececco}@unitn.it
[2] German Research Center for Artificial Intelligence (DFKI), Saarbrücken, Germany
fabrizio.nunnari@dfki.de

Abstract. In this paper, we present a method to synthetically generate the training material needed by machine learning algorithms to perform human action recognition from 2D videos. As a baseline pipeline, we consider a 2D video stream passing through a skeleton extractor (Open-Pose), whose 2D joint coordinates are analyzed by a random forest. Such a pipeline is trained and tested using real live videos. As an alternative approach, we propose to train the random forest using automatically generated 3D synthetic videos. For each action, given a single reference live video, we edit a 3D animation (in Blender) using the rotoscoping technique. This prior animation is then used to produce a full training set of synthetic videos via perturbation of the original animation curves. Our tests, performed on live videos, show that our alternative pipeline leads to comparable accuracy, with the advantage of drastically reducing both the human effort and the computing power needed to produce the live training material.

Keywords: Action recognition · Random forest ·
Augmentation animation · Virtual Environment

1 Introduction

The mutual understanding and the collaboration between men and machines lead to a new concept of the relation between users and tools in several fields, such as automotive [1], medical [2,3], industry 4.0 [4], home assistance [5] or even gaming [6]. In these contexts, a more suitable way of managing the relation between man and machines relies on action recognition. The constraints introduced by joysticks, joypads, mice or any other physical controller will disappear. It is therefore important to study action recognition to fulfill the increasing need for teaching robots and make them 'understand' human movements. Augmented reality devices available in the market, like the Microsoft's Hololens[1] and

---

[1] https://www.microsoft.com/en-us/hololens – Feb 8th, 2019.

© Springer Nature Switzerland AG 2019
L. T. De Paolis and P. Bourdot (Eds.): AVR 2019, LNCS 11614, pp. 23–42, 2019.
https://doi.org/10.1007/978-3-030-25999-0_3

Fig. 1. Left: OpenPose applied on a live recorded video. Right: The same action recreated on a Virtual Human in the Virtual Environment.

the Meta 2[2], already allow for an interaction based on free hand movements, like pointing and grabbing[3].

Teaching to a machine how to recognize an action is performed by Machine Learning algorithms, which needs to be trained in order to produce a classifier. The classifier is then able to recognize recurring patterns. The training process is generally accomplished providing datasets. In the case of action recognition, a dataset consists of a number of videos of real persons performing actions, each video associated to information about the action performed. A person can perform several actions, or the same action several times, or a combination of the two. The datasets have to be properly organized and include a large amount of data to allow the artificial intelligence to discern and classify successfully different gestures and attitudes. However, this approach suffers of two main limitations. First, it is slow and expensive, due to the (usually) manual collection of data. Second, the data acquisition can not be automatized, and it is difficult to control parameters' evolution as long as the gestures are usually performed by actors.

We propose to analyze a new trend based on Virtual Environment (VE) data generation. Given an action, instead of collecting live videos of many real subjects performing the same action, we collect a single video of one subject performing a prototype of an action (See Fig. 1, left). Then, the prototype action is converted into a 3D animation of a virtual character (See Fig. 1, right) through Rotoscoping. Rotoscoping is a widely used technique in digital computer animation that consists of tracing over real videos images, frame by frame, to recreate a realistic replica.[4] Finally, the animation data of the prototype 3D animation are copied and altered, augmented, in order to generate a desired number of

---

[2] https://www.metavision.com/ – Feb 8th, 2019.

[3] https://www.ted.com/talks/meron_gribetz_a_glimpse_of_the_future_through_an_augmented_reality_headset?language=it.

[4] https://en.wikipedia.org/wiki/Rotoscoping.

variations. The resulting animation videos are used for training in place of the real ones.

The advantage of this approach is that the VE allows to control the parameters and automatize the data acquisition. In this way, it is possible to noticeably reduce the time and the effort required to collect the data.

This paper is structured as follows. Section 2 describes the approaches pursued in literature. Section 3 returns a general overview of our approach. Section 4 illustrates the procedures for VE and animation creation, data acquisition, and dataset generation. Sections 5 and 6 present the testing phase and discuss the results achieved. Finally, Sect. 7 concludes the paper by commenting on the limitations that affect the pursued methods and introduces future direction.

2 Related Work

Traditional approaches on gesture recognition are based on data collection from the real environment. Dataset generation has, usually, been pursued manually recording actors during the gesture performance [7–9]. However, machine learning requires often a large amount of data, but datasets are difficult to provide because of the need, together with actors, of a proper equipment, a post-production, and an accurate annotation phase. Eventually, it results to be an expensive and slow process.

It is also possible to rely on already available datasets, such as the KTH human motion dataset [10], the Weizmann human action dataset [11], INRIA XMAS multi-view dataset [12], UCF101 sport dataset [13]. These datasets are often used as reference for comparing different action recognition approaches. They are quite useful, because they allow to save a lot of time that usually data acquisition unavoidably takes. Unfortunately, they are frequently limited to specific cases [14]. This leads to use them mainly for different machine learning methods comparison and performance estimation.

Generally, both methods do not allow to control parameters as long as even the best actor can not introduce controlled variability thousand of times. Therefore, variability of the gesture and pattern distribution is populated repeating the action over and over, hoping to cover all the possible cases.

Training on synthetic data allows to manage these limitations. First of all, relying on a VE, data collection can be easily automatized. This relevant aspect permits to save time, funds, and effort, as long as it does not require neither actors, real cameras or equipment. Second, the procedure can control accurately all parameters, which leads to a more effective data generation: several versions of the same action can be achieved without repeating redundant information.

Similar research, which exploits a VE for Machine Learning, has been pursued on different classifications. For example, related to the autonomous driving, VE has been used to simulate a urban driving conditions. Data acquired from the synthetic world were used to generate a Virtual Training dataset for pedestrian detection [15].

Previous research gave us a reference guideline to follow as well as a benchmark for the results comparison. In addition, often they have been taken in to

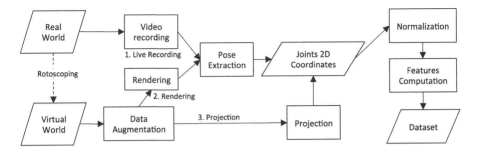

Fig. 2. Three different pipelines to produce the dataset for an action recognition system.

account during features selection. In our case, we focused the attention in establishing the performances of a VE dataset generation, obtained from synthetic data. In many research projects, the acquisition of the skeletal pose of a human can be obtained via affordable acquisition devices that can be found on the consumer market, like the Microsoft Kinect. Even if 3D skeleton acquisition provide much more spatial information [16], the Kinect is still an expensive solution with respect to a simple 640×480 2D camera [17]. For this reason, a 2D video-stream analysis has been pursued with Open Pose [18]. In this way, real and synthetic videos were processed in a similar way and in a reasonable amount of time.

Concerning the choice of a classification algorithm, a Random Forest [19] have been preferred with respect to Deep Neural Networks to limit the time needed for training and focus on the study of real vs. synthetic training material.

From the application point of view, gestures have been selected considering previous research projects in the field of rehabilitation [2,5,20]. In this context, the main goal of the application is to assist a mild cognitive patient while he or she is cooking, reconizing a few common kitchen actions such as: grabbing, pouring, mixing, and drinking.

3   Method Overview

Figure 2 outlines our approach for generating the dataset for gesture recognition. It shows the plots of three generation strategies, organized in three *paths* sharing the same starting and ending points.

Starting from top-left, *path 1* shows the most traditional approach, widely used in previous works [7,9]. It involves data collection from the Real Environment followed by a Pose Extraction step performed with Open Pose. The result of the pose extraction is a dataset (Joints 3D coordinates) containing the evolution in time of the 2D coordinates of the joints of the human skeleton. Obviously, the skeleton is not accurate, but only the approximation inferred by Open Pose through the analysis of 2D color video frames (See Fig. 1, left, for an example).

The two other paths, two alternative Virtual Training approaches, share a first step where a single live animation is transferred to a 3D editor (in our case

Blender) through rotoscoping (left side of the figure). Rotoscoping is a common animation technique consisting of drawing over a (semi-)transparent layer positioned over some reference animated material, which is visible in the background. This is a manual operation which leads to the Virtual Environment: a simplified representation of the real world where a virtual character performs a prototype version of some actions (See Fig. 1, right). The Virtual world provide the synthetic environment for the Data Augmentation process, which procedurally alters the animation data in order to produce a multitude of variations of the prototype animations. The augmented data is then used by paths 2 and 3.

Following *path 2*, the augmented animations are rendered as 2D videos, and the analysis proceed as for path 1, using OpenPose for pose detection. Differently, *path 3* skips the rendering and the joint coordinates are directly, and precisely, transformed into 2D coordinate via a computationally inexpensive world-to-screen projection.

Regardless of the generation path, the time-evolving projected 2D coordinates proceed into a normalization process, followed by a feature extraction step, to become the reference training datasets.

4 Experiment: Action Recognition in a Kitchen

We applied our proposed approach in a real-world scenario set in a kitchen. The scenario accounts for a monitoring device which has to check if a user is correctly performing four actions: *grabing*, *pouring*, *mixing* or *drinking*. The goal of the assistant is to classify which of the four actions is being currently performed. The following subsections describe each of the phases needed to configure the classifier.

4.1 Virtual Avatar and Virtual Environment

In order to recreate an action which is feasible by a human, we edited a 3D environment composed of a virtual human, a kitchen, and three interactive objects: a coffe-pot, a coffe-cup, and a spoon (See Fig. 3). All the editing was performed using the Blender[5] 3D editor. The virtual human was generated using a freely available open-source add-on for Blender called MB-Lab[6]. This choice allowed us to save a conspicuous amount of time and rely on a well-made character. The VE creation takes the cue from a real environment. The reference objects have been crafted from scratch by one of the authors. Both the virtual human and the objects were disposed in the environment with the goal to realize animations for the four above-mentioned actions.

---

[5] https://www.blender.org/ – Apr 30, 2019.
[6] https://github.com/animate1978/MB-Lab – Feb 9th, 2019.

Fig. 3. Virtual Environment recreated in Blender 3D

4.2   Animation Generation Through Rotoscoping

As already anticipated in Sect. 3, a limited amount of reference animations where recorded live and meticulously replicated in the 3D environment through roto-scoping. We call these the *prototype* animations.

Two classes of animation have been generated. The first class of animations was created for *calibration* purposes. We identified five key locations in front of the body, with the right arm extended centrally, up, down, left, right, with respect to the related shoulder. For each position, we edited an animation where the hand is waving around its reference location. In this way, a simple check on features and armature comparison was easily pursued between real world and virtual one.

The second class of animations consists of four kitchen-related actions: grab-bing, pouring, mixing, and drinking. These actions involve the interaction with the three objects in the environment: a coffee pot, a spoon, and a coffee cup. While the first class of animations was edited completely manually, the second class of animations went through a data augmentation process.

To ease the rotoscoping process, inverse kinematic (IK) controls were con-figured on the hands of the virtual character. The author of the rotoscoping procedure must only move the bone representing the palm of the hand. The IK routine will automatically compute the rotation values of all the joins of the arm, up to the spine. This is a standard time-saving technique widely used in the digital animation industry. Consequently, the animation curves defining an action affects only the IK controller, while the rotation of the intermediate bones are computed in real-time during the playback of the animation.

An *animation curve* is a function mapping the time domain to the value of an element of the VE. In our context, an animated element can be one of the six degrees of freedom (3 for position and 3 for rotation) of an object or skeletal joint. Hence, for gesture or posture recognition, animation curves are useful to track the evolution over time of the position and rotation of IK controllers in the 3D space, as well as the evolution of the rotation of the joints. The Animation curves are strictly related with the animation generation. During the animation authoring (in our case, through rotoscoping) the author stores a set of reference body poses as *key frames*. A key frame maps a specific time point to a value for each of the character's animation curves. Afterward, during animation playback, all the animation curves are generated by interpolating between the key-points saved along the animation timeline. In Blender, animation curves can be modified programmatically. This allows us to generate and modify automatically several animations with a routine.

4.3   Data Augmentation: Procedural Variation of the Prototype Animations

It is important to remember that the introduction of a significant variability that populate the feature distribution returns often better training outcome. For this reason, we implemented four data augmentation methods using the Blender's internal scripting system, which is based on Python.

Randomization of Objects' Position. This first method is designed to change, randomly, within certain intervals, the disposition of the interactive objects (coffee pot and cup) in front of the avatar. In this way, the bone structure is re-arranged by the inverse kinematic in order to reach, grab, and move these object in different locations the 3D space during the action development.

Camera Rotation. This method takes care of moving the camera in different positions and rotations around the vertical axis of the avatar. The function requires as input the total number of variations – always an odd number, in order to keep the central framing – and the angular step between two consecutive views. Eventually, the procedure takes as reference the frontal perspective, which is located at 90°, and moves the camera along a circumference of fixed radius. The animation videos were augmented with a combination of five different points of view and an angular step of 18°, leading to positions of the camera respectively at 54°, 72°, 90°, 108°, 126° (Fig. 4).

Time Scaling. The time scaling method controls the duration of the animation sequence, shrinking or expanding the time interval between two consecutive key-frames. This augmentation technique takes into consideration that different actors perform the same action along different time intervals. The transformation function takes as input the animation curve and the scaling factor. The process does not modify the first frame, which is kept as the reference starting point.

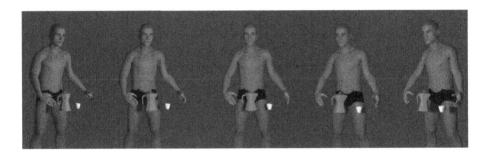

Fig. 4. Rendering of the same action frame using different camera viewing angles.

Perturbation of the Hand Trajectory. This method modifies the animation curves of the hand IK controller. The animation curves keep the track of the position and rotation changing along the animation.

This method requires as input the three animation curves covering the translation of the hand IK controller and introduces, in between the main key-frames, an extra key-frame which "perturbs" the interpolation connection, imposing a passage by a different point. This method aims to simulate the human feedback control while an object reaching is pursued. The introduced perturbation is randomized in a range with the same order of magnitude of the other key points.

Combining these methods as nesting dolls a large number of variations of the same gesture took place. These methods produced already a good amount of variability to appreciate the potentiality of the augmentation approach.

4.4   Joint Coordinate Collection

This phase consists of storing in a data frame the evolution of the movement of the body joints. Each line of the data frame will be associated to a time stamp. Each column of the data frame is one cartesian component (x or y) of the join 3D position after a projection on the 2D image space. Table 1 shows an example.

Table 1. Example of the data frame resulting from the projection of the joint coordinates in time.

Time (ms)	hand.x	hand.y	wrist.x	wrist.y	elbow.x	elbow.y	...
0.00	0.123	0.234	0.345	0.345	0.762	0.324	...
16.66	0.124	0.232	0.346	0.344	0.760	0.328	...
33.33	0.125	0.230	0.348	0.342	0.759	0.329	...
...	...	...	...	...	...	...	

Given an input animation, the joint coordinates collection was performed using two different methods: (i) pose estimation using OpenPose, and (ii) direct

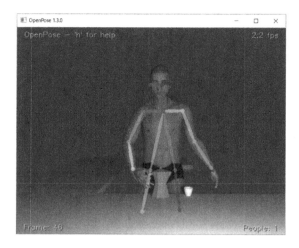

Fig. 5. OpenPose applied on a frame of a rendered video

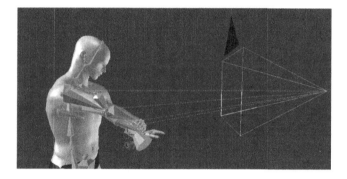

Fig. 6. Joint coordinates projection on the camera view plane

projection (in Blender) of the 3D joint coordinates into the 2D camera plane. With reference to Fig. 2, the first method is used by paths 1 and 2, while the second method is used in path 3.

As already mentioned, the first method exploits Open Pose and reflects the more traditional approach, as it is possible to point out from Fig. 5. Open Pose extrapolates the skeleton animation from 2D videos and returns the coordinates on a pixel-based metric. The result of the processing on synthetic videos is very similar to the one on real videos. With this method, the processing is highly computational expensive because the augmented animations must be rendered.

Differently, the second method skips the rendering step and uses the virtual camera embedded into the virtual environment to collect the 3D joint coordinates and projects them on the 2D virtual screen (See Fig. 6). These coordinates are expressed in a meter-base metric.

Fig. 7. Comparison between Open Pose (red) and projected coordinates (green) skeletal structures on the same frame. (Color figure online)

This strategy emulates the Open Pose video analysis with the difference of avoiding any error in the estimation of the joint positions. The absence of error in the projection creates a discrepancy between the perfectionism of the data set with respect to the estimation errors introduced by Open Pose (See Fig. 7). In this case, perfectionism is not an advantage, because Open Pose will be anyway used to analyse real world video during the action recognition on human subjects. This implies that training data and test data will come from different distributions.

Even if the second method results to be faster with respect to the first one, it returns a skeleton that appears to be slightly different from the Open Pose standard structure. The main difference being the presence, in the Open Pose structure, of a central upper-spine bone always positioned as mid-point between the clavicles. Hence, we aligned the two structures by computing, in the Blender skeleton, the coordinates of a fictional spine bone by averaging the coordinates of the clavicles.

Regardless the used method (pose extraction or projection), the coordinates obtained as output are then collected, normalized, and used for computing Classification Features.

4.5 Coordinates Normalization

The projected joint coordinates have to be pre-processed before features computation takes place. As anticipated, this occurs because the two methods use different coordinate systems: the open pose output is in a pixel-based metric, while the projected coordinates are expressed in meters (unit of reference used in the VE). See Fig. 8, left.

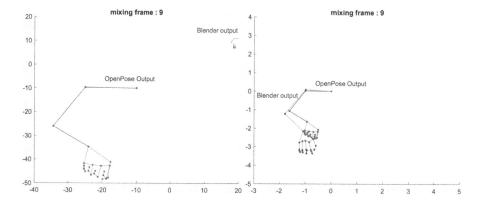

Fig. 8. Comparison between the coordinates projected from Blender skeleton structures and Open Pose skeleton structure. Left: before normalization. Right: after normalization.

Eventually, the midpoint between the two shoulder sets the common origin of the reference system, which represents, from now on, every other couple of joints coordinates. Therefore, every joint coordinate was translated to have the midpoint as origin.

After that, the two representations needed to be scaled. The clavicle extension appeared to be the most suitable length to choose as reference. In fact, it remains constant enough during the entire gesture performance. This is due to the fact that the analysis has been conducted mainly with respect to a frontal camera point of view, which did not change remarkably over time. Another advantage of this strategy is that farther and closer subjects are normalized and analyzed as they were at the same distance.

Hence, every distance is divided by the reference length in order to represent every skeleton structure with the same dimensions. In this way it has been possible to overlap accurately the Open Pose skeleton and the Blender one (See Fig. 8, right).

4.6   Features Selection

The feature selection was the most critical step of this work, as it can deeply affect the final result and jeopardize the output performances. Literature and academic paper of reference mainly suggest two approaches. The first one, based on a Kinect data acquisition or 2D skeleton extraction, compute the features from geometrical relations between joints, such as distances, angles or relative velocities [7,9,21]. The second approach, based on 2D images analysis, suggest a Deep Learning approach where pixels are provided to the algorithm and the best features combinations are selected by the network itself [1,17].

For this work, we used random forests as classifiers, hence features were selected manually. Different choices have been pursued and performances were compared with respect to each other to find out the best combination.

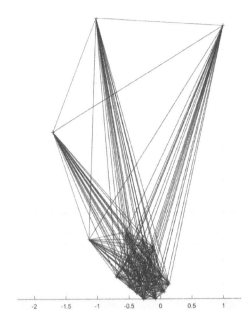

Fig. 9. Feature set one: all the possible combination of relative distances between two non-consecutive joints.

The computation of the features dataframe uses a temporal sliding window of 5 frames. The first 5 rows of the coordinates dataset produces the first row of the features dataframe. The procedure iterates by shifting the time window of 2 frames at each step, until the end of the coordinates dataset is reached.

The time windows size (5) and sliding step (2) were empirically chosen considering the high frame rate (compared with the gesture duration), and with the goal of reducing the similarities with respect to the previous five frames and the consecutive ones. Therefore reducing the chance of computating two identical samples. This approach has also the advantage of acting as low-pass filter against the high frequency jittering of the coordinates returned by Open Pose.

The computed features (columns of the features dataframe) belong to four categories: raw normalized coordinates, distances between joints, joint movement speed (computed over the 5 frames of the time window), and angles (of the armpit and elbow).

Data analysis and features computation have been accomplished focusing on the right arm, which involves 23 bones, from the clavicle and down to the fingers. The results reported in the remainder of this paper come from the two feature sets which returned the best results. The best choice is a list of features computed as all the possible combinations of distances between two non-consecutive joints (Fig. 9), while the second selection, simpler and more intuitive, collects the normalized $<x, y>$ joints coordinates.

The two choices of features returned different results on different groups of actions. In fact, for the first group, composed by the five different hand position – mainly accomplished as a calibration experiment – the x, y joint coordinates returned the best result. Differently, the list of relative distances between all the non consecutive joints returned a better outcome in classifying the grab, pour, mixing, and drinking actions.

Moreover, further comparison and focused analysis led us to introduce also angles measurements of the shoulder and the elbow joint and to exclude the relative distance combination between fingers – due to Open Pose low accuracy in locating them – in order to increase the performances.

5   Training and Testing

Once the features are computed and filled up the dataset, the Machine Learning method receive them as input in order to proceed with the training. When training is done, the performance of the algorithm are evaluated using a test dataset.

It is worth now reminding that the purpose of this study is mainly to assess the advantages (or disadvantages) of using a virtual environment, rather than a real one, to train a classifier. Hence, the classifiers were trained with two different type of video sources: the first made of live-recorded videos of humans (path 1), the second with the video material generated synthetically (paths 2 and 3). However, in order to assess the behaviour on real scenarios, the test set was always entirely composed of live videos.

Training Material. For each action, each prototype animation has been augmented into nine videos of roughly one second each; for a total of 36 videos. At the render rate of 60 frame per second, we collected around 28 feature samples per video. The same amount of videos have been collected from the real environment in order to generate a balanced training dataset using the traditional approach.

Similar criteria applied for testing material collection from the Real Environment. Four different actors performed nine times each action, for a total of 144 videos. The four actors accomplished each gesture in different amount of time. Consequently, even if the same number of videos have been exploited, for each action there is a slightly different amount of samples.

Table 2 reports on the time needed to create a dataset for each of the three generation paths.

Environment setup refers to the time needed to prepare the live scene, or the 3D editing. Data acquisition refers to the time needed for live shooting, rendering, or coordinates projection. Skeleton Extraction is the time taken by OpenPose. Data Conversion refers to the time needed to transform the skeleton extracted coordinates in matrix form, which is more suitable for Matlab operations. Converting Open Pose data is slower because the coordinates of each frame are saved in a separate JSON file. Finally, Feature Computation refers to

Table 2. Measured time for the creation of the training material

Training method	Environment setup	Data acquisition per single action	Skeleton extraction per single action	Data conversion per single action	Feature computation per single action
Traditional (path 1)	30 s (per single action)	1 min	3 min	6 s	4 min
VE rendering (path 2)	30 min	10 s	3 min	6 s	4 min
VE projection (path 3)	30 min	3 s	n/a	2 s	4 min

the conversion from coordinates to features dataset, as described above in this section.

With this timings, the creation of the 36 videos of the training set took a total of 292.1 min for path 1, 291.6 min for path 2, and 177 min for path 3. The latter, represents a 40% time saving compared to path 1. It is possible to appreciate, that the virtual training approach results to be much faster if the training require a large dataset composed by many actions. In different conditions of video length, computational procedures or feature selection, timings would proportionally scale. The traditional approach would eventually remain the slowest, while the coordinates projection based would always result to be the fastest.

The Training and Testing processes exploit two different methods. The first one is the Classification Learner of Matlab, available as toolbox already implemented and accessible from the toolbar of the main window. The second one is an experimental Random Forest developed by the University of Trento [19].

MATLAB Classification Learner. The training procedure of the Matlab Classification Learner asks the user, at first, to select a validation strategy. Due to the large amount of samples – rows of the dataset – the "Cross out Classification" and the "No Classification" options have been discarded, while the "Hold Out Selection"[7] appeared to be the most suitable one. This choice allows selecting the percentage of samples that the trained classifier uses for the validation. This subgroup of samples does not take part to the training, but provides the entries for the validation procedure.

Following, the user is further asked to select the list of features to use for the training. This is useful to experiment with different combination of features or to simply discard a subset if is already known that some features are less relevant with respect to the others, and therefore can be neglected.

Once the Classification Learner has loaded the dataset, it is possible to choose the Classification Algorithm, selecting it from a list of available ones. We chose to use the Fine Tree algorithm, because we wanted to compare a basic decision Tree

---

[7] https://www.mathworks.com/help/stats/select-data-and-validation-for-classificatio n-problem.html – Apr 30, 2019.

based classifier with the sigma-z Random Forest developed by the University of Trento [19].

After selecting the classifier algorithm, the training process begins, followed by validation and testing. The training uses the features measurements to learn how to classify the action. The validation set is used during the training to help adjusting the hyper parameters and improve the performance of the classifier. Finally, the real performance of the classifier is measured on the test set.

Using Matlab Classification Learner, Validation and Testing process are quite different to each other. During our tests, the validation percentage returned often very high accuracy levels, between 98% and 100%. This means that the classification algorithms works extremely well on video of the same type. However, even when training on synthetic videos, test is performed on live videos. For this reason, testing process results to be far more significant because it represent how good the VE data emulate the Real Environment ones.

Random Forests. The Sigma-z Random Forest, proposed by the University of Trento, works differently. First of all, it is important to introduce that this Random Forest does not process data in the same way pursued by traditional Random Forests [19]. In fact, it relies on a probability computation analyzing two main aspects. The first one is the population uncertainty, computed considering the variability over the whole dataset, the second one is the sampling uncertainty, which reflects the uncertainty related to the new entry measurement (through the variance computed over the five frames of the feature computation). These two aspects, combined together, determine the probability for a certain new entry to belong to a specific class. In this way, it is possible to provide also a confidence level for each new entry classification in addition to the accuracy level, which is usually reported after every testing procedure. This remarkable aspect allows knowing the percentage of confidence the classifier is providing while every new entry is classified, in other words "how much it is sure". The Random Forest does not accomplish the validation on a subgroup of samples of the training dataset, but it directly pursues the validation using the testing dataset. Hence, in the Random Forest case, Validation and Testing are actually the same process.

Results. We decided to report the results of the Random Forest proposed by the University of Trento, because it returned an outcome which is slightly better with respect to the Classification Learner of Matlab. Each one of the following tables is an average result of multiple testing sessions conducted separately with different actors. Tables 3, 4 and 5 report the confusion matrix and the overall accuracy achieved during the testing procedure over the four cooking actions. Table 3 presents the results obtained from classifier trained with real environment videos (Traditional approach, path 1). Tables 4 and 5 report the results obtained training the classifier with a virtual environment through video rendering (path 2) and avatar joint coordinates projection (path 3), respectively.

Table 3. Confusion matrix of the classifier trained with the **traditional approach**, using real environment videos

True\Predicted	Drinking	Mixing	Grabbing	Pouring
Drinking	94.27%	1.67%	1.32%	2.74%
Mixing	13.38%	68.79%	16.36%	1.47%
Grabbing	0.96%	23.45%	71.47%	4.12%
Pouring	10.16%	1.87%	2.71%	85.26%
Average accuracy	79.94%			

Table 4. Confusion matrix of the classifier trained with the **proposed approach, path 2**, using video rendering.

True\Predicted	Drinking	Mixing	Grabbing	Pouring
Drinking	96.02%	1.3%	0.25%	2.43%
Mixing	1.03%	59.02%	37.15%	2.80%
Grabbing	0.91%	17.28%	81.16%	0.65%
Pouring	1.1%	1.7%	3.08%	94.12%
Average accuracy	82.58%			

It is possible to state that the reached accuracy is averagely over 80%. During the multiple training and testing trials the average accuracy mainly fluctuated between 71% and 85%.

Both the approaches based on VE (Tables 4 and 5) returned results performances comparable with the traditional approach. Moreover, there is not a conspicuous difference between the two VE based approaches.

6   Discussion

Even if average performances of the proposed method results to be definitely promising, sometimes the classifier has difficulties in classifying a specific class, lowering the session performances around 70%, while usually state of art accuracy values results to be higher than 80% [2,6]. Higher accuracy resulted difficult to pursue mainly due to several limitations which occurred during data acquisition and feature selection. Even if Open Pose skeleton mismatching has been compensated through data filtering – selection based on the joint acquisition confidence value, provided by Open Pose – the 2D spatial characterization between actions has been hardly achieved.

The cooking group of actions characterizes each gesture with geometrical features, such as the rotation of the wrist or the hand opening/closure tracking through the relative position between fingers. However, these intuitive features are quite difficult to monitor properly from a 2D flat screen. This problem is further increased due to the prospective point of view. Hence, while the five hand

Table 5. Confusion matrix of the classifier trained with the **proposed approach, path 3**, using projected coordinates.

True\Predicted	Drinking	Mixing	Grabbing	Pouring
Drinking	92.14%	0.74%	1.15%	5.97%
Mixing	1.79%	57.98%	38.28%	1.95%
Grabbing	1.0%	16.19%	82.16%	0.65%
Pouring	1.5%	1.2%	2.97%	94.33%
Average accuracy	81.65%			

position group of gesture – just used for calibration purposes and, therefore, not widely reported – has been better classified by x, y features, the second group (of cooking actions) returned better results comparing a much larger number of features. These features, as mentioned before, are the relative distances between non consecutive joints, angles between joints, and their average difference between two consecutive frames (within the interval of five frames).

In fact, the lack of a depth component affected considerably this second group testing procedure, as soon as cooking actions discernment relied on features less distinct and less trackable in a 2D context.

It is intuitive to understand–and possible to observe–that the accuracy level is strictly related to the feasibility of the Avatar gestures features with respect to the real ones that the actor performs. For this reason, a meticulous iterative analysis has been performed over the gesture evolution in order to understand, once for all, how to realize a natural movement. The group of the five hand position helped remarkably in this process, because of the simplicity of the action allowed to focus more deeply on the joint representation aspects. Furthermore, translation and scaling augmentation procedures played an important role in coordinate feasibility, forcing the similarity between the skeleton structures of the real and the virtual environments. As it is possible to state from the images of Fig. 10, all along the four cooking actions performance, the sequence of the avatar joints render a skeleton structure indistinguishable from the actors' ones.

7   Conclusions

This paper presented a novel approach to train machine learning systems for the classification of actions performed by humans in front of video cameras. Instead of training on a multitude of live-recorded real-world videos, we proposed to realize a limited amount of live videos and manually translate them in to 3D animations of a virtual character. The few hand-crafted 3D animations are then augmented via procedural manipulation of their animation curves.

Tests on real, live videos showed comparable results between the two training approaches. Hence, despite the action recognition accuracy (81%–84% with the classifiers used in this study) being far from human-level, this research allowed us to show the versatility and the performances of the virtual training.

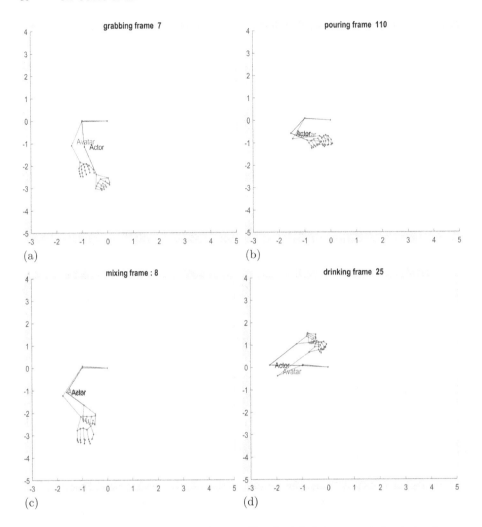

Fig. 10. Comparison between the Avatar (Synthetic) skeleton and the Actors' one.

Applied on a large scale set of different actions, the proposed approach can contribute in significantly cutting the time needed for the collection of training material and, consequently, lower production costs, yet with no loss in performances.

As a matter of improving the classification accuracy, it is important to remember that the entire work has been pursued based on a 2D analysis. It is thus reasonable to believe that performances could further increase if an extension to the 3D analysis – for example using a Microsoft Kinect – is accomplished. This aspect would lead to a more accurate similarity between the real and the virtual skeletons and to more precise information regarding the joint coordinates position and rotation.

It is also important to remark that the basic animation relies on the rotoscoping of one single person, while the testing procedure has been conducted over a sample of four people. A more general approach – which should involve several different people, each one with different personal movement style – may encounter more difficulties in discriminating gestures with similar posture patterns, such as mixing and pouring, and would lead to a lower accuracy estimation, independently from the followed path (live or synthetic).

As a possible improvement, it would be also possible to skip skeleton recognition and rely on deep convolutional neural networks for an end-to-end mapping between frame pixels and action class. This would include, for example, an object recognition algorithm. The object recognition would support the gesture recognition one, helping the selection of the candidate action in relation to the detected object.

References

1. Nilsson, M.: Action and intention recognition in human interaction with autonomous vehicles (2015)
2. D'Agostini, J., et al.: An augmented reality virtual assistant to help mild cognitive impaired users in cooking a system able to recognize the user status and personalize the support. In: 2018 Workshop on Metrology for Industry 4.0 and IoT, pp. 12–17. IEEE (2018)
3. Dariush, B., Fujimura, K., Sakagami, Y.: Vision based human activity recognition and monitoring system for guided virtual rehabilitation. US Patent App. 12/873,498, 3 March 2011
4. Gorecky, D., Schmitt, M., Loskyll, M., Zühlke, D.: Human-machine-interaction in the industry 4.0 era. In: 2014 12th IEEE International Conference on Industrial Informatics (INDIN), pp. 289–294. IEEE (2014)
5. Mizumoto, T., Fornaser, A., Suwa, H., Yasumoto, K., De Cecco, M.: Kinect-based micro-behavior sensing system for learning the smart assistance with human subjects inside their homes. In: 2018 Workshop on Metrology for Industry 4.0 and IoT, pp. 1–6. IEEE (2018)
6. Bloom, V., Makris, D., Argyriou, V.: G3D: a gaming action dataset and real time action recognition evaluation framework. In: 2012 IEEE Computer Society Conference on Computer Vision and Pattern Recognition Workshops (CVPRW), pp. 7–12. IEEE (2012)
7. Papadopoulos, G.T., Axenopoulos, A., Daras, P.: Real-time skeleton-tracking-based human action recognition using kinect data. In: Gurrin, C., Hopfgartner, F., Hurst, W., Johansen, H., Lee, H., O'Connor, N. (eds.) MMM 2014. LNCS, vol. 8325, pp. 473–483. Springer, Cham (2014). https://doi.org/10.1007/978-3-319-04114-8_40
8. Wang, C., Wang, Y., Yuille, A.L.: An approach to pose-based action recognition. In: Proceedings of the IEEE Conference on Computer Vision and Pattern Recognition, pp. 915–922 (2013)
9. Li, W., Zhang, Z., Liu, Z.: Action recognition based on a bag of 3D points. In: 2010 IEEE Computer Society Conference on Computer Vision and Pattern Recognition Workshops (CVPRW), pp. 9–14. IEEE (2010)

10. Blank, M., Gorelick, L., Shechtman, E., Irani, M., Basri, R.: Actions as space-time shapes, pp. 1395–1402. IEEE (2005)
11. Schuldt, C., Laptev, I., Caputo, B.: Recognizing human actions: a local SVM approach. In: Proceedings of the 17th International Conference on Pattern Recognition, ICPR 2004, vol. 3, pp. 32–36. IEEE (2004)
12. Weinland, D., Ronfard, R., Boyer, E.: Free viewpoint action recognition using motion history volumes. Comput. Vis. Image Underst. **104**(2–3), 249–257 (2006)
13. Soomro, K., Zamir, A.R., Shah, M.: Ucf101: a dataset of 101 human actions classes from videos in the wild. arXiv preprint arXiv:1212.0402 (2012)
14. Poppe, R.: A survey on vision-based human action recognition. Image Vis. Comput. **28**(6), 976–990 (2010)
15. Marin, J., Vázquez, D., Gerónimo, D., López, A.M.: Learning appearance in virtual scenarios for pedestrian detection. In: 2010 IEEE Conference on Computer Vision and Pattern Recognition (CVPR), pp. 137–144. IEEE (2010)
16. Abbondanza, P., Giancola, S., Sala, R., Tarabini, M.: Accuracy of the microsoft kinect system in the identification of the body posture. In: Perego, P., Andreoni, G., Rizzo, G. (eds.) MobiHealth 2016. LNICST, vol. 192, pp. 289–296. Springer, Cham (2017). https://doi.org/10.1007/978-3-319-58877-3_37
17. Lie, W.-N., Le, A.T., Lin, G.-H.: Human fall-down event detection based on 2D skeletons and deep learning approach. In: 2018 International Workshop on Advanced Image Technology (IWAIT), pp. 1–4. IEEE (2018)
18. Simon, T., Wei, S.-E., Joo, H., Sheikh, Y., Hidalgo, G., Cao, Z.: OpenPose (2018)
19. Fornaser, A., De Cecco, M., Bosetti, P., Mizumoto, T., Yasumoto, K.: Sigma-z random forest, classification and confidence. Meas. Sci. Technol. **30**(2), 025002 (2018)
20. Fornaser, A., Mizumoto, T., Suwa, H., Yasumoto, K., De Cecco, M.: The influence of measurements and feature types in automatic micro-behavior recognition in meal preparation. IEEE Instrum. Meas. Mag. **21**(6), 10–14 (2018)
21. Yoo, J.-H., Hwang, D., Nixon, M.S.: Gender classification in human gait using support vector machine. In: Blanc-Talon, J., Philips, W., Popescu, D., Scheunders, P. (eds.) ACIVS 2005. LNCS, vol. 3708, pp. 138–145. Springer, Heidelberg (2005). https://doi.org/10.1007/11558484_18

Debugging Quadrocopter Trajectories in Mixed Reality

Burkhard Hoppenstedt[1]([✉]), Thomas Witte[2], Jona Ruof[2], Klaus Kammerer[1], Matthias Tichy[2], Manfred Reichert[1], and Rüdiger Pryss[1]

[1] Institute of Databases and Information Systems, Ulm University, Ulm, Germany
burkhard.hoppenstedt@uni-ulm.de
[2] Institute of Software Engineering and Programming Languages, Ulm University, Ulm, Germany

Abstract. Debugging and monitoring robotic applications is a very intricate and error-prone task. To this end, we propose a mixed-reality approach to facilitate this process along a concrete scenario. We connected the Microsoft HoloLens smart glass to the Robot Operating System (ROS), which is used to control robots, and visualize arbitrary flight data of a quadrocopter. Hereby, we display holograms correctly in the real world based on a conversion of the internal tracking coordinates into coordinates provided by a motion capturing system. Moreover, we describe the synchronization process of the internal tracking with the motion capturing. Altogether, the combination of the HoloLens and the external tracking system shows promising preliminary results. Moreover, our approach can be extended to directly manipulate source code through its mixed-reality visualization and offers new interaction methods to debug and develop robotic applications.

Keywords: Quadrocopter · Mixed Reality · Robot Operating System · 3D trajectories

1 Introduction

In recent years, quadrocopters became affordable for a wider audience and were successfully used for commercial use cases, such as delivery [7,22]. The autonomous operation of quadrocopters requires complex approaches, such as planning of *flight paths* [13], *coordinated actions* [20], or *collision avoidance* [10]. A similar development took place in the field of augmented reality. Especially smart glasses provide more features and offer a better interaction with the real world. The development efforts for quadrocopter software are very high and we see potential in utilizing augmented reality to assist the development of quadrocopter software. Augmented reality has proven to be a suitable approach for complex tasks, such as machine control [19]. In this work, we connected a quadrocopter via the *Robot Operating System* (ROS) to a Microsoft HoloLens. In the smart glass, we can see all the trajectories of the flight plan. Using the proposed

© Springer Nature Switzerland AG 2019
L. T. De Paolis and P. Bourdot (Eds.): AVR 2019, LNCS 11614, pp. 43–50, 2019.
https://doi.org/10.1007/978-3-030-25999-0_4

framework, it is possible to receive an instant 3D feedback for the programming of a planned path and the method can be seen as *visual 3D programming* [12].

The remainder of this paper is structured as follows: Sect. 2 discusses related work, while Sect. 3 introduces the background for quadrocopters, Mixed Reality, and the Robot Operating System. In Sect. 4, the developed prototype is presented, in which the mixed-reality application and the processing pipeline are presented. Threats to validity are presented in Sect. 5, whereas Sect. 6 concludes the paper with a summary and an outlook.

2   Related Work

Quadrocopters have been already tested in the context of Mixed Reality [14]. Hereby, the simplification of debugging, the elimination of safety risks, and the easier development of swarm behavior by complementing physically existing robots with simulated ones were named as outstanding advantages. Furthermore, a study for the programming of a small two-wheeled robot [16] revealed no impact on the learning performance when using Augmented Reality (AR). The debugging of quadrocopters with a fixed camera in combination with LEDs is presented by [11]. In this approach, autonomously driving robots are operated in a testing environment, and a camera stream is overlapped with AR data. Finally, the use of mobile LED projectors for easily understandable visualizations of robot data is presented by [15]. However, the projection is limited to 2D data and might be insufficient for complex flight paths of quadrocopters. However, to the best of our knowledge, a combination of technologies as shown in this work, has not been presented in other works so far.

3   Fundamentals

3.1   Robot Operating System

The Robot Operating System (ROS) [18] is a *meta operating system* designed for the development of robot-specific applications. It is not classified as a standard operating system, it is rather a collection of programs. ROS is basically designed as a *peer-to-peer* architecture and supports the *publish and subscribe pattern* (see Fig. 1). A central instance, denoted as master, provides a central lookup directory and organizes the connections between different programs. The clients of ROS are called *nodes*, and communicate by publishing messages in a YAML-like [8] syntax. All messages are send to so-called *topics*, which can be seen as a black board for messages. This concept allows the asynchronous transfer of large amounts of data.

3.2   Mixed Reality

We used the Microsoft HoloLens to utilize mixed-reality concepts. Mixed reality, in turn, is a concept in which the real environment and the virtual environment

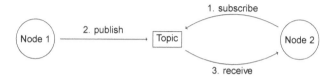

Fig. 1. Publish and subscribe principle of the Robot Operating System

are combined. Concerning the *Virtuality Continuum* [17], mixed reality has the higher intersection of reality and virtuality. More specifically, the HoloLens uses *spatial mapping* to create a virtual model of the real environment. Therefore, interactions of holograms and real-world objects become possible. Mixed Reality is widely used in the contexts of industrial maintenance, medicine, and architecture. The HoloLens is equipped with various sensors, such as a depth sensor, a RGB Camera, and an ambient light sensor. Infinite projections are not possible, neither to the distance nor to the proximity. With a weight of 579 g, the HoloLens may cause dizziness when used for a longer period of time. In an intensive use, the battery lasts for about 2.5 h, which inhibits a long-time usage.

3.3 Tracking

In our architecture, the HoloLens is tracked with a *motion capturing system*, which uses an array of infrared emitting and receiving cameras. The HoloLens is equipped with a set of three (at least) reflective markers (see Fig. 2, right side). The markers are placed asymmetrically, since a symmetric marker placement would yield an ambiguous tracking position. Furthermore, the HoloLens provides its own tracking, which is relative to the starting pose. The motion capturing system, in turn, provides an accuracy on millimeter level and is not susceptible to drifts due to *dead reckoning*, which outperforms the internal HoloLens tracking (see Fig. 2, left). After a calibration, the pose data is sent to the ROS as a *virtual reality peripheral network stream* [23].

4   System Overview

Using a motion capturing system with 16 cameras, the system can detect, identify, and track objects with high precision. The motion capturing software then makes this information available to the ROS through a continuous stream of pose data. Inside the ROS network, the decentral and dynamic nature of ROS makes it possible to connect components and other applications, such as the control of a quadcopter swarm, in a flexible way. The motion data can, for example, be processed to estimate the velocity and acceleration of objects, which is of paramount importance for the motion planning and control of quadcopters. This amount of different data sources and objects might be complex to grasp or understand, and is therefore often visualized using tools such as rviz [4]. We use the same data format as common rviz marker messages to feed our mixed-reality

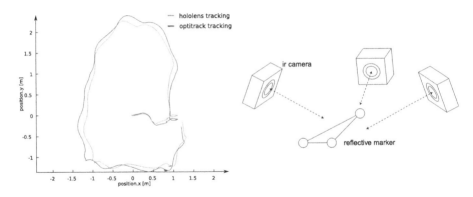

Fig. 2. Tracking accuracy (left), Marker reflection (right)

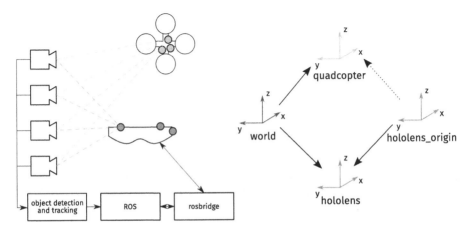

Fig. 3. Architecture of the system (left). Coordinate systems and transformations (right).

overlay. This enables to use the HoloLens as a plug and play replacement for rviz and connect the 3D visualizations directly to the objects that provide the data. The communication between the HoloLens and the ROS is realized through a rosbridge [5]. The latter translates the TCPROS [6] message passing protocol to a websocket-based protocol. The messages are then decoded using a HoloLens rosbrige [1], interpreted, and eventually displayed on the HoloLens.

The HoloLens uses its own motion tracking [2], including internal cameras and its Inertial Measurement Unit (IMU) to localize itself and keep the visual overlay stable with respect to the environment. The reference frame of this internal tracking is not directly available to the ROS components of our system. We can only observe the tracked position of the HoloLens itself (in the static *world* frame of the motion capturing system) and communicate the internal (estimated) position back to the ROS network (see Fig. 3). Using these two coordinate transformations, we are able to calculate the position of arbitrary tracked objects with respect to the internal HoloLens reference frame *hololens_origin*.

This setup is highly sensitive to timing. The transmission of data through different hosts over a network as well as inside the ROS network introduces a significant delay to the pose data that is used to transform the object positions into the mixed-reality overlay. An additional delay is related to the tracking of objects and the HoloLens itself. Figure 4 shows a time difference in the tracking data, received from the HoloLens and the motion capturing system in ROS of at least 0.1s. This results in a shifted or delayed positioning of the overlay, which is especially visible during fast head turns or fast movements. We currently do not estimate or compensate this delay as it would be necessary to synchronize clocks across multiple systems. Instead, we update the *HoloLens_origin* frame only sporadically during periods of low head movements. This way, the internal tracking of the HoloLens can keep the overlay stable, and the positioning of the overlay is corrected when timing is less critical.

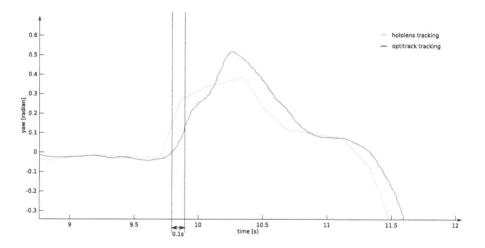

Fig. 4. Time difference of the received pose data in ROS.

The system was tested in the quadrocopter lab at Ulm University. An area of approximately $9.5 \times 5.5 \times 4\,\mathrm{m}$ is used as the room for the quadrocopters to operate and make maneuvers. The *Optitrack* [3] tracking system is installed closely to the ceiling of the lab. The HoloLens has been made trackable with markers, so that the motion capturing system can deliver a continuos stream of pose information. This stream is sent to a computer in the lab, running an instance of ROS. Finally, a ROS visualization is sent back to the HoloLens. This message contains all the information about the current visualization. It further includes lists of all objects with their location and required changes (e.g., modify, delete, insert). The HoloLens parses this messages and transfers its content into the internal object representation. The visualisation message can be seen as a toolbox to build arbitrary complex scenes from the provided basic components (see Fig. 5). A flight path can then be visualized using, e.g., a line strip or a list of points, while arrows can be used to visualize the desired acceleration and velocity vectors at a waypoint.

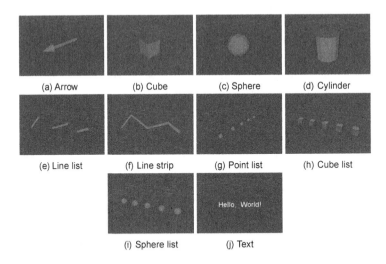

Fig. 5. Basic 3D components

5   Threats to Validity

The following limitations of the presented approach need to be discussed. First, the approach has not been tested in a study yet. Thus, it is unclear if users benefit from the immersive representation of quadrocopter flight paths in terms of speed or usability. Second, the approach relies heavily on a robust network connection, therefore network problems will negatively influence the application behavior. In general, network limitations might reduce the performance of animations and the general application framerate. Third, the system is designed under laboratory conditions and cannot be applied outside unless a marker-based environment is established.

6   Summary and Outlook

We presented a prototype for visualizing data of a quadrocopter system (e.g., quadcopter trajectories) in Mixed Reality. The communication is realized using the Robot Operating System. Different 3D objects serve as a toolkit to build visualizations in the context of quadrocopter flight plan programming. In a next step, we will extend the approach from the monitoring to the interaction level.

Using the concept of *source origin tracing*, as used, e.g., by [9], it is possible to directly manipulate constants in source code through the visualization. For example, when using a tap gesture, the position of an object in the room could be modified as well as the attached source code. Therefore, the user of the application may program his or her quadrocopter through immersive interactions with the displayed holograms. As the proposed approach contains novel user interaction patterns, it should be evaluated in a usability study (cf. [21]) to identify the mental load during its use. In this context, the speed advantage compared to a traditional approach should be evaluated. Altogether, this work has shown that the combination of the Robot Operating System and Mixed Reality is a promising strategy.

References

1. HoloLens-Rosbridge. https://github.com/roastedpork/hololens-rosbridge. Accessed 13 Feb 2019
2. HoloLens Tracking-System. https://docs.microsoft.com/en-us/windows/mixed-reality/enthusiast-guide/tracking-system. Accessed 13 Feb 2019
3. Optitrack. http://optitrack.com. Accessed 15 Feb 2019
4. ROS Vizualization rviz. http://wiki.ros.org/rviz. Accessed 13 Feb 2019
5. ROSBridge. http://wiki.ros.org/rosbridge_suite. Accessed 13 Feb 2019
6. TCPROS. http://wiki.ros.org/ROS/TCPROS. Accessed 15 Feb 2019
7. Bamburry, D.: Drones: Designed for product delivery. Des. Manag. Rev. **26**(1), 40–48 (2015)
8. Ben-Kiki, O., Evans, C., Ingerson, B.: Yaml ain't markup language (yaml^{TM}) version 1.1. Technical report, p. 23 (2005). yaml.org
9. Breckel, A., Tichy, M.: Live programming with code portals. In: Workshop on Live Programming Systems - LIVE 2016 (2016)
10. Gageik, N., Müller, T., Montenegro, S.: Obstacle detection and collision avoidance using ultrasonic distance sensors for an autonomous quadrocopter. University of Wurzburg, Aerospace information Technologhy (Germany) Wurzburg, pp. 3–23 (2012)
11. Ghiringhelli, F., Guzzi, J., Di Caro, G.A., Caglioti, V., Gambardella, L.M., Giusti, A.: Interactive augmented reality for understanding and analyzing multi-robot systems. In: 2014 IEEE/RSJ International Conference on Intelligent Robots and Systems (IROS 2014), pp. 1195–1201. IEEE (2014)
12. Green, T.R.G., Petre, M., et al.: Usability analysis of visual programming environments: a 'cognitive dimensions' framework. J. Vis. Lang. Comput. **7**(2), 131–174 (1996)
13. Hehn, M., D'Andrea, R.: Quadrocopter trajectory generation and control. In: IFAC World Congress, vol. 18, pp. 1485–1491 (2011)
14. Hoenig, W., Milanes, C., Scaria, L., Phan, T., Bolas, M., Ayanian, N.: Mixed reality for robotics. In: 2015 IEEE/RSJ International Conference on Intelligent Robots and Systems (IROS), pp. 5382–5387. IEEE (2015)
15. Leutert, F., Herrmann, C., Schilling, K.: A spatial augmented reality system for intuitive display of robotic data. In: Proceedings of the 8th ACM/IEEE International Conference on Human-Robot Interaction, pp. 179–180. IEEE Press (2013)

16. Magnenat, S., Ben-Ari, M., Klinger, S., Sumner, R.W.: Enhancing robot programming with visual feedback and augmented reality. In: Proceedings of the 2015 ACM Conference on Innovation and Technology in Computer Science Education, pp. 153–158. ACM (2015)

17. Milgram, P., Takemura, H., Utsumi, A., Kishino, F.: Augmented reality: a class of displays on the reality-virtuality continuum. In: Telemanipulator and Telepresence Technologies, vol. 2351, pp. 282–293. International Society for Optics and Photonics (1995)

18. Quigley, M., et al.: ROS: an open-source robot operating system. In: ICRA Workshop on Open Source Software, Kobe, Japan, vol. 3, p. 5 (2009)

19. Ralston, S.E.: Augmented vision for survey work and machine control, US Patent 6,094,625, 25 July 2000

20. Ritz, R., Müller, M.W., Hehn, M., D'Andrea, R.: Cooperative quadrocopter ball throwing and catching. In: 2012 IEEE/RSJ International Conference on Intelligent Robots and Systems (IROS), pp. 4972–4978. IEEE (2012)

21. Schobel, J., Pryss, R., Probst, T., Schlee, W., Schickler, M., Reichert, M.: Learnability of a configurator empowering end users to create mobile data collection instruments: usability study. JMIR mHealth uHealth **6**(6) (2018)

22. Stolaroff, J.: The need for a life cycle assessment of drone-based commercial package delivery. Technical report, Lawrence Livermore National Laboratory (LLNL), Livermore, CA (United States) (2014)

23. Taylor II, R.M., Hudson, T.C., Seeger, A., Weber, H., Juliano, J., Helser, A.T.: VRPN: a device-independent, network-transparent VR peripheral system. In: Proceedings of the ACM Symposium on Virtual Reality Software and Technology, pp. 55–61. ACM (2001)

Engaging Citizens with Urban Planning Using City Blocks, a Mixed Reality Design and Visualisation Platform

Lee Kent[(⊠)], Chris Snider, and Ben Hicks

University of Bristol, Bristol, UK
lee.kent@bristol.ac.uk

Abstract. This paper looks at city scale visualisation during concept generation as a method to engage citizens with city planning. A Virtual Reality based platform is proposed, prototyped and piloted. The platform, City Blocks, utilises an abstracted physical design kit and a generated, explorable digital twin to visualise and analyse concepts. It is found that through the Virtual Reality based platform, later design stage visualisation of ideas can be brought to concept development and that citizens are able to reason and refine designs via an abstracted physical twin. The paper presents an analysis of the 149 city designs, including a comparison between citizens and experienced urban planners.

1 Introduction

During any design activity with integrated stakeholder involvement, it is crucial to foster the conversation element, utilising appropriate tools to decipher concepts and translate terminology. An accessible and deskilled design activity empowers end-users and stakeholders to contribute meaningfully much earlier in the process [1, 2]. Urban planning is a complex domain with many uncompromising constraints, stakeholders and requirements.

'Inclusive and sustainable urbanisation and capacity for participatory, integrated and sustainable human settlement planning and management' is recognised by the UN and is a Sustainable Cities and Communities target [3]. With 54% of the global population currently living in urban areas, a percentage expected to increase to 66% by 2050 [4], there is a global drive to empower citizens to contribute to the planning of their communities.

Digital Twinning is the synchronous pairing of a physical asset to a virtual model to simulate and predict through-life behaviours [5]. This relationship can be one to one or one to many, as a CAD model can be physically manufactured several times with each instance subject to intrinsic variances introduced in the manufacturing process. The object will be subject to different uses, loads, environments, and the properties of the object will change as a result. These variations can be captured using a variety of modern metrological techniques and used to produce a digital, through-life counterpart that evolves in tandem with the physical object. It has been proposed to scale digital twins up to factory floor [6, 7], University campus [8] and even city wide with several

© Springer Nature Switzerland AG 2019
L. T. De Paolis and P. Bourdot (Eds.): AVR 2019, LNCS 11614, pp. 51–62, 2019.
https://doi.org/10.1007/978-3-030-25999-0_5

smart city initiatives attempting to interconnect digital and physical data at a city scale [9–11].

There are several initiatives utilising design by play as a method for designing public spaces and iterative city development. The innately creative video game Minecraft has been successfully demonstrated as a tool to reimagine spaces. In Nairobi, Block by Block [12] helped players design a space that they would feel safe walking through at night. On a larger scale, the entirety of Stockholm was virtualized without all of its buildings in the project Blockholm [13]. The public could then reimagine the city, one city block at a time. During a series of workshops in London, a community was invited to reimagine their local pedestrian space using items made by local school children that were scanned and placed into a model of the park using the Unreal Engine [14]. The designs were explored using Virtual Reality to understand the use of space as they designed the park and as a platform to present ideas to their peers

By combining the benefits of modern visualisation techniques, tangible objects, and immersive Virtual Reality, a novel platform, City Blocks, has been created to support citizen engagement with city planning. Using Lego; a city can be designed where the colour of each brick represents common elements of a city. The design is scanned, and a three-dimensional city is generated at a city scale providing a virtualised likeness seen in Fig. 1 or early stage digital twin, of the physical concept that can be explored in Virtual Reality. The city is analysed, with tangible and numerical requirements extracted without participant technical understanding of the city planning process or expert terminology. Utilising the platform, the following proposition can be addressed, and is broken up into three key discussion points.

Fig. 1. Left: The City Blocks template transitioning to Right: In simulation visualisation

Is the digital-physical paradigm able to support city planning from concept generation and is there any value in having a digital twin of an idea?

1. Can quantifiable and beneficial data be obtained from the digital twin?
2. Does the platform encourage citizens to engage with city planning?

3. Is Virtual Reality an appropriate technology to help a non-expert designer to visualise a concept?

The pilot of the platform, City Blocks, was scheduled to be exhibited at a local housing festival, a community-driven initiative supported by local council, and open to members of the public. Due to the nature of the exhibition, players could spend as much or as little time interacting with the platform, able to disengage with the activity at any point. Figure 2 presents the player experience, with decision points in which the citizen could choose to what extent they engaged with the platform, if at all.

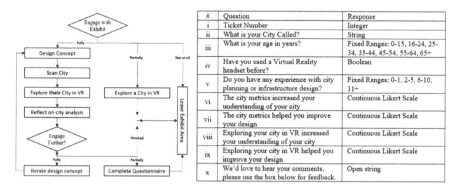

#	Question	Response
i	Ticket Number	Integer
ii	What is your City Called?	String
iii	What is your age in years?	Fixed Ranges: 0-15, 16-24, 25-34, 35-44, 45-54, 55-64, 65+
iv	Have you used a Virtual Reality headset before?	Boolean
v	Do you have any experience with city planning or infrastructure design?	Fixed Ranges: 0-1, 2-5, 6-10, 11+
vi	The city metrics increased your understanding of your city	Continuous Likert Scale
vii	The city metrics helped you improve your design	Continuous Likert Scale
viii	Exploring your city in VR increased your understanding of your city	Continuous Likert Scale
ix	Exploring your city in VR helped you improve your design	Continuous Likert Scale
x	We'd love to hear your comments, please use the box below for feedback.	Open string

Fig. 2. Left: The user engagement pathway. Right: Questionnaire items and response options

2 City Blocks

The City Blocks platform consists of two separate pieces of software, working in parallel on the same machine. The first, a Python application, uses a webcam and image recognition algorithms to track the citizens Lego configuration. The second, an application created using the Unreal Engine, generates a 3D environment to be explored in Virtual Reality. To explore the potential of City Blocks, a study was conducted with members of the public as participants. On engagement with the activity, the citizen becomes a 'player' of City Blocks. Figure 3 shows the complete system diagram, with each of the following subsections corresponding to the outlined groups. An optional questionnaire was also available for City Blocks players, with items shown in Fig. 2.

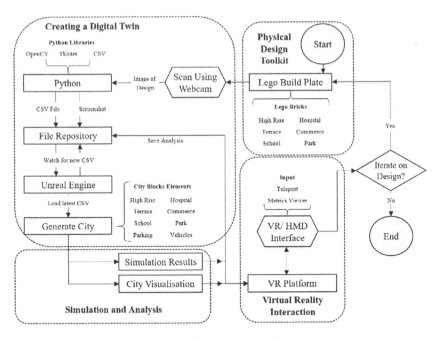

Fig. 3. City Blocks system diagram

2.1 Physical Design Toolkit

Using physical design kits, spatial constraints can be investigated [15], systematically or exploratively. Lego, seeming a heavily constraining design tool, can provide a rich, tangible interface with six standard, four by two stud Lego bricks having 915,103,765 build permutations [16]. This playful interface, when combined with a set of rules, has been linked to having a significant effect on design variation [17]. Players created a single block high city on a 16×16 stud two-dimensional board. The pilot study utilised six colours with a seventh element being parking in the absence of a block. The grid had a pink Lego brick placed at each corner, outside of the build area, used for tracking and localising of the build plate.

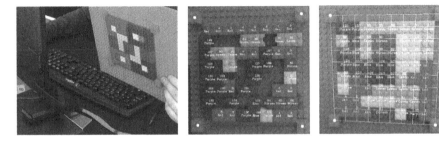

Fig. 4. Left: A city design being scanned Centre/Right: Colour recognition algorithm output

2.2   Creating the Digital Twin

On completion, the design was scanned using a high definition webcam and analysed using the Python library OpenCV [18] to determine the brick configuration. This was achieved by finding the four pink bricks placed at the corners outside of the build area. Once the four corners had been found an internal grid was laid over the build area that was skewed, scaled and rotated to match the relative transform of the physical city design. The grid was looking for patterns of 2 × 2 square bricks, meaning 64 measurements were taken, as opposed to finding every stud. This was a practical constraint due to the Lego pieces obstructing visibility to the ground plate. At each intersection of the grid, the average colour was taken and compared against the colour palette and a two-dimensional array of colours was formed. The array is saved locally as a CSV file. The algorithm would not compensate if the player placed Lego pieces at odd intervals, i.e. starting at the second stud or using 3 × 2 sized bricks. This can be seen in the top row of the centre image in Fig. 4. Due to this, there is some margin of error with the scanning process and a more robust scanning algorithm is required for future versions.

The application built with the Unreal Engine, detects the CSV of a new city design, the game engine generates a three-dimensional city reflecting the player's concept. Each element in the scanned two-dimensional array is assigned to its equivalent space in the virtual world and is contextualised depending on its colour. Each city block has procedurally generated elements, reducing the likelihood of repeated blocks looking and feeling the same.

2.3   Simulation and Analysis

Each digital city can be analysed both individually in real time and collectively after the activity. The initial algorithms for the pilot were based on some basic principles and were deliberately obfuscated from players, and were agreed on by the festival

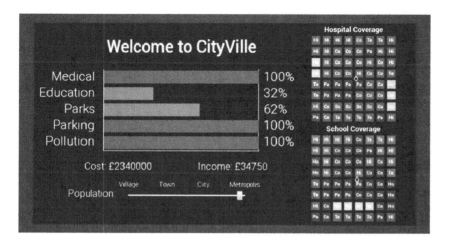

Fig. 5. Analysis and metrics provided to the player from within the simulation (Color figure online)

organisers hosting the pilot. This was to instil an element of discovery into understanding how their changes to their design would affect their results. The key metrics were given as percentage provision and covered medical, education, parks, parking, pollution, cost, income and population, as shown in Fig. 5. Education and medical coverage was analysed in more detail. Each school or hospital had a fixed range that it could provide to. Different buildings had different requirements, e.g. Parks did not need school coverage, and was presented to the user as a traffic light system, with problem areas highlighted in red.

2.4 Virtual Reality Interaction

There are two key interactions with the city simulation from within Virtual Reality. The first is travel via teleportation. Holding the teleport button on the Virtual Reality tracked motion controller projects an arc of significant reach from the controllers' position in the virtual environment. On release of the button, the player is transported instantly to the point of intersection of the arcs first collision with an object. This can be used to scale buildings, climb lampposts and even jump onto vehicles and can be used to explore the city from a variety of perspectives. The second key interaction is the metrics viewer. When a player looks at their wrist like looking at a watch, a screen is projected in from of them displaying the real-time analysis metrics. Figure 5 shows the information provided to the player through the metrics interactions.

3 Results

In total 149 cities were designed by 125 unique players during the pilot, with several choosing to iterate several times on their initial design. Of these 125, 31 City Blocks players also filled out the associated questionnaire and can be considered very engaged. Cities designed by questionnaire respondents can be analysed further, utilising the metadata gained about the player, for example being able to draw comparisons between inexperienced and expert city planners. Moving forward, results using the extra metadata will be referred to as respondents cities; alternatively it will be all of the city designs.

45% of questionnaire respondents were under 24, with 55% over. Five of the cities were designed by experienced city planners, four of whom had greater than six years' experience, four chose not to disclose their experience and the remaining 22 had zero to one year of experience. Previous Virtual Reality use was also considered an important factor in how players reacted to being able to explore their designs. The result could also point to how much the technology contributed to levels of engagement. It was found that 45% had not tried any form of Virtual Reality before while 48% had. The remaining players were unsure.

Players were asked to mark on a line how influential the digital twin analysis and use of Virtual Reality was to them and their designs. Table 1 shows how useful each player found the analysis and visualisation of their city as part of the (U) understanding of their design and (I) supported the improvement of their design. The continuous Likert scale option asked players to denote their answer with a cross on a 6.5 cm line from "Not at all" to "Significantly". The mark was then measured and given as a result from 0 to 6.5 with 6.5 being "Significantly". A slice of the results are presented in Table 1.

Table 1. Questionnaire results (U) = Understanding, (I) = Improve

Questionnaire item		Metrics		Virtual reality	
		(U)	(I)	(U)	(I)
Previous VR use	Yes (14)	4.4	4.1	5.2	4.6
	No (15)	4.6	4.9	5.0	4.6
	Blank(2)	4.6	4.8	5.6	5.2
Planning experience	0–1 (22)	4.5	4.5	5.0	4.6
	2+ (5)	4.8	5.4	5.0	4.5
	Blank(4)	4.8	5.4	5.0	4.5
All respondents	(31)	4.5	4.5	5.1	4.6

Using Content Analysis of the qualitative feedback section, (x) of the 31 questionnaires, the following themes were elicited; 5x blank, 13x positive experience, 6x platform improvement ideas, 9x acknowledgement of benefits to planning and 6x positive comments on the technology.

The average city had bricks covering 80% of the ground plate; spaces where no brick was detected defaulted to a small open-air car park in the visualisation. The centrepiece of the city, a 2×2 brick canvas, provides an interesting insight into what players consider important in their city centres. Table 2 shows a count of each time a brick was used as the centre of a city. Due to the scanning process being independent of orientation, data can only be compared when the cartesian transform of the brick is insignificant.

Table 2. Lego brick colours, assigned attributes and how often they were used in all 149 designs

Brick colour	Brick type	Avg per city	Centre usage	STDEV
Light Green	Park	20.0%	30.7%	12.7
Red	High Rise	12.8%	8.4%	8.2
Blue	Commercial	13.8%	13.7%	8.8
Purple	Terrace	14.2%	7.4%	9.1
Yellow	School	10.0%	10.8%	6.4
Dark Green	Hospital	10.0%	11.5%	6.4
No Brick	Parking	19.2%	17.4%	12.3

Two metrics, medical and education provision, were given as percentage scores. Every city block was given a requirement for whether it needed to be within a hospital or a school catchment area. This was relayed back to the player as a minimap from within the Virtual Reality headset. An example city is shown in Fig. 5. The black squares do not require hospital coverage, the green require medical provision and have it, the red squares require provision but are too far away, and the yellow are the hospitals themselves. The citizen players averaged 89% medical provision compared to the experienced planner average of 92.4%. Education, with schools having a much smaller catchment area, was more difficult to get a higher score and players got an average of 65.2% compared to the experienced planner average of 92.6%. Each brick colour has a large standard deviation (STDEV in Table 2), with Schools and Hospital having similar usage. Using the metadata from respondents in future refined studies outliers and correlated designs can be sought to define reasoning and requirements from the activity.

4   Discussion

The platform, City Blocks, was created to explore the potential application of early stage digital-physical twinning in the context of city planning. On completion of the public study, a series of conclusions can be drawn about digital-physical twinning, citizen engagement with city planning, Virtual Reality as an enabling technology and the platform itself as a planner's tool. Three components were proposed as discussion points to answer the following question:

Is the digital-physical paradigm able to support city planning from concept generation and is there any value in having a digital twin of an idea?

4.1   Analysing a Digital Twin of a Concept

The abstracted digital twin provided the opportunity to analyse physical designs, both in real time as part of the simulation and after the event, with each scan saved in several formats. This has been demonstrated through the use of the in-simulation metrics provided to the player and the city centre analysis provided in the results section. Table 2 shows analysis run after the exercise. The results from the questionnaire, as shown in Table 1, highlight that experienced city planners saw more use from the simulations and metrics provided to the player.

From the simulations, user requirements and statistics can be gathered directly or inferred. From the pilot there is evidence for players wanting parks for their city centres, followed by a need for commercial spaces. Comparisons between citizen players and experienced urban planner designs can be utilised to verify algorithms or be used to gather quantifiable requirements. The algorithms used in the pilot were for proof of concept purposes, to ensure meaningful and viable analysis can be obtained from a digital twin. Moving forward these will need to be tailored and developed on a case-by-case basis.

4.2 Encouraging Citizens to Engage with City Planning

The average city had 80% brick coverage, implying that the players completed their design. There was no minimum number of bricks or exemplar cities and players could disengage from the activity at any time they felt inclined. The gamification elements keeping citizens designing were the opportunity to obtain a better city with higher scores and the fact there was freedom to fail. These elements motivated some players to iterate and improve their designs with 13 choosing to create at least 2 iterations of their city concept. 29% of the open question (x) responses were positive comments about using City Blocks as a platform to support planning activities, with one user commenting "So fun! It made me think - I'd never have put myself in the shoes of a city planner except through play! Thank you". Other users saw the benefits for experienced planners, "Really good fun and great way for planners to create a perfect city". It should be noted there were no negative comments about the platform in the comments, leading towards potential demand characteristic bias, something to be addressed in further studies using the platform.

4.3 Virtual Reality to Visualise Concepts

It was found that having experience with Virtual Reality before engagement with the platform did not influence how players felt about the platform in a city planning context. The questionnaire results, Table 2, show that all of the respondents saw value in using Virtual Reality to help them both understand and improve their designs as opposed to the given metrics. This may stem from the ability to experience and explore scale and depth within a Virtual Reality environment, something that can only be represented monoscopically on standard computer monitors. One player commented with a desire to share the experience to help them understand what others thought of their design. "The ability to understand the city in a very real way was great. Maybe other people in the city would show how they interact with the city". Multiplayer and collaborative exploration would be an interesting addition to the platform, with designers able to give virtual tours of their digital twin, with discussion happening within the simulation as opposed to after the exploration within Virtual Reality.

The interactivity afforded by using Virtual Reality also has benefits over traditional visualisations. The player has the freedom to explore the digital twin simulation as they wish, able to teleport to the tops of moving vehicles, trees or buildings in search of new and novel perspectives of their city design. Being able to summon the analysis results using a simple gesture also added to the immersive experience. Players could look at their stats and observe their surroundings to deduce why they were given the results they were (Fig. 6). Due to the scale of the design, it was important to provide indicators to ensure players recognised the link between their physical Lego-based design and the digital twin visualisation they were immersed in. Each city block is a prefabricated digital space with procedurally placed exterior components and a floor matching the colour of the physical contextualised Lego brick. There was also a small arrow indicating the position and orientation of the player within the simulation.

There was an even split between those that had and had not experienced Virtual Reality before, implying the activities key engagement factor was not to try the Virtual

Reality hardware. As a result, conversations could be focused around community planning, with the technology enabling communication of and discussion around concepts. The hardware is perceived less as a novel technology and more as another tool in the visualization toolbox.

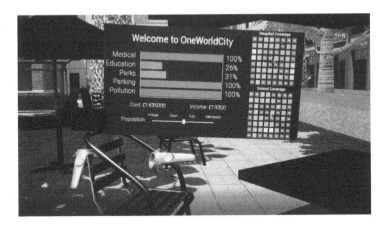

Fig. 6. A visualisation of a player reviewing their concept analysis from within City Blocks

4.4   Value in Creating a Digital Twin of an Idea

City Blocks provides an abstracted, physical design environment that can be scanned using any standard webcam. From this scan, a digital twin is procedurally generated, indicating a possible real-world likeness that can be visualised, communicated and provides early feasibility simulations. The digital-physical twinning paradigm on a city scale is being realised, with City Blocks being a proof of concept that it can be expanded to the ideation and innovation phases of design. Being able to run an analysis on the concept has been shown to be valuable to the players.

Of the five experienced urban planners respondents, those with 2+ years of experience, acknowledged the value of the provided metrics to help them improve their cities, justifying the value of a digital twin. One experienced designer commented, "Great tool that would be excellent for early stage community engagement in the planning process". Through discussion of the three supporting questions, there is demonstrated value, acknowledged by city planners to creating a digital twin of a concept, particularly in a city planning context.

4.5   Limitations of System

Whilst City Blocks has demonstrated value as an engagement tool, there are inherent limitations associated with the platform and methodology. The abstraction of the process may be too significant and simplified from the initial goal of collaborative and open urban planning. If citizens engaging with the activity treat it solely as a playful endeavour without considerations of practicality, requirements drawn may be skewed.

With the current methodology, players cannot experience their design until they have completed the first iteration. This means a lot is left to imagination until the second iteration. A suggestion to resolve this is to use a feed through camera on the front of the VR headset, with the player able to quickly switch between viewing the build plate in front of them and the virtual simulation, with the simulation updating in real time as the city is designed. The build plate recognition algorithm also requires some improvement, as known, systematic errors are induced as discussed in Sect. 2.2. The design is currently only scanned on completion, but the link should be moved to real-time if practicable.

5   Conclusions and Future Work

The creation and conversation around City Blocks open up a myriad of research opportunities. The platform approaches the possibility of early stage visualisation and analysis of concepts, utilising the digital-physical twinning paradigm. The prototype platform, City Blocks, was piloted at a local festival with members of the public opting to engage with the exhibit to the extent of their own volition. 149 unique cities were designed physically using the Lego design kit with contextualised bricks by 125 individual players. Each design was scanned and a digital twin was created, the digital twin was analysed and visualised using the platform and explored using Virtual Reality with various metrics communicated to the player from within the simulation. Players had the option to iterate and improve their cities, fill in a questionnaire or disengage with the exhibit. It was found that citizens with urban planning experience found the most use from the metrics, particularly with supporting design iteration.

Virtual Reality is an ideal tool for large visualisations where scale is difficult to communicate. Design using play should also be explored further, with emphasis on understanding the impact of rulesets, how abstracted the Lego bricks can be whilst retaining value, and how different types of play could be prescribed to different design activities. Integrating a concept with a real city digital twin would be the ideal outcome, mirroring, integrating and simulating the concept to be explored in Virtual Reality.

In summary, the paper has sought to explore the concept of digital-physical twinning in the context of the early stages of urban planning and wider stakeholder engagement. In extending the concept of twinning we have created a Lego-based physical environment that acts both as an accessible and inclusive design tool and the interface for the Virtual Reality digital twin. What separates this work from purely interface design is that the physical and digital are twinned and the affordances of each leveraged. The physical domain was a playful design interface for non-experts and the digital domain provided Virtual Reality led simulation, immersion and feedback. The platform demonstrated benefits in a city planning context for both experienced and novice participants.

References

1. Goudswaard, M., Hicks, B., Nassehi, A.: Towards the democratisation of design: exploration of variability in the process of filament deposition modelling in desktop additive manufacture. In: Transdisciplinary Engineering Methods for Social Innovation of Industry 4.0 (2018)
2. Al-Kodmany, K.: Visualization tools and methods in community planning: from freehand sketches to virtual reality. J. Plan. Lit. **17**(2), 189–211 (2002)
3. United Nations: Cities - United Nations Sustainable Development Goal 11: Make cities inclusive, safe, resilient and sustainable. https://www.un.org/sustainabledevelopment/cities/. Accessed 28 Nov 2018
4. United Nations: World Urbanization Prospects (2014)
5. Qinglin, Q., Fei, T., Ying, Z., Dongming, Z.: Digital twin service towards smart manufacturing. In: 51st CIRP Conference on Manufacturing Systems (2018)
6. Grieves, M.: Digital Twin: Manufacturing Excellence through Virtual Factory Replication A Whitepaper. In: Digital Twin White Paper (2014)
7. Hamer, C., Zwierzak, I., Eyre, J., Freeman, C., Scott, R.: Feasibility of an immersive digital twin: the definition of a digital twin and discussions around the benefit of immersion (2018)
8. Dawkins, O., Dennett, A., Hudson-Smith, A.: Living with a Digital Twin: Operational management and engagement using IoT and Mixed Realities at UCL's Here East Campus on the Queen Elizabeth Olympic Park (2018)
9. National Research Foundation: Virtual Singapore (2018). https://www.nrf.gov.sg/programmes/virtual-singapore. Accessed 28 Nov 2018
10. Bristol Transit Platform. https://portal-bristol.api.urbanthings.io/#/home. Accessed 28 Nov 2018
11. MappingGM. https://mappinggm.org.uk/. Accessed 28 Nov 2018
12. Mojang: A Mojang, Microsoft, and UN-Habitat Collaboration — Block by Block. https://www.blockbyblock.org/about/. Accessed 28 Nov 2018
13. Blockholm - Ett Stockholm I Minecraft. http://blockholm.se/index.html. Accessed 28 Nov 2018
14. Thamesmead: Claridge Way Virtual Reality Project (2018). https://www.thamesmeadnow.org.uk/claridgeway/workshops-and-events/claridge-way-virtual-reality-project-october-november-2018/. Accessed 28 Nov 2018
15. Shelley, T., Lyons, L., Shi, J., Minor, E., Zellner, M.: Paper to Parameters: Designing Tangible Simulation Input (2010)
16. Davidson, K., Junge, D.: A Lego Brickumentary (2014)
17. Mathias, D., Boa, D., Hicks, B., Snider, C., Bennett, P.: Design variation through richness of rules embedded in lego bricks. Hum. Behav. Des. **8**(21), 99–108 (2017)
18. OpenCV team: OpenCV library (2018). https://opencv.org/. Accessed 28 Nov 2018

Convolutional Neural Networks for Image Recognition in Mixed Reality Using Voice Command Labeling

Burkhard Hoppenstedt[1]([✉]), Klaus Kammerer[1], Manfred Reichert[1],
Myra Spiliopoulou[2], and Rüdiger Pryss[1]

[1] Institute of Databases and Information Systems, Ulm University, Ulm, Germany
burkhard.hoppenstedt@uni-ulm.de
[2] Faculty of Computer Science, Otto-von-Guericke-University,
Magdeburg, Germany

Abstract. In the context of the Industrial Internet of Things (IIoT), image and object recognition has become an important factor. Camera systems provide information to realize sophisticated monitoring applications, quality control solutions, or reliable prediction approaches. During the last years, the evolution of smart glasses has enabled new technical solutions as they can be seen as mobile and ubiquitous cameras. As an important aspect in this context, the recognition of objects from images must be reliably solved to realize the previously mentioned solutions. Therefore, algorithms need to be trained with labeled input to recognize differences in input images. We simplify this labeling process using voice commands in Mixed Reality. The generated input from the mixed-reality labeling is put into a convolutional neural network. The latter is trained to classify the images with different objects. In this work, we describe the development of this mixed-reality prototype with its back-end architecture. Furthermore, we test the classification robustness with image distortion filters. We validated our approach with format parts from a blister machine provided by a pharmaceutical packaging company in Germany. Our results indicate that the proposed architecture is at least suitable for small classification problems and not sensitive to distortions.

Keywords: Mixed Realiy · Image recognition ·
Convolutional Neural Networks

1 Introduction

Image recognition [2] has become an important factor in the digitalization of industrial factories. Camera systems support the industrial production, e.g., by automatically detecting faulty parts of a machine. The current development of smart glasses [13] offers the possibility to utilize them as mobile cameras, with reduced resolution and expected noise due to the users' movements. Smart

© Springer Nature Switzerland AG 2019
L. T. De Paolis and P. Bourdot (Eds.): AVR 2019, LNCS 11614, pp. 63–70, 2019.
https://doi.org/10.1007/978-3-030-25999-0_6

glasses offer the further possibility of location independent image classification. Interestingly, a paradigm change in the field of image classification could be observed in the last years. The excellent classification rates of *convolutional neural networks* (CNNs) outperformed traditional approaches in many use cases [7]. The traditional approaches rely on the explicit definition of image features, while CNNs offer a more generic approach and are able to find complex relationships in images. In the broader context of supervised learning approaches like CNNs, each image needs a *label* to be classified. To tackle the labeling problem, in this work, we generate the labels by mapping voice commands to the smart glasses video stream. More specifically, three technical parts of a machine from an industrial company are classified, whereas the classification process is afterwards tested by using image distortion filters (i.e., blur, noise, and overexposure filters) and measuring the effect on the classification accuracy. In general, our approach tries to provide a simplified image recognition from scratch, for which a user scans his or her environment and all objects during a *calibration/labeling phase*. This input is processed in a machine learning pipeline using CNNs and, eventually, presented to the user through a web service for live classification.

The remainder of the paper is structured as follows: Sect. 2 discusses related work, while Sect. 3 introduces relevant background information for image recognition, mixed reality, and convolutional neural networks. In Sect. 4, the developed prototype is presented, in which the mixed-reality application, the processing pipeline, and the classification algorithm are presented. The results of the distortion algorithms are shown in Sect. 5. Threats to validity are presented in Sect. 6, whereas Sect. 7 concludes the paper with a summary and an outlook.

2   Related Work

Convolutional neural networks (CNN) are widely used in the field of image recognition. CNNs have been successfully tested in the context of face recognition with a high variability in recognizing details of a face [8]. Even though CNNs are widely used for image recognition, they can also be applied to other use cases, such as speech recognition and time series prediction [9]. Since the training of neural network is very time intensive, but the execution time is rather low, CNNs are also suitable for real-time object recognition [10]. Object recognition for augmented reality is mostly performed in a marker-based manner [14], which means that markers (e.g., barcodes) support the recognition process. Standard architectures for CNNs have been proposed, e.g., by AlexNet [7], GoogleNet [16], or InceptionResnet [15]. In large scale scenarios, deep convolutional neural networks incorporate millions of parameters and hundreds of thousands neurons [7], and therefore need an efficient GPU implementation. Furthermore, an evolving topic in the field of image recognition using CNNs is denoted as *transfer learning* [12]. Hereby, image representations from large-scale data sets are transferred to other tasks to limit the necessary training data. In general, not only the content of the image can be learned, but also the image style [3]. The latter offers the possibility of high level image manipulation. The aforementioned techniques denote a promising extension level of our approach. However, these techniques are, in our opinion, not suitable for a small scenario, as a larger computational power would be necessary. To the best of our knowledge, existing works do not combine image recognition, mixed reality, and voice commands as we have realized for the solution presented in the work at hand.

3 Fundamentals

3.1 Convolutional Neural Networks

In general, neural networks are mathematical models for optimization problems, for which the influence of each neuron is expressed with a *weight*. The network constitutes a construct build from neurons that receive an input and compute its output via an *activation function* (e.g., *sigmoid*). A stack of neurons in a single line is denoted as a layer. The first layer constitutes the input, the last layer is called output and all layers in between are denoted as *hidden layers*. In the case of a Convolutional Neural Network (CNN), the neurons form convolutional layers. The most important parameter in a convolutional layer is the filter size, which denotes the window size of the convolution. Each convolution reduces the input's size, so that we use - for the example of image recognition - a padding at the image borders to keep the images dimensions. To reduce the spatial size, a pooling layer is applied. The most common pooling operation is denoted as *max pooling*, where a filter applies the maximum function on the image. The combined information of the neural network is denoted as model. Hereby, the prediction accuracy represents the quality of the model. To keep the computation simple, not all training data is loaded into the network at once. Instead, small batches with a predefined *batch-size* are used in each training iteration. In one *epoch*, the model is fed with all the training images. In general, three types of data sets exist: *Training data, validation data* and the *test set*. The model is trained with the training data and tested with the validation data. The test set consists of images from a separate data set to test the generalization of the network and prevent *overfitting* [4]. CNNs are classified as a supervised learning method, which means that they need a *label* that assigns the correct output for each input. In our approach, we try to simplify this labeling process via voice commands. Finally, the speed of the learning progress can be influenced via the *learning rate*, where a high learning rate enables the model to adapt the weight of each neuron quickly.

3.2 Mixed Reality

Mixed Reality tries to achieve the highest overlapping of reality and virtuality in the reality-virtuality continuum [11]. When using the Microsoft HoloLens, then, the latter performs spatial mapping [5] to generate a virtual model. Therefore, virtual objects can be placed in the real world and stay in a fixed position through tracking features. The HoloLens is equipped with various sensors, including a

RGB camera, a depth sensor, and a Mixed Reality capture feature. Furthermore, the HoloLens offers the usage of individual speech commands based on natural language processing.

4   Prototype

4.1   Workflow of the Approach

In general, our approach (see Fig. 1) aims at a simple labeling for the image recognition. The first step is to define all names of the objects to be recognized. When starting the mixed-reality application, the HoloLens loads these names from the database and defines speech commands for these terms. Then, the *labeling phase* starts by recording a video. When an object enters the user's field of view, the user says its name. Thereby, the HoloLens logs the current timestamp and the name of the object into a file. The same procedure takes place when the objects leaves the user's field of view.

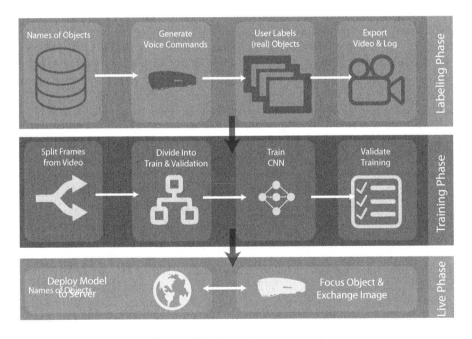

Fig. 1. Workflow of the approach

As the next step, we generate a *mapping* of images to objects, defined by the period of time between the start and the end voice command. At the end of the labeling phase, the resulting video and log file are sent to an offline application. This latter divides the video into frames and separates all frames into the corresponding folders with images for each object or background. The images are chosen randomly to have the same number of images for each object class. Using the deep learning framework *Tensorflow* [1], the neural network is being trained (i.e., learning phase). As a very simple architecture is used, the network can be trained by using a normal CPU. The image input is automatically split up into 80% training data and 20% validation data. After the learning process is finished, the network is accessible via a RESTful API using a python server. Practically, the user operates with the smart glass, puts the focus to an object, and says the voice commands *classify*. The image is then sent to the server, predicted by the use of the network, and the prediction result is eventually sent back. Note that we needed to include a manual correction into the labeling process. Theoretically, the timestamp t of a voice command should fit exactly to the video frame where the user has seen the object. Unfortunately, there is a calculation time before the voice command gets recognized. We measured this delay and calculated a mean difference between timestamp and frame of 1.16 s with a variance of 0.13 s. Therefore, we included this delay as a static threshold in the processing pipeline. Altogether, the following technologies were used to realize this approach. As a database for all possible objects, we chose the document-based NoSQL database *MongoDB*. The web interface is provided by the python webserver *Flask*. All machine learning operations are provided by *Tensorflow*, which uses the image library *OpenCV* for image processing. We developed the mixed-reality application in *Unity* and used the Java library *JCodec* to split the video and map the recorded timestamps. Finally, all distortion filters were generated using the software *Matlab*.

4.2   Convolutional Neural Network

The network is implemented using a simple CNN structure (see Fig. 2). The input consists of a *4D tensor*, with the dimensions number of images, width, height, and number of color channels. The weight of the neurons, that will be adapted during the training through *back propagation*, are initialized with a random normal distribution. As an optimizer, we use the *Adam* algorithm [6] for gradient calculation and weight optimization. We choose 0.0001 as a learning rate and a batch size of sixteen. After each convolution step with 32 filters, a max pooling is applied to the result.

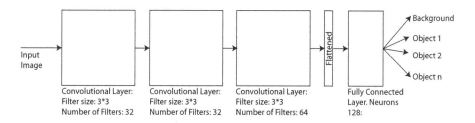

Fig. 2. Used CNN architecture

4.3 Distortion Filters

To test the effects of bad image quality in our approach, we tested the three distortion filters blurring, noise, and overexposure (see Fig. 3). We applied one filter each time and tested the resulting images with our model in terms of accuracy. For the blurring, we used a box filter with dimensions 11 × 11. Moreover, a *salt & pepper* noise with a density of 0.2 is applied and, lastly, the brightness effect is achieved by increasing the RGB value by fifty.

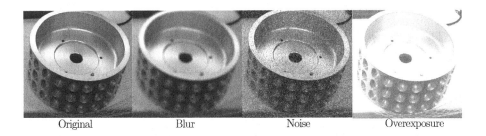

Original	Blur	Noise	Overexposure

Fig. 3. Distortion filters

5 Results

The prototype was tested with 500 images per class (i.e., 2000 images in total). Each object to be detected (three in total) and the background are represented by a training class. In general, the training process was stopped after five epochs to measure the accuracy. The training process of the images without any distortion revealed a validation accuracy of 90.6% (see Fig. 4). The noise on the image led to an accuracy of 86.2%, the blurring lowered the accuracy to 85.6% and, lastly, the images with an increase brightness were classified with an accuracy of 81.0%. Therefore, when performing image recognition in Mixed Reality, attention should be paid to a good illumination. The blurring effect, likely caused by fast head movements, was not critical for the classification. In general, the distortion filters did not disrupt the classification significantly.

6 Threats to Validity

Our approach is tested in only one room and with a low number of objects. The higher the number of objects is, the more likely it is that the classification accuracy will decrease. Moreover, as every user is responsible for the labeling process him- or herself, the classification will fail if the objects were not focused precisely or the voice commands are not correctly synchronized with the gaze. Furthermore, neural networks are not a transparent method of machine learning. Therefore, it will be hard to find failure reasons in case of a low classification rate. Despite these limitations, we consider our approach as an easy-to-use object recognition process with high accuracy rates on small data sets.

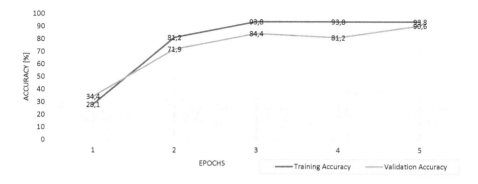

Fig. 4. Learning progress

7   Summary and Outlook

We provided an approach in Mixed Reality that allows users to train objects by labeling frames in the recorded video via voice commands. The generated output is then processed and put into a convolutional neural network. The classification of an image during the use of the HoloLens is achieved by sending the image to a web server. Here, the image is classified with the previously trained model and the response is sent back to the HoloLens. This information can further on be used to monitor additional information for the recognized object. New types of mixed-reality glasses might introduce new possibilities for object recognition (e.g., better image resolution) and could improve this approach. Furthermore, the approach could be tested versus approaches, for which the objects are labeled manually. Moreover, the scalability of this approach should be further investigated. The neural network architecture is conceived in such a way that everyone can provide the computational power for the training phase. When tackling more complex problems, more convolutional layers could be introduced. Currently, the workflow demands that the user names all objects at the beginning. However, the user may consider some objects as more important than others and concentrate on them first. Hence, a future step could be to have the user add labels gradually. This would turn the static learning task into a stream learning task, in which the CNN must be adapted to new classes. In conclusion, we consider convolutional neural networks in combination with a labeling based on voice commands in Mixed Reality as an appropriate approach for object detection, especially for scenarios in the context of the Industrial Internet of Things (IIoT).

References

1. Abadi, M., et al.: TensorFlow: large-scale machine learning on heterogeneous systems (2015). Software https://www.tensorflow.org/
2. Fu, K.S., Young, T.Y.: Handbook of Pattern Recognition and Image Processing. Academic Press (1986)
3. Gatys, L.A., Ecker, A.S., Bethge, M.: Image style transfer using convolutional neural networks. In: Proceedings of the IEEE Conference on Computer Vision and Pattern Recognition, pp. 2414–2423 (2016)
4. Hawkins, D.M.: The problem of overfitting. J. Chem. Inf. Comput. Sci. **44**(1), 1–12 (2004)
5. Izadi, S., et al.: Kinectfusion: real-time 3D reconstruction and interaction using a moving depth camera. In: Proceedings of the 24th Annual ACM Symposium on User Interface Software and Technology, pp. 559–568. ACM (2011)
6. Kingma, D.P., Ba, J.L.: Adam: a method for stochastic optimization. In: Proceedings of the 3rd International Conference on Learning Representations (2014)
7. Krizhevsky, A., Sutskever, I., Hinton, G.E.: Imagenet classification with deep convolutional neural networks. In: Advances in Neural Information Processing Systems, pp. 1097–1105 (2012)
8. Lawrence, S., Giles, C.L., Tsoi, A.C., Back, A.D.: Face recognition: a convolutional neural-network approach. IEEE Trans. Neural Netw. **8**(1), 98–113 (1997)
9. LeCun, Y., Bengio, Y., et al.: Convolutional networks for images, speech, and time series. handb. Brain Theory Neural Netw. **3361**(10), 1995 (1995)
10. Maturana, D., Scherer, S.: Voxnet: a 3D convolutional neural network for real-time object recognition. In: 2015 IEEE/RSJ International Conference on Intelligent Robots and Systems (IROS), pp. 922–928. IEEE (2015)
11. Milgram, P., Takemura, H., Utsumi, A., Kishino, F.: Augmented reality: a class of displays on the reality-virtuality continuum. In: Telemanipulator and Telepresence Technologies, vol. 2351, pp. 282–293. International Society for Optics and Photonics (1995)
12. Oquab, M., Bottou, L., Laptev, I., Sivic, J.: Learning and transferring mid-level image representations using convolutional neural networks. In: Proceedings of the IEEE Conference on Computer Vision and Pattern Recognition, pp. 1717–1724 (2014)
13. Rauschnabel, P.A., Ro, Y.K.: Augmented reality smart glasses: an investigation of technology acceptance drivers. Int. J. Technol. Mark. **11**(2), 123–148 (2016)
14. Rekimoto, J.: Matrix: a realtime object identification and registration method for augmented reality. In: Proceedings of the 3rd Asia Pacific Computer Human Interaction, pp. 63–68. IEEE (1998)
15. Szegedy, C., Ioffe, S., Vanhoucke, V., Alemi, A.A.: Inception-v4, inception-resnet and the impact of residual connections on learning. In: AAAI, vol. 4, p. 12 (2017)
16. Szegedy, C., et al.: Going deeper with convolutions. In: Proceedings of the IEEE Conference on Computer Vision and Pattern Recognition, pp. 1–9 (2015)

A Framework for Data-Driven Augmented Reality

Georgia Albuquerque[1,3](✉), Dörte Sonntag[2], Oliver Bodensiek[2],
Manuel Behlen[1], Nils Wendorff[1], and Marcus Magnor[1]

[1] Computer Graphics Lab, TU Braunschweig, Braunschweig, Germany
{georgia,behlen,wendorff,magnor}@cg.cs.tu-bs.de
[2] Institute for Science Education Research, TU Braunschweig,
Braunschweig, Germany
{doerte.sonntag,o.bodensiek}@tu-bs.de
[3] Software for Space Systems and Interactive Visualization, DLR,
Braunschweig, Germany

Abstract. This paper presents a new framework to support the creation of augmented reality (AR) applications for educational purposes in physics or engineering lab courses. These applications aim to help students to develop a better understanding of the underlying physics of observed phenomena. For each desired experiment, an AR application is automatically generated from an approximate 3D model of the experimental setup and precomputed simulation data. The applications allow for a visual augmentation of the experiment, where the involved physical quantities like vector fields, particle beams or density fields can be visually overlaid on the real-world setup. Additionally, a parameter feedback module can be used to update the visualization of the physical quantities according to actual experimental parameters in real-time. The proposed framework was evaluated on three different experiments: a Teltron tube with Helmholtz coils, an electron-beam-deflection tube and a parallel plate capacitor.

Keywords: Augmented Reality · Physics education ·
Real-time interaction

1 Introduction

Augmented Reality (AR) is expected to play a major role in future educational technology since it offers a high potential to integrate physical, digital and social learning experiences in hybrid learning environments. Some of the observed effects of AR applications in education are: improved learning gains, higher motivation and better interaction and collaboration [2]. It is, however, still subject to current research to identify the specific moderating variables for these positive effects [10]. The knowledge about these variables and how to address them in practice, summarized in evidence-based frameworks for using

© Springer Nature Switzerland AG 2019
L. T. De Paolis and P. Bourdot (Eds.): AVR 2019, LNCS 11614, pp. 71–83, 2019.
https://doi.org/10.1007/978-3-030-25999-0_7

AR in education [3], would facilitate goal-oriented instructional design of AR applications. In addition, there is still a lack of authoring tools enabling users without deep programming skills to create own AR learning applications for head-mounted displays.

Especially in natural sciences, AR offers the potential to close the gap between theory- or calculus-related learning and experimental learning in a single hybrid environment by visual augmentation of experiments. In addition to the possibility of superimposing – measured but yet invisible – physical quantities on real experiments as presented by Strzys et al. [21] and Kapp et al. [14], the visual overlay of simulation results and their real-time dynamic coupling to experimental parameters is another benefit of AR. Ultimately, visualizations of measured data and different model simulations can directly be compared.

Besides applications themselves, it is also important to develop frameworks and authoring tools for educators and students which allow for an easy implementation and application of AR-experiments without programming skills. In this paper we propose a framework to support the creation of experimental physics AR applications in which virtual visualizations of physical entities are overlaid on real experiment setups. Using our framework, the user only needs to generate simulation input data on the one hand and to provide a CAD-model of the experiment on the other hand. Our main contributions are:

- A framework to support the creation of AR applications for physics experiments.
- A user interface to facilitate the configuration of the AR applications without requiring programming skills.
- An evaluation of how AR applications created with our framework can affect learning processes in physic lab courses.

2   Related Work

An increasing amount of AR applications with educational purpose have been presented in the last years, including medical engineering [5], cultural history [22] and other various areas [2]. Yuen et al. [23], Radu [20] and Bacca et al. [2] presented an overview and future directions for AR in education as a whole. Although most existing AR approaches and applications were developed for students with a more experienced background like high school and university students, applications for younger students were also proposed [2,15]. The app LifeLiqe [17] is an example of a well-established AR application for educational purpose that provides many models extracted from fields as geography, physics and anatomy, allowing to walk around projected objects and parts to be observed in detail.

Similar to our work, Farias et al. [7] presents an AR authoring tool that allows teachers to create general AR presentations without requiring programming skills. Our framework also provides an authoring tool, but with the difference that our approach offers real-time parameter interaction and feedback in the AR visualizations. AR contributions that allow the visualization of invisible physical quantities were introduced by Li et al. [16], who presented an AR approach to visualize gravity forces. Applications to teach basic concepts of electromagnetism and evaluation of their defectiveness were also presented, using a simple mocked magnet [19], and AR markers [13]. Similarly, Strzys et al. [21] presented AR applications to augment a thermal flux experiment using an infrared camera to measure spatially distributed temperature distributions visualizing them with a HoloLens. Although their implementation moved augmentation of experiments towards real-time data driven AR, their approach using an infrared camera cannot be extended to scalar or vector fields in electromagnetism. Along these developments, Bodensiek et al. [6] presented an AR experiment to help students to study physics of charged particles in magnetic fields, and Kapp et al. [14] proposed an AR experiment for high school students to study Kirchhoff's circuit laws in electrical DC circuits using real-time measurement data. Following this prior work, we go a step further in this paper and propose a framework to ease the creation of data-driven AR applications for physics education. Furthermore, we perform an exploratory evaluation on how applications created with our framework can improve the learning processes.

3   Framework

Besides the Microsoft HoloLens, our hardware resources consist of a physical experiment setup with power supplies, cabling and a measuring module. Figure 1 shows our fundamental hardware resources. To allow for a real-time parameter feedback, our setup includes a measuring module composed by multimeters and a RaspberryPi single-board computer that is used as a simple web server to transmit the current measurements to the HoloLens application. For high voltage experiments, a voltage divider is used together with the respective multimeter to transform the necessary values.

Figure 2 gives an overview of the four main modules that compose our framework: input data, object positioning, parameter feedback, and augmented visualization. The framework was developed using the DirectX SDK, OpenCV and C# in the Universal Windows platform for the HoloLens. For each application, two main components need to be externally created: a CAD model file of the central experimental setup (CAD-Model), such as an electron deflection tube including base but without cables, power supply or measurement tools in *.obj* format, and the simulation data for the experiment as vector data for each physical field in text format containing field and position vectors. CAD models and simulation data can be created using modeling, and Finite-Element Method (FEM) tools or using tabular data of analytical models, respectively.

To ease the implementation of new applications, we developed a simple user interface to facilitate the configuration of new AR applications without requiring programming skills, as can be seen in Fig. 3. The user interface receives three general components as input: The configuration field *Application Name* determines the name of the AR App that will be built. After finishing the creation of the application this is the name that will appear in the HoloLens' main menu. The next component is the name of the *CAD model* file of the central experimental setup, as described above. The third general setting is the *IP address*

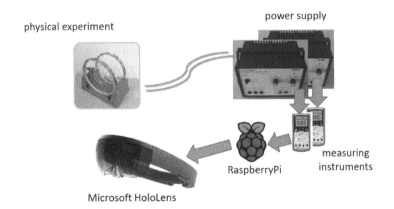

Fig. 1. The typical hardware resources setup consists of the HoloLens, RaspberryPi, measuring devices and the physical experiment itself with its power supply.

of the web server that will be used to connect the application to the correct server. Finally, we allow users to add up to three different visualizations which can be simultaneously observed in the application. For each visualization, the user can check the *enable* checkbox and fill up the respective fields. Our Framework supports three different visualization types that can be selected in the *Data Type* drop-down menu: vector fields, particle beams and density fields, as further described in Sect. 3.3. The remaining configuration fields describe the the simulation data for each of the visualizations. The *Data scale* field can be used to correct a scale factor of the CAD model that was used to create the simulation data. The HoloLens assumes a metric unit system, which means that the value 1 is equivalent to one meter. If the simulation data is provided in another scale, for example in millimeters, it can be corrected using this field. The *Simulation Data Folder* field determines the folder that contains the simulation data in the web server for each visualization. The last configuration options are the parameters of each experiment, as currents and voltages that are described in the simulation data. It includes a parameter identifier and the file name of the file that contains the simulation data of the respective parameter in the web server. Depending on the experiment, up to three parameters are used. Section 4 shows three applications examples and their respective parameters.

3.1 Input Data

In order to solve the partial differential equations describing the involved electrical and magnetic fields we use the Finite Element Method (FEM). More precisely, Comsol MultiPhysics software is used as an external tool to compute the numerical data. In principle, any other numerical tool can be used to create the input data provided that the exported data text-file follows the expected structure. For simple geometries even analytical methods or other tools are possible. The exported data text-file must contain tab-separated columns with the corresponding experimental parameters such as voltage or current and a column each of the

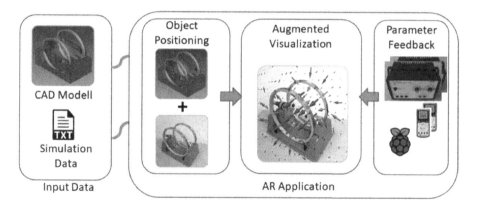

Fig. 2. Framework pipeline including our four main modules: input data, object positioning, parameter feedback, and augmented visualization.

components of both position and field vectors. This data is pre-computed for a range of experimental relevant parameters and accessed by the AR-application from local storage during run-time. If the actual parameter value from the real experiment falls in between constant computed parameters stored in the input files, the data is linearly interpolated.

3.2 Object Positioning

Once the input data is supplied and the application generated, the first step to visualize new virtual elements is to determine where these elements should be placed. We choose a coarse-to-fine approach to determine the position of the virtual elements. We first compute a coarse position of the new elements by comparing an image sequence of the physical setup obtained from the Hololens camera, with the supplied CAD model. Specifically, we generate multiple views of the CAD model, and apply a sobel edge detection algorithm [11] to these views. The resulting edge images are then used as hypothesis and are compared to the edge image of the real sequence using a particle filter-based approach [4] implemented for the HoloLens GPU. The algorithm keeps choosing the best hypothesis, and generating new ones based on them until the user executes a *Tap* gesture and the current best hypothesis for the position is taken. Figure 4 shows an example of hypothesis generated for an experiment setup and its respective CAD model. Using an image-based approach instead of relying on the infrared sensors of the HoloLens has the advantage of being better suitable to detect metallic surfaces as they do not reflect infrared rays as well as non-metallic surfaces [9]. Some of the experiments we describe in this paper have detailed metallic structures. Finally, our user interface provides positioning arrows that can be used for any needed correction or fine tuning.

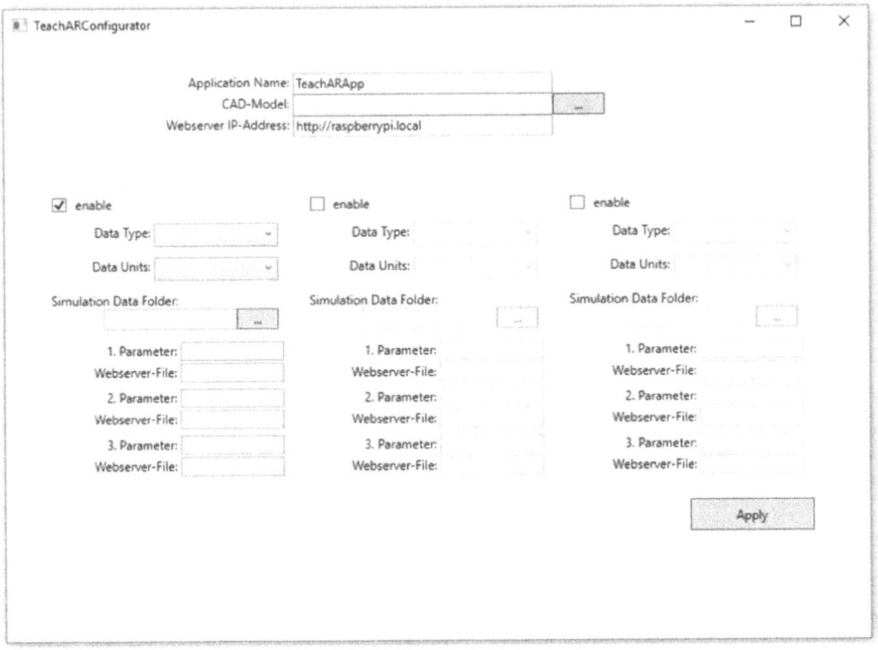

Fig. 3. The user interface provides different setting options to generate a new application without the need for any knowledge of programming languages.

3.3   Parameter Feedback and Augmented Visualization

From an educational perspective, the parameter feedback module is an essential part of the AR applications as it relates experimental actions to the behavior of a mathematical model, exemplifying the bridge between theory and experiment. As aforementioned, we use multimeters to measure the current parameters of the experiments connected through USB ports to a RaspberryPi single-board computer. A web server is implemented for the RaspberryPi which makes the current values measured by the multimeters available and can be accessed by the HoloLens application through the local network. The web service is automatically initialized when a multimeter is connected to the RaspberryPi. Once the service is on-line, the application can request the desired parameter values from the web server and use them to update the respective visualization modules. Our framework includes three different physical augmented visualization: vector fields, particle beams and density fields. Additional visualization types can be easily added to the framework is necessary. The first kind of augmented visualization is a vector field that is used to represent electrical or magnetic fields in the experiment. In a vector field, each point of the space is assigned to a vector that characterizes the alignment and intensity of the field in that point. The simulation input data described in Sect. 3.1 contains the positions and their respective vectors for the actual field. For each data point, an arrow

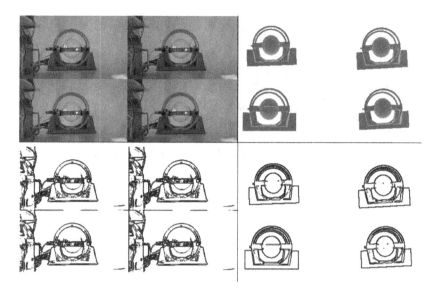

Fig. 4. Example of object positioning. Top left: repeated camera frame; Top right: example of rendered hypothesis; Bottom left: edge features of the camera frame; Bottom right: edge features of the rendered hypothesis.

representing the corresponding vector is rendered. The field intensity in each point is represented by the width, length and color of the arrow. Smaller vectors are displayed in green and larger vectors in red as can be seen in Fig. 5.

The second kind augmented visualization is a line that is used to represent the appearance of electron beams. Real electron beams cannot be well visualized when the experiment takes place in bright environments. A virtual electron beam can therefore be used to improve the visualization of a real electron beam in general light conditions. The simulation file that underlies the electron beam visualization is structurally different from the files of the vector field, containing a chronological list of the positions of an electron. The positions are stored in a dynamic vertex buffer and updated according to current parameters. Since the graphics card does not have to process triangle data but points, the primitive type *line strip* is used to connect the individual points with lines. Figure 6, shows an example of electron beam visualization.

The last augmented visualization is a density plot that is used to visualize the energy distribution of electrical and magnetic fields. In this visualization, the intensity of the field is displayed as a two-dimensional area between the sources in form of color gradients. In this case, the orientation of the field is not clearly visible, but the intensities of the field can be clearly observed. Either kind of simulation data can be used as parameter for our density field visualization: a vector field or a point cloud describing the field in two dimensional areas perpendicular to the field sources. In both cases, the vertices of the central area are selected to generate the visualization. Finally, we interpolate these selected

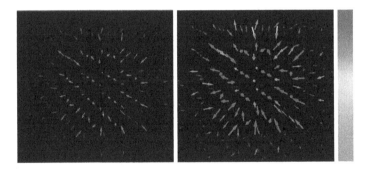

Fig. 5. Depending on the intensity of the field, an arrow is colored and scaled at the specific position.

vertices using a Delauney' triangulation [8]. The vertex color is computed based on the length of the vector, if available. Figure 7 shows an example of density field visualization.

4   Application Examples

In this section we show three AR applications that were created for different electrodynamics experiments using our framework: a Teltron Tube, a Parallel Plate Capacitor and an Electron Beam Deflection Tube. In the Teltron Tube and the Electron-Beam-Deflection-Tube experiments an electron beam is observable. Its shape is influenced by the magnetic and electric fields present and a visualization of the latter can help understanding the interplay of charged particles and fields. The Parallel Plate Capacitor experiment is typically used to determine its electrical capacity varying both the plate distance and the voltage. As the electrical capacity is directly related to the energy density of the electric field and the squared field strength, respectively, it is useful to visualize this scalar field along with the electric vector field.

4.1   Teltron Tube with Helmholtz Coils

The first AR experiment we describe in this section is a *Teltron tube with Helmholtz coils*, which is further detailed in [6]. The aim of this experiment is to determine the specific charge of an electron, as well as to investigate the deflection of electrons in a magnetic field. A Helmholtz coil pair creates an almost homogeneous magnetic field in between the coils. The amplitude of the magnetic field can be altered by changing the current intensity I_C in the coils. To show the influence of the magnetic field on a moving charge, a bundled electron beam is guided vertically towards a glass flask filled with hydrogen molecules under pressure. The beam passes through the magnetic field and is guided towards a circular path as a result of the Lorentz force. In this experiment students can adjust two main parameters:

– I_C, the current in the Helmholtz coils
– U_A, the anode voltage to accelerate the electron beam

Our AR application renders a dynamic visualization of the magnetic field and the electron beam according to I_C and U_A parameters, respectively, as described in Sect. 3.3. Figure 6 shows the augmented visualization of a Teltron Tube with Helmholtz Coils using our application from the HoloLens view.

Fig. 6. Teltron Tube with Helmholtz Coils: Electron beam and the magnetic field are visually overlaid on the real experiment.

4.2 Electron-Beam-Deflection Tube

The electron beam deflection tube is an experimental setup that illustrates the behavior of moving charges in magnetic and electric fields. Like the *Teltron tube with Helmholtz coils* experiment, the magnetic field is generated by a pair of coils which creates an almost homogeneous field inside the glass tube. The voltage U_C on a capacitor plate pair installed in the tube, generates an electric field perpendicular to the magnetic fied. A heated cathode releases electrons which are accelerated by the anode voltage U_A in the horizontal direction. The resulting electron beam is then deflected by both the electric and magnetic field. As in the previous experiment, the current I_C in the coils is roughly proportional to the magnetic field force in the center. In summary the following parameters can be modified during this experiment:

– U_C, the voltage on the capacitor plates
– I_C, the current in the coil pair
– U_A, the acceleration voltage of electrons

Fig. 7. A parallel plate capacitor with a movable plate to study the relationship between the distance of the plates and the intensity of the electric field.

In this experiment both the electric and magnetic field are visualized as a vector plot, the corresponding energy densities are visualized as a semi-transparent density plot and the electron beam can also be visualized.

4.3   Parallel Plate Capacitor

The last application that we describe in this section was created for a parallel plate capacitor. In this experiment, an electric field is generated by a voltage U_C between the capacitor plates. Its main goal is to teach how changes on U_C and on the distance between parallel capacitor plate d influence the electric capacity and to determine it. As noted above, the capacity is proportional to the energy density of the electric field which is hence visualized along with the electric field itself and updated along with changes of the two experimental parameters:

- U_C, the voltage on the capacitor plates
- d, the distance between the capacitor plates

The reference CAD model which is used to detect the experiment is constructed with a constant distance between the plates of the capacitor, such that this distance should be similar to the initial position of the plates during the real experiment to facilitate the detection. The CAD model is only important during the initial positioning of the virtual experiment. The dynamic parameter d is measured using a infrared sensor mounted near the mobile plate of the experiment and connected to the RaspberryPi. Figure 7 shows the augmented visualization of a Parallel Plate Capacitor using our application from the HoloLens view.

5   Evaluation

We conducted a perceptual study to investigate in which aspect AR applications created with our framework can affect learning processes in physic lab courses.

We performed the study with 14 participants from a physics lab course for undergraduate teacher students with similar background knowledge. To evaluate the effect of the AR application, the perceptual study was conducted in two parts: The first part was composed by 6 participants without any augmented reality support, as a control group. In the second part, 8 participants used augmented reality applications created with our framework in the HoloLens. During both parts, the participants worked in pairs. They were asked to perform one of the lab experiments described in Sects. 4.1, 4.2 and 4.3. Before starting each experiment, we explained to the participants how to carry out the experiment, and we performed a pre-interview with each participant about basic conceptions of electric and magnetic fields related to the experiment. Additionally, in the second part of the study, where the participants used AR, we also explained to them how to use the respective AR applications with the HoloLens. A post-interview, based on the same basic concepts as the first interview, was carried out at the end of each experiment.

Table 1. Percentage of changes between the pre-defined concepts for control and AR Group. Positive links describe linkages between two correct basic conceptions. False links describe linkages between and with misconceptions. Representation links describe linkages between correct representation of electric or magnetic fields and basic concepts.

	Positive links	Negative links	Representation links
Control group	+8%	−27%	+22%
AR group	+51%	−26%	+68%

To evaluate how the applications can affect the learning processes, we compare the results of the pre- and post-interviews of the control and AR groups. For the evaluation of the pre- and post-interviews we transcribed them and created a deductive category system based on both basic concepts of electric and magnetic fields and educational research on misconceptions [1,12,18]. The resulting category system had 14 main concepts and 76 sub-concepts that represent expected learning concepts of the subject. For each participant, we compare how the execution of the experiment increased the number of correct connections between basic concepts as well as correct field representations and how it decreased the number of wrong connections between misconceptions and correct concepts. Two concepts are considered to be linked if the participant places them in a logical context on their own. If a participant declares that, e.g., *an electrical field has a negative and a positive pole, and it can be described with field lines from the positive to the negative pole* and that *an electrical field can be created with a capacitor, for example* this contains three different (sub-) concepts: *negative and positive pole* (cause), a *capacitor* (example), *field lines between negative and positive poles* (field description), and the correct connection between them. We specifically compare changes between connections deducted from the pre-interview to the connections deducted from the post-interview. Table 1 shows the average changes in connections between concepts for both, control and AR groups. Note that the group of participants that executed the experiments using our AR applications shows a general improvement of 51% correct connections between the concepts after the experiment, while for the control group only an improvement of 8% can be observed. The number of wrong connections similarly decreased in both, control (-27%) and AR group (-26%). Finally, the number of connections related to the representation of the invisible quantities, e.g. electrical and magnetic fields, was enhanced by $+68\%$ using our applications, while an increase of only $+22\%$ could be observed in the control group.

6   Conclusion

In this work, we presented a flexible framework that can be used to support the creation of AR applications in physics education, implementing visualization for invisible physical quantities like electrical and magnetic vector fields or the corresponding energy-density fields to be overlaid on real-word experiment setups. Moreover, our framework includes an authoring tool to facilitate the configuration of the AR applications without the need of programming skills. It establishes an AR work-flow that allows for generating applications for physics experiments including visualizations with real-time parameter feedback. To evaluate the framework and its functionality, applications for three different experiments were created: a Teltron tube with Helmholtz coils where the electron beam and the magnetic field can be visualized, an electron-beam-deflection tube where the electron-beam, magnetic and electric field can be visualized, and a parallel plate capacitor where the electric field can be superimposed to the experiments. An exploratory evaluation of applications created with our framework, regarding an educational added value, was positive and evenly more evident when considering representation aspects of the invisible fields.

Acknowledgements. This work was supported in part by the German Science Foundation (DFG MA2555/15-1 *Immersive Digital Reality* and DFG INST 188/409-1 FUGG ICG Dome) and in part by the German Federal Ministry of Education and Research (01PL17043 *teach4TU*).

References

1. Albe, V., Venturini, P., Lascours, J.: Electromagnetic concepts in mathematical representation of physics. J. Sci. Educ. Technol. **10**(2), 197–203 (2001)
2. Bacca, J., Baldiris, S., Fabregat, R., Graf, S., et al.: Augmented reality trends in education: a systematic review of research and applications. J. Educ. Technol. Soc. **17**(4), 133 (2014)
3. Bacca, J., Baldiris, S., Fabregat, R.: Kinshuk: framework for designing motivational augmented reality applications in vocational education and training. Australas. J. Educ. Technol. **53**(3), 102 (2019)
4. Blake, A., Isard, M.: The condensation algorithm-conditional density propagation and applications to visual tracking. In: Advances in Neural Information Processing Systems, pp. 361–367 (1997)

5. Blum, T., Kleeberger, V., Bichlmeier, C., Navab, N.: mirracle: an augmented reality magic mirror system for anatomy education. In: 2012 IEEE Virtual Reality Short Papers and Posters (VRW), pp. 115–116. IEEE (2012)
6. Bodensiek, O., Sonntag, D., Wendorff, N., Albuquerque, G., Magnor, M.: Augmenting the fine beam tube: from hybrid measurements to magnetic field visualization. Phys. Teach. **57**, 262 (2019)
7. Farias, L., Dantas, R., Burlamaqui, A.: Educ-AR: a tool for assist the creation of augmented reality content for education. In: 2011 IEEE International Conference on Virtual Environments Human-Computer Interfaces and Measurement Systems (VECIMS), pp. 1–5. IEEE (2011)
8. Fortune, S.: Voronoi diagrams and delaunay triangulations. In: Handbook of Discrete and Computational Geometry, pp. 377–388. CRC Press Inc. (1997)
9. Garon, M., Boulet, P.O., Doironz, J.P., Beaulieu, L., Lalonde, J.F.: Real-time high resolution 3D data on the hololens. In: 2016 IEEE International Symposium on Mixed and Augmented Reality (ISMAR-Adjunct), pp. 189–191. IEEE (2016)
10. Garzon, J., Pavon, J., Baldiris, S.: Virtual Reality, Special Issue: VR in Education (2019)
11. Gonzalez, R.C., Woods, R.E., et al.: Digital image processing (2002)
12. Guisasola, J., Almudi, J.M., Zubimendi, J.L.: Difficulties in learning the introductory magnetic field theory in the first years of university. Sci. Educ. **88**(3), 443–464 (2004)
13. Ibáñez, M.B., Di Serio, Á., Villarán, D., Kloos, C.D.: Experimenting with electromagnetism using augmented reality: impact on flow student experience and educational effectiveness. Comput. Educ. **71**, 1–13 (2014)
14. Kapp, S., et al.: Augmenting Kirchhoff's laws: using augmented reality and smartglasses to enhance conceptual electrical experiments for high school students. Phys. Teach. **57**(1), 52–53 (2019)
15. Kaufmann, H., Papp, M.: Learning objects for education with augmented reality. In: Proceedings of EDEN, pp. 160–165 (2006)
16. Li, N., Gu, Y.X., Chang, L., Duh, H.B.L.: Influences of AR-supported simulation on learning effectiveness in face-to-face collaborative learning for physics. In: 2011 11th IEEE International Conference on Advanced Learning Technologies (ICALT), pp. 320–322. IEEE (2011)
17. Lifeliqe inc. (2019). https://www.lifeliqe.com/
18. Maloney, D.P., O'Kuma, T.L., Hieggelke, C.J., Van Heuvelen, A.: Surveying students' conceptual knowledge of electricity and magnetism. Am. J. Phys. **69**(S1), S12–S23 (2001)
19. Matsutomo, S., Miyauchi, T., Noguchi, S., Yamashita, H.: Real-time visualization system of magnetic field utilizing augmented reality technology for education. IEEE Trans. Magn. **48**(2), 531–534 (2012)
20. Radu, I.: Augmented reality in education: a meta-review and cross-media analysis. Pers. Ubiquit. Comput. **18**(6), 1533–1543 (2014)
21. Strzys, M., et al.: Augmenting the thermal flux experiment: a mixed reality approach with the hololens. Phys. Teach. **55**(6), 376–377 (2017)
22. Sylaiou, S., Mania, K., Karoulis, A., White, M.: Exploring the relationship between presence and enjoyment in a virtual museum. Int. J. Hum Comput Stud. **68**(5), 243–253 (2010)
23. Yuen, S.C.Y., Yaoyuneyong, G., Johnson, E.: Augmented reality: an overview and five directions for ar in education. J. Educ. Technol. Dev. Exch. (JETDE) **4**(1), 11 (2011)

Usability of Direct Manipulation Interaction Methods for Augmented Reality Environments Using Smartphones and Smartglasses

Alexander Ohlei[✉], Daniel Wessel[✉], and Michael Herczeg[✉]

University of Lübeck, 23562 Lübeck, SH, Germany
{ohlei,wessel,herczeg}@imis.uni-luebeck.de

Abstract. This contribution presents a study examining five different interaction methods for manipulating objects in augmented reality (AR). Three of them were implemented on a smartglass (virtual buttons, swipe pad of a smartglass, remote control via the touchscreen of a smartwatch) and two of them on a smartphone (virtual buttons, touch interaction). 32 participants were asked to scale and rotate a virtual 3D object. We studied the usability of the interaction methods by measuring effectiveness, efficiency, and satisfaction of the users. The results of the study showed that smartphone interaction is superior to any of the studied smartglass interaction methods. Of the interaction methods implemented for the smartglass, the interaction via smartwatch shows the highest usability. Our findings suggest that smartwatches offer higher grades of usability when interacting with virtual objects rather than using the swipe pad of the smartglass or virtual buttons.

Keywords: Augmented reality · Gesture control · User study ·
Usability study · Smartglass · Google Glass · Smartwatch

1 Introduction

Commercially available augmented reality (AR) glasses that are able to display digital 3D objects have become more common in recent years. These devices allow for free movement during use, but differ in their specifications and use cases. Differences include monocular (e.g., Google Glass or Vuzix M300) or stereoscopic binocular devices (e.g., Microsoft HoloLens or Magic Leap), weight, or battery life. Regarding use cases, in this paper we focus on using AR objects in educational settings like museums. In these settings, AR glasses can be used, e.g., to augment physical exhibits, provide additional information, or to more closely examine a virtual representation without actually touching the original work.

A crucial part of dealing with AR objects is the appropriate orientation, position, and scaling of these virtual objects in the physical environment. Once a creator (e.g., museum curator or educator) creates a 3D object used for AR, it is usually necessary to change the scaling, translation, and rotation of the object to fit adequately into the physical space. Likewise, recipients of AR objects might want to see the object from all sides or in more detail, so they need to be able to rotate or scale the object as well.

© Springer Nature Switzerland AG 2019
L. T. De Paolis and P. Bourdot (Eds.): AVR 2019, LNCS 11614, pp. 84–98, 2019.
https://doi.org/10.1007/978-3-030-25999-0_8

While these tasks can be done via touch gestures on a smartphone, using AR glasses might allow for easier and more comfortable placement, orientation, and scaling.

Fig. 1. A study participant using Glass, (left) interaction with the connected smartwatch, (right) interaction with virtual buttons.

It is still unclear, how novices can best interact with AR objects on smartglasses and how different interaction methods compare to using touch gestures on a smartphone.

To examine which interaction method with virtual AR objects works best on Google Glass (Glass), we present a comparative usability study (Fig. 1) of three different interaction methods on Glass and compare them against two interaction methods on a smartphone.

2   Related Work

Several methods of interaction in AR environments have been developed and studied in recent years. Interaction methods with wearable sensors, data from head tracking, eye tracking, camera-based finger tracking, tangible AR interaction [1] and touch interaction [2] can be used.

Laviola [3] showed that when interacting with virtual objects, the error rate decreases when using stereo vision and head tracking. Hürst et al. [4] investigated finger tracking for gesture-based interaction. They used a smartphone camera to recognize finger gestures to move a 3D object on a 2D plane. The results were compared to direct input on the smartphone touchscreen. Results showed that the gestures in midair had lower accuracy. Using touch input on a plane surface was the fastest and most accurate form of interaction. Lee et al. [5] examined virtual buttons, i.e. using a matrix of markers onto paper with an app recognizing when markers were covered by a hand. Participants interacted with a calculator played a ball game and changed the color of an object. The study has shown that users found this interaction technique easy and intuitive to apply.

Smartwatches have been researched for either presenting AR content [6] or as devices for controlling head mounted displays (HMDs) [7]. However, so far, little research has focused on usability studies comparing interaction methods for AR object manipulation with Glass. Based on related work, we expected that interaction methods on smartphones show lower error rates and higher accuracy than interaction methods on Glass. Specifically, our corresponding research question:

Which of the interaction methods for Glass is best suited for AR interaction regarding rotation and scaling?

3 Selection and Implementation of Interaction Methods

Based on the use case "AR exhibits", we selected and implemented five different interaction methods, including rotation and scaling of virtual objects. Three were implemented on Glass and two on a smartphone (see below). For Glass we chose interaction methods that do not require extra devices carried in the hand of the user, using either the device itself or a connected smartwatch. We decided against implementing speech input, as this would not fit into contexts like a museum or a school. Each interaction technique was implemented as a separate app with a high level of code reuse.

The application is based on the app InfoGrid [8, 9], which is intended to be used to digitally enriched learning spaces [10] like museums and schools. InfoGrid is primarily used on users' mobile devices and uses image recognition to detect physical objects. AR content is delivered by a web-connected backend framework the Network Environment for Multimedia Objects (NEMO) [11], which stores and handles all data in a cloud-based semantic database. InfoGrid enables creators and editors (e.g., curators in museums) to use a web-based application to place AR content in physical environments, and it allows consumers (e.g., museum visitors or students) to view the AR content through a mobile app for smartphones, tablets or smartglasses.

3.1 Glass Interaction Methods

All interaction methods on Glass use an extended build of InfoGrid, called Info-Grid4Glass. It has been developed using the Unity [12] development environment and the Vuforia image recognition framework. The Vuforia [13] framework provides the functionality to detect visual markers and supports the implementation of virtual buttons. The three interaction methods implemented on Glass were:

1. **TouchPad on Glass (TG)**
The touchpad of Glass was used for input. A swipe forward caused a rotation to the left and a swipe back to the right (Figs. 1 and 2). A double tap on the swipe pad triggered a change from the rotation to scaling mode and vice versa. A swipe forward caused an increase in object size and a swipe backward caused a decrease in object size.

Fig. 2. Interaction gestures on the swipe pad of Glass.

2. Virtual Buttons through Glass (VG)

Four virtual buttons were displayed (Figs. 3 and 1 right) on Glass. Users can select a button by covering it while holding up their fingers in midair.

Fig. 3. Screenshot of the 3D object of the head displayed in a starting orientation in the Unity development environment. The blue icons are virtual buttons. (Color figure online)

3. Smartwatch Connected to Glass (SG)

A smartwatch was paired to Glass via a Wi-Fi network. To accelerate the device discovery, we set up the smartwatch to use a static IP address. The AR object was displayed on Glass along with a symbol indicating the active interaction method

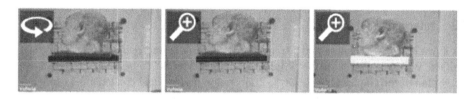

Fig. 4. Screenshots of InfoGrid4Glass running with the connected smartwatch, (left) rotation mode, (center) resizing mode, (right) 3D Object is correctly aligned.

(Fig. 4). A swipe to the right or left on the smartwatch caused a rotation to the corresponding side. A double tap was used to switch the mode from the rotation to scaling mode. In scaling mode, a swipe from left to right causes a zoom in and a swipe to the left causes a zoom out.

3.2   Smartphone Interaction Methods

On the Smartphone, two interaction types were implemented:

1. Touchscreen on Smartphone (TS)
Interaction by the user on the touchscreen of the smartphone was used for rotation and scaling (Fig. 5).

Fig. 5. Interaction gestures on the touchscreen of the smartphone.

2. Virtual Buttons Through Smartphone (VS)
Virtual buttons were used, activated by pointing the finger into defined areas of the camera view of the smartphone (VS) (Fig. 6).

Fig. 6. Interaction gestures in midair to activate the virtual buttons.

3.3   Interaction Object

As a test object for rotation and scaling, we used a digital representation of a physical artifact. We took photos of the physical sculpture "Heads" (by Günter Grass) in the exhibition of the Günter Grass-Haus (a cooperation partner) and created a 3D Model of it using photometrical methods. We attached a rectangular stand to the model to indicate the current orientation and scaling (Fig. 3). As the visual background, we used an image of a wall. Printed on paper, the image served as a recognition marker for the AR app. A virtual shelf was displayed additionally for guidance regarding the correct orientation of the AR object: the head had to be oriented in a particular direction and the object stand had to be aligned with the shelf.

4 Study: Methods

4.1 Study Design

A within-subjects design was used to conduct the study. Participants tried out the different interaction methods to rotate and scale the virtual object.

4.2 Study Participants

32 participants invited via email and direct communication took part in the study. Participants received a chocolate bar as a compensation for their time and cooperation. Age varied between 19 and 62 years ($M = 30.88$, $SD = 11.06$). Participants were recruited on the campus of our university as well as inside the museum (cooperation partner). 22 participants were male and 10 were female. 17 participants were wearing glasses. 32 participants owned a smartphone and four a smartwatch. 22 participants did already know AR apps and 15 did know about Google Glass.

4.3 Study Instruments

In this section, we present the hardware, tasks, and questionnaire used for the study.

Hardware
As hardware, we used a Google Glass Explorer Edition V2 with a 640×360 px display, a Samsung Galaxy S5 smartphone with a 5.1" Full HD display and a Sony SWR3 smartwatch with a 1.6" 320×320 px display. To ensure low network latency, a wireless access point was used to provide a private network for the devices.

Rotation and Scaling Task
The task for the user was to align the object stand with the shelf, with the head looking to the left. To ensure comparability of the conditions, we held the deviation between starting position and final position constant at 110°. Likewise, for scaling, a constant start deviation of 35% was used. Interaction time and error rates were logged and saved onto the SD-card of the device. The apps continuously checked the current rotation and size of the object. To assess deviations during use, if a participant rotated the object further than necessary, the deviation from the correct position to the selected position was summed up. The app did not count the degrees back to the correct position as an error. The same applied to the scaling error. The task finished when the user rotated and scaled the object until it matched the correct position. Based on our pre-study tests, we used a tolerance of 5% in size and an error rate of 2.22% in rotation as correctly aligned in all conditions.

Usability Questionnaire

To measure the usability of AR applications, several questionnaires exist. Many of these questionnaires determine the usability of handheld AR systems for performing multiple tasks [14–16]. In our study, we focus on assessing detailed information on the interaction task that is used either with handheld or with HMD devices. Thus, we adapted a usability questionnaire, which has already been successfully used to measure usability in similar contexts [2, 17, 18].

The following list shows the items of the questionnaire (translated from German; 7-point Likert scale ranging from 1: strongly disagree to 7: strongly agree):

1. I was able to use the interaction method well.
2. The object was easily visible.
3. The interaction method was well suited for rotation.
4. The interaction method was well suited for scaling.
5. Switching between scaling and rotation was easy.
6. The interaction method was quick to master.
7. The interaction method was intuitive.
8. The interaction method required a lot of dexterity.
9. The interaction method required a lot of concentration.

Preferred Glass Interaction Method

To force a decision between the Glass-based interaction methods, participants were asked which Glass-based interaction method they liked best.

Setting

The study was carried out in the usability lab of our research institute as well in the use context of a museum (Günter Grass-Haus). Two markers were placed on the wall at the eye level of the participant (one with, the other without, buttons for the virtual button condition, (see Fig. 1), with another image showing the final orientation and final size of the virtual object.

Procedure

Study participants were welcomed and the tasks and procedures were explained to them. The markers were aligned to the participants' eye level, Glass was cleaned and the position of the prism of Glass adjusted for the participant. The participants tried out the touch input on the smartphone to familiarize themselves with the system.

Table 1. Order and initial values of the interaction methods.

No.	Interaction method	Starting rotation	Initial size
1	TS: touch interaction on the smartphone	−110°	135%
2	VG: virtual buttons on Glass	−110°	65%
3	TG: touch interaction on the swipe pad of Glass	110°	135%
4	VS: virtual buttons on the smartphone	110°	65%
5	SG: touch interaction on the smartwatch connected to Glass	−110°	135%

Participants were then asked to perform the tasks in the five conditions. To familiarize the participants with the task and to avoid overheating of Glass after continued use, the order of the interaction methods was fixed in the following order (Table 1):

After completing each task, the participants were asked to fill out the usability questionnaire for that interaction method. Filled in pages were removed to ensure the participants could not review their previous answers when filling in the next questionnaire. After all conditions were assessed, the participants were asked to choose which Glass-based interaction method they liked best and answered demographic questions.

5 Study: Discussion

Statistical calculations were made with GraphPad Prism (GraphPad 3.5.1) and R (with RStudio 1.1.453).

5.1 Interaction Duration

Mean and standard errors of the interaction times are shown in Fig. 7.

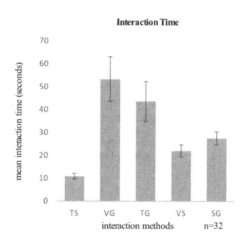

Fig. 7. Time in seconds the participants needed to correctly align the virtual 3D object. TS: touch interaction on the smartphone; VG: virtual buttons on Glass; TG: touch interaction on the swipe pad of Glass; VS: virtual buttons on the smartphone; SG: touch interaction on the smartwatch connected to Glass.

A D'Agostino-Person Omnibus K2 Test [19] did show that the data is not equally distributed, so we used the non-parametric Friedman Test [20]. As significant differences did exist, we used the non-parametric post-hoc-test of Dunn [21] to determine which interaction methods did differ in duration. Significant differences were found between the interaction times of the touch interaction on the smartphone (TS) and all other interaction methods; with the smartphone being significantly faster (all p-values are adjusted: TS-VG: $p < .001$; TS-TG: $p < .001$; TS-VS: $p < .0443$; TS-SG: $p < .001$). We also found significant differences between virtual buttons on Glass (VG) and virtual buttons on the smartphone (VS) with the smartphone being significantly faster (adjusted p-value: VG-VS: $p < .005$). We repeated the tests for the recorded errors of the rotation (Fig. 8) and the scaling (Fig. 9).

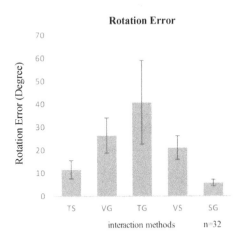

Fig. 8. Rotation errors (y-axis). TS: touch interaction on the smartphone; VG: virtual buttons on Glass; TG: touch interaction on the swipe pad of Glass; VS: virtual buttons on the smartphone; SG: touch interaction on the smartwatch connected to Glass.

The data was not equally distributed and the non-parametric post hoc test of Dunn [21] did show a significant difference between the scaling errors using the virtual buttons on Glass (VG) and the touch interaction on the swipe pad of Glass (TG); the

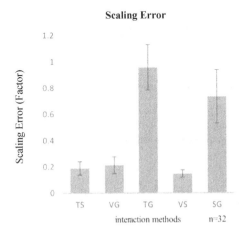

Fig. 9. Scaling errors (y-axis). TS: touch interaction on the smartphone; VG: virtual buttons on Glass; TG: touch interaction on the swipe pad of Glass; VS: virtual buttons on the smartphone; SG: touch interaction on the smartwatch connected to Glass.

touch interaction on the smartphone (TS) and the touch interaction on the swipe pad of Glass (TG); the touch interaction on Glass (TG) and the virtual buttons on the smartphone (VS) (all p-values are adjusted: TS-TG: $p < 0.001$; VG-TG: $p < 0.001$; TG-VS: $p < 0.001$).

In addition, significant differences between the rotation errors of the touch interaction on the swipe pad of Google Glass (TG) and the touch interaction using the connected smartwatch with Google Glass (SG) as well as between the virtual buttons on the smartphone (VS) and the smartwatch connected to Glass (SG) were found (all p-values are adjusted: TG-SG: $p < .0103$; VS-SG: $p < .0344$).

5.2 Usability Questionnaire

Mean and p-values of the results of the questionnaire are shown in Table 2.

Table 2. Results of the questionnaire. Mean values of the questions of a 7-item Likert scale. The higher the values, the better the results (the item values of Q8 and Q9 were for this purpose). The last column shows the p-values of the Friedman test.

Question	TS	VG	TG	VS	SG	p
1	6.78	5.44	5.75	5.63	6.44	<0.01
2	6.59	4.50	4.69	6.53	5.19	<0.01
3	6.69	4.78	5.97	5.22	6.41	<0.01
4	6.56	4.94	5.66	5.34	6.28	<0.01
5	6.81	6.31	5.03	6.16	6.13	<0.01
6	6.84	5.53	6.09	5.88	6.38	<0.01
7	6.57	5.07	5.50	5.37	5.67	<0.01
8	5.19	3.44	4.19	3.78	4.78	<0.01
9	5.47	3.19	4.44	4.22	5.00	<0.01
Total	6.39	4.80	5.26	5.35	5.81	

Internal consistency (Cronbach's Alpha) was assessed for the usability questionnaire in all five conditions. Items 7 and 9 were dropped in all conditions to increase internal consistency (see Table 3, values ranging from questionable in TS and SG, acceptable in TG, to good in VG and VS).

Table 3. Results of the usability questionnaire ($n = 32$). Values could range from 1 to 7, with higher values meaning higher usability (item 8 was reverse-coded and was thus inverted; items 7 and 9 were dropped).

Condition	M(SD)	Min-Max	Cronbach's Alpha
TS	6.50(0.47)	5.14–7.00	.69
VG	4.99(1.17)	2.57–7.00	.88
TG	5.50(1.08)	2.50–7.00	.77
VS	5.50(1.08)	2.57–7.00	.89
SG	5.94(0.62)	4.86–7.00	.64

Figure 10 shows the boxplots of the usability questionnaire values of the five conditions. A one-way repeated measures ANOVA was conducted to compare mean usability ratings in the five conditions (means and standard deviations, see Table 2). There was a significant effect for condition, $F (4, 124) = 21.75$, $p < .001$. Bonferroni adjusted post hoc tests did show significant differences between TS and all other conditions (TS higher usability) and between SG and the other Glass-based conditions TG and VG (SG higher usability).

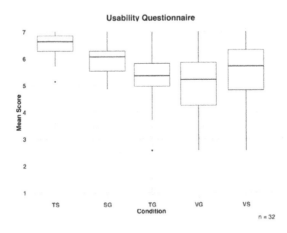

Fig. 10. Mean usability scores in the five conditions: TS: touch interaction on the smartphone; SG: touch interaction on the smartwatch connected to Glass. TG: touch interaction on the swipe pad of Glass; VG: virtual buttons on Glass; VS: virtual buttons on the smartphone

5.3 General and Demographic Questions

All participants owned a smartphone; therefore, no effects could be calculated regarding smartphone ownership. The correlation between the participants that have a smartwatch ($n = 4$) and the participants that do not have a smartwatch ($n = 28$) was not calculated, because of the inequality of the number of members between the groups. All other correlations were calculated with the Spearman Rank correlation [22]. The Spearman correlation values were assessed after Cohen [23] for strong correlations ($r > 0.5$). Three strong correlations were found between the age and the interaction time on the swipe pad of Glass ($r = 0.628$, $p < 0.001$; Fig. 11), the age and the virtual buttons on the smartphone ($r = 0.648$, $p < 0.001$; Fig. 12), the age and the smartwatch connected to Glass ($r = 0.623$, $p < 0.001$; Fig. 13).

Participants who had previous experience with AR ($r = -0.536$, $p = 0.002$) or had experience with Google Glass ($r = -0.545$, $p < 0.002$) made less scaling errors than the participants that did not have previous experience. No other statistically significant correlations were found.

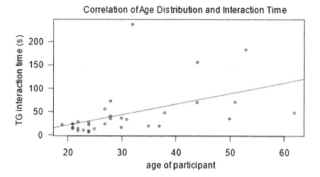

Fig. 11. Correlation of interaction time for the touch interaction on the swipe pad of Glass (y-axis) and the age of the participant. The line represents the linear increase of the interaction time depending on the age and was calculated with linear regression.

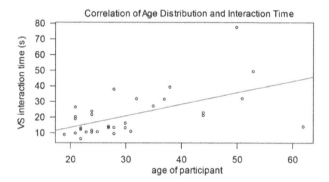

Fig. 12. Correlation of interaction time for the virtual buttons smartphone (y-axis) and the age of the participant. The line represents the linear increase of the interaction time depending on the age and was calculated with linear regression.

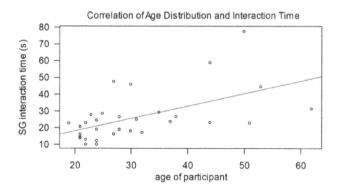

Fig. 13. Correlation of interaction time for the smartwatch with Glass (y-axis) and the age of the participant. The line represents the linear increase of the interaction time depending on the age and was calculated with linear regression.

5.4   Preferred Interaction Method

Asked, which interaction method on Glass they liked best, of 32 participants, 20 preferred the interaction on the smartwatch in connection with Glass, nine the touch interaction with the swipe pad of Glass, and three the virtual buttons on Glass.

5.5   User Observations

Multiple participants of the study mentioned that they do not know the diopter strength of their glasses. Therefore, we did not test for these values as initially planned. One participant mentioned that it is difficult to use the virtual buttons on Glass because of the need to lift the arm during the interaction. One participant mentioned that it is more exhausting using the virtual buttons because of the need to switch their focus from the buttons to the object multiple times during interaction.

6   Discussion

Participants needed the least time to solve the task with a smartphone. Both smartphone conditions are superior to using any Glass-based interaction method. All interaction methods on Glass had the same FPS.

Regarding time used, no statistically significant differences were found within the Glass-based conditions. Thus, no Glass-based interaction method is preferable when it comes to the time used.

Looking only at the Glass-based interaction methods, Glass with virtual buttons (VG) led to the least scaling errors and Glass with a smartwatch (SG) to the lowest rotation errors. The results of the usability questionnaire indicate that of the Glass-based interaction methods, using the smartwatch (SG) was rated best (albeit not as well as touch interaction on the smartphone, TS).

Regarding preference, asking just about the Glass-based interaction methods, participants also preferred the combination with the smartwatch.

Thus, when it comes to Glass-based interaction, using a smartwatch seems to be the best interaction method. However, using Glass with a smartwatch was also the last condition tested with a fixed order of conditions, so training effects cannot be ruled out. An experimental design with a random order of conditions would further increase confidence in the results. As for the comparison between Glass- and smartphone-based interactions, general learning effects likely also play a role, as all participants used smartphones in their everyday life, while there is no such prior experience for Glass. As the focus of the paper is novices (currently the typical museum visitors interacting with AR exhibits), the lack of familiarity with Glass is not that relevant.

7 Conclusion and Future Work

In this paper, we presented the evaluation of three different interaction methods for a smartglass and two interaction methods on a smartphone for manipulating objects in augmented reality (AR), assessing usage times, error rates, usability rating and preference.

As expected, the touch interaction method on the smartphone is significantly faster than all other interaction methods. It also received the highest scores in the usability questionnaire. The users also made the least scaling errors with the smartphone, but not the least rotation errors. The smartglass combined with a connected smartwatch had the highest usability scores of all three smartglass-based interaction methods and is most frequently preferred by the participants.

Further studies should increase internal validity and effects of different device designs and assess the usability of similar interaction devices and stereoscopic AR glasses like the Microsoft HoloLens.

The distribution of monocular AR smartglasses has recently shifted from the consumer to the enterprise market. Google distributes its former Google Glass now only as Google Glass for enterprise edition. Companies like Focalmax with the smartglass Scati S1 and Vuzix with the smartglass Vuzix M300 also focus mainly on enterprise applications in the fields of manufacturing, field service, warehousing, work instructions and others. Our research supports this trend for, in first order, professional usage of smartglasses. We think it makes sense for museum professionals as well to be able to place their virtual content with smartglasses and to see detailed information of the exhibits hands-free. For museum visitors, the highest usability can currently be achieved by using their own smartphones or tablets to perceive and interact with AR content.

References

1. Damala, A., Hornecker, E., van der Vaart, M., van Dijk, D., Ruthven, I.: The loupe: tangible augmented reality for learning to look at ancient Greek art. Mediterr. Archaeol. Archaeom. **16**, 73–85 (2016). https://doi.org/10.5281/zenodo.204970
2. Budhiraja, R., Lee, G.A., Billinghurst, M.: Using a HHD with a HMD for mobile AR interaction. In: 2013 IEEE International Symposium on Mixed and Augmented Reality, ISMAR 2013 (2013). https://doi.org/10.1109/ismar.2013.6671837
3. Laviola, J.J.: The influence of head tracking and stereo on user performance with non-isomorphic 3D rotation. In: 14th Eurographics Symposium on Virtual Environments, Virtual Environments 2008, pp. 111–118 (2008). https://doi.org/10.2312/EGVE/EGVE08/111-118
4. Hürst, W., Van Wezel, C.: Gesture-based interaction via finger tracking for mobile augmented reality. Multimed. Tools Appl. **62**, 233–258 (2013). https://doi.org/10.1007/s11042-011-0983-y
5. Lee, G.A., Billinghurst, M., Kim, G.J.: Occlusion based interaction methods for tangible augmented reality environments. In: Proceedings of the 2004 ACM SIGGRAPH International Conference on Virtual Reality Continuum and Its Applications in Industry, pp. 419–426. ACM, New York (2004)

6. Wenig, D., Schöning, J., Olwal, A., Oben, M., Malaka, R.: WatchThru: expanding smartwatch displays with mid-air visuals and wrist-worn augmented reality. In: Proceedings of the 2017 CHI Conference on Human Factors in Computing Systems, pp. 716–721 (2017). https://doi.org/10.1145/3025453.3025852

7. Serrano, M., Hasan, K., Ens, B., Yang, X., Irani, P.: Smartwatches + Head-Worn Displays: the 'New' Smartphone. Mobile Collocated Interactions: From Smartphones to Wearables (2015)

8. Ohlei, A., Bouck-Standen, D., Winkler, T., Herczeg, M.: InfoGrid: an approach for curators to digitally enrich their exhibitions. In: Mensch und Computer 2018 - Workshopband (2018)

9. Ohlei, A., Bouck-Standen, D., Winkler, T., Herczeg, M.: InfoGrid: acceptance and usability of augmented reality for mobiles in real museum context. In: Mensch und Computer 2018 - Workshopband (2018)

10. Winkler, T., Scharf, F., Hahn, C., Herczeg, M.: Ambient learning spaces. In: Education in a Technological World: Communicating Current and Emerging Research and Technological Efforts, pp. 56–67 (2011)

11. Bouck-Standen, D., Ohlei, A., Winkler, T., Herczeg, M.: An approach to auto-enhance semantic 3D media for ambient learning spaces. In: Kropf, J., Berntzen, L. (eds.) AMBIENT 2018 - The Eight International Conference on Ambient Computing, Applications, Services and Technologies. IARIA, Athens, Greece (2018)

12. unity3d.com. https://unity3d.com/de

13. www.vuforia.com, https://www.vuforia.com/

14. Brooke, J.: SUS-A quick and dirty usability scale. In: Usability Evaluation in Industry. CRC Press (1996)

15. Finstad, K.: The usability metric for user experience. Interact. Comput. **22**, 323–327 (2010). https://doi.org/10.1016/j.intcom.2010.04.004

16. Santos, M.E.C., Polvi, J., Taketomi, T., Yamamoto, G., Sandor, C., Kato, H.: Toward standard usability questionnaires for handheld augmented reality. IEEE Comput. Graph. Appl. **35**, 66–75 (2015). https://doi.org/10.1109/MCG.2015.94

17. Bai, H., Lee, G.A., Billinghurst, M.: Freeze view touch and finger gesture based interaction methods for handheld augmented reality interfaces, p. 126 (2013). https://doi.org/10.1145/2425836.2425864

18. Yang, Z., Weng, D.: Passive haptics based MR system for geography teaching, pp. 23–29 (2016). https://doi.org/10.1145/3013971.3013995

19. D'Agostino, R.B.: Goodness-of-Fit-Techniques. Taylor & Francis (1986)

20. Friedman, M.: The use of ranks to avoid the assumption of normality implicit in the analysis of variance. J. Am. Stat. Assoc. **32**, 675–701 (1937). https://doi.org/10.1080/01621459.1937.10503522

21. Dunn, O.J.: Multiple comparisons among means. J. Am. Stat. Assoc. **56**, 52–64 (1961). https://doi.org/10.1080/01621459.1961.10482090

22. Zar, J.H.: Significance testing of the spearman rank correlation coefficient. J. Am. Stat. Assoc. **67**, 578–580 (1972). https://doi.org/10.1080/01621459.1972.10481251

23. Cohen, J.: A power primer. Psychol. Bull. **112**, 155–159 (1992). https://doi.org/10.1037/0033-2909.112.1.155

An Augmented Reality Tool to Detect Design Discrepancies: A Comparison Test with Traditional Methods

Loris Barbieri$^{(\boxtimes)}$ (ID) and Emanuele Marino (ID)

Department of Mechanical, Energy and Management Engineering (DIMEG),
University of Calabria, Via P. Bucci, 87036 Arcavacata di Rende, CS, Italy
{loris.barbieri,emanuele.marino}@unical.it

Abstract. Augmented Reality (AR) is an innovation accelerator for Industry 4.0 that supports the digitalization and improves the efficiency of the industrial sector by providing powerful tools able to enhance the workers' visual perception by combining the real world view with computer-generated data.

In this context, the paper presents a new AR tool and an exploratory test in order to examine how well it supports the user's tasks for the detection of design discrepancies. In particular, the test aims to evaluate the effectiveness and efficiency of the proposed tool and how it compares to other instruments traditionally adopted for this end, such as technical drawings and CAD systems.

The experimental findings show that the proposed AR tool presents similar results with the other instruments in term of effectiveness and very encouraging results about its efficiency.

Keywords: Augmented reality · Assistive tool · Design variations · Industry 4.0

1 Introduction

In the industrial field, it is not uncommon to find discrepancies between the design defined by the technical office and the correspondent product. The causes have different origins and in the majority of cases are due to unexpected issues that arise during the manufacturing and prototyping activities because of the technology adopted, operators' capacities or environmental context. These design discrepancies are quite common in the Oil&Gas sector, and in particular, in the production departments for the manufacturing and assembly of small lot sizes of tubing and piping. In fact, because of the complexity of the piping networks and systems, they often require modifications and improvements to be made on-site by workers which intentionally perform a number of design changes and adjustments in order to minimize the number of bendings, reduce overall dimensions, streamline pipe routes, and increase the visibility and manoeuvrability space for easier assembly of the components. These design variations, if not properly and promptly detected, unavoidably lead to a significant increase in the production cycle's time and can cause slowdowns and inefficiencies both in the design, production and maintenance areas.

© Springer Nature Switzerland AG 2019
L. T. De Paolis and P. Bourdot (Eds.): AVR 2019, LNCS 11614, pp. 99–110, 2019.
https://doi.org/10.1007/978-3-030-25999-0_9

At the moment, a very common solution for the detection of these design changes consists of a paper-based approach in which workers write down a few notes about the discrepancies on papers or on technical drawings. These annotations, sometimes accompanied by photographic material, are then gathered and send to the technical office that will take care of the updating of the CAD models on the basis of which the next technical drawings will be generated. Clearly, this is not an efficient and reliable approach, in fact, the hand-written notes may be lost or may not be exhaustive and consequently lead to miss-understandings. In any case, it is not possible for the technical office infers from 2D sources, such as technical drawings and pictures, a precise determining of the position in 3D space of the design changes.

From the perspective of Industry 4.0, Augmented Reality (AR) represents a key technology that can overcome these limitations and inefficiencies. In fact, in the last decades, AR technology has been successfully and efficiently applied in many sectors of the industrial field to help workers to accomplish several tasks [1–5] thanks to its capability to superimpose virtual data directly on the real environment. In particular, the display technology adopted for the development of AR-based solutions typically relies on head-mounted displays (HMDs) [6, 7], projectors [8, 9], and mobile devices [10, 11] such as smartphones and tablets. To the detriment of these varieties of technological solutions, the question about which device or display solutions fit better the industrial context of the future smart factories remains still open. This is due to the fact that there is not an ultimate technology that is useful for all types of environments with sufficient accuracy and reasonable cost, but the kind of solution to adopt it is strictly dependent by the specific context of the application, contingent users' needs and industrial constraints.

About the abovementioned industrial context, the specific AR application should allow workers to conduct the detection of the design discrepancies at the workplace while watching the superimposed images on the real scenario. To this end, digital projectors offer the advantage to superimpose virtual data directly on the real environment, but this requires to be placed in fixed positions and are characterized by limited areas of usage, surface-based distortions, lower color fidelity and visibility of projection. HMDs devices provide more freedom of movement, and the user can use both hands while watching the superimposed images but they suffer from bad ergonomics and excess of weight. Furthermore, both projectors and HMDs are expensive devices and require a workstation for the running of software applications. As a consequence, mobile devices could be more adequate for the specific application field. If, on the one hand, mobile handheld devices require both hands for the interaction, on the other hand, users can bring them at the workplace and interact with the AR application by moving and rotating the standalone device in the real space.

On the basis on these considerations, the paper presents a novel AR tool that provides support at the workplace to easily detect design changes by augmenting virtual 3D models, as defined in the project plan, on the actual design. The proposed AR tool requires no external hardware but a smartphone device. In this manner, the user can easily hold the device with one hand and interact with the dominant hand. The paper focuses on an exploratory and comparison test in which the AR tool is compared to technical drawings and CAD systems, traditionally adopted for the individuation of design variations, in order to investigate and quantify the impact and the validity of the proposed tool over the other two.

This first exploratory lab test activity has been carried out before proceeding with field testing in order to collect a preliminary set of usability data that will be helpful to quantify its potentials and improve its design in terms of functionalities and ease of usage. This first lab test is part of the development process of a novel instrument that should support workers at the workplace. Then further lab and field, assessment and validation, tests will be carried out within the context of a user-centered design approach, with usability testing conducted throughout the whole designing process in order to develop a usable handheld AR-based tool.

2 Research Aim and Test Materials

As stated in Sect. 1, an exploratory study has been carried out to evaluate the effectiveness and efficiency of a new AR tool that allows to detect design discrepancies between real parts and their associated construction data. In particular, the main goal of the testing is: compare the proposed AR tool with traditional instrument adopted for performing design discrepancy check by measuring user performances while completing the task.

The test sample has been specifically designed in order to simulate a simple real case study in which the physical item presents some variations with respect to the design defined in the project plan. In particular, Fig. 1a depicts the 3D model of a set of five tubes that consists in single socket pipes, of the diameter of 40 mm, assembled by means of 45° and 90° bend and equal branch connection pipes.

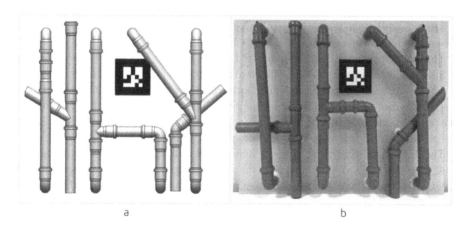

a   b

Fig. 1. 3D model of the piping (a) and its physical prototype with ten design discrepancies (b).

The correspondent full-size physical prototype (Fig. 1b) has been made with push-fit lightweight PVC pipes that have been placed by means of hot glue on a thin stiff sheet of transparent plexiglass with the approximate dimensions of 83 cm (length) × 70 cm (height). On each pipe of the physical mock-up (Fig. 1b) two design changes have been intentionally introduced, for a total of ten design variations as compared to the 3D model (Fig. 1a). The design changes are limited to length and inclination of pipes, with the exception of a missing component on the first pipe on the left.

3   AR Tool

The AR tool consists of an application developed with Google Project Tango Development Kit [12]. This software development technology, introduced by Google in 2014, has been preferred to other solutions because it offers hybrid tracking techniques that combine vision-based and sensor-based methods to calculate device's motion and orientation in 3D space in real-time. In particular, the application takes advantage of both marker-based and natural feature-based techniques, and combine these ones with a sensor fusion technique that uses the various sensors (motion tracking camera, 3D depth sensor, accelerometer, ambient light sensor, barometer, compass, GPS, gyroscope) equipped on the device to remember areas that it has travelled through and localize the user within those areas to up to an accuracy of a few centimeters.

Between the years 2014 and 2017, the Tango technology has been enabled on a limited number of consumer Android devices: the Peanut mobile phone and the Yellowstone tablet produced by Google; the Zenfone AR smartphone produced by ASUS and the Lenovo Phab 2 Pro [13] smartphone. All these devices have in common the hardware architecture required to enable Tango technology. In particular, the motion tracking is achieved using a high-performance accelerometer and gyroscope sensors, depth perception happens thanks to a proprietary RGB-IR camera, and area learning entails the recording of it all for real-time or rendered data. Because of the need of these special sensors recently Google has introduced a new augmented reality system, known as ARCore [14], that works by exploiting the hardware already on board on the consumer Android smartphones.

The developed application runs on a Lenovo Phab 2 Pro smartphone. It is an Android device equipped with a Qualcomm Snapdragon 652 (1.80 GHz) processor, a 16MP camera, and integrated depth and motion tracking sensors (accelerometer, digital compass, gyroscope, proximity sensor, ambient light sensor). Since all the computations are carried out on the device itself, there are no other external hardware components required for the data input and processing.

The following illustration (Fig. 2) depicts the software architecture of the AR application that takes as input images of the real scenario framed by the camera and 3D models in order to combine this information to create the augmented reality visualization provided to the user by means the display of the handheld device.

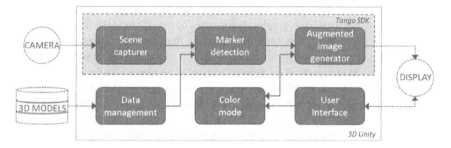

Fig. 2. Software architecture of the AR tool.

The different modules, depicted in Fig. 2, have been programmed in 3D Unity [15] and each one is dedicated to one or more specific operations.

A real-world live video is given as an input from the Android cell phone camera to the "Scene capturer" module. Then the "Marker detection" module gets access to this data in real time by means of the Tango SDKs in order to recognize the presence of a marker in the real world scenario.

When a marker is detected, the "Data management" module gets access to an external file system in which the 3D models are stored and sends this data to the "Marker detection" module. This last module calculates the relative position and orientation of the detected marker and assigns to it a local reference frame on which the 3D models' coordinate system is aligned. All these data are sent to the "Augmented image generator" module which performs a precise superimposition of the virtual objects on the real world scenario.

The "Marker detection" module is responsible also for motion tracking capabilities. In particular, it takes advantage of the visual-inertial odometry algorithm, embedded in the Tango device, to combine the information provided by the inertial sensors and camera of the device and then track its own movement and orientation through 3D space.

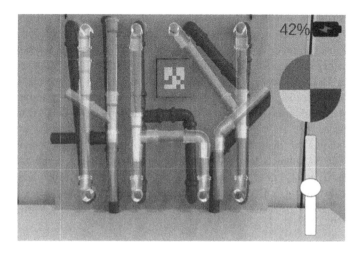

Fig. 3. 3D models augmented on the physical prototype.

The "Color mode" module uses functions to set material properties and change, from opaque to transparent, the graphics properties of the 3D models. These properties are made available by the "User Interface" module that displays a menu by means of which the user can perform a real-time customization of the graphic appearances of the 3D models (Fig. 3). In particular, default setting provides a shaded display of the augmented virtual models in a light gray color but the user can change this color by means of a color palette displayed on the right side of the screen and set transparency level through a slider. The customization of the visual style of the 3D models makes easier the detection of the design discrepancies because it allows to clearly distinguish between the augmented virtual models and the real world background.

4   Usability Testing

As abovementioned in the previous sections, a usability testing, consisting in an exploratory and comparison test, has been carried out in order to collect empirical data while observing users using three different instruments to perform a realistic task for the detection of design discrepancies.

The test has been carried out in the laboratories of the Department of Mechanical, Energy and Management Engineering (DIMEG) of the University of Calabria. All the participants were students enrolled in the first year of Masters degree in mechanical engineering. In particular, for this comparison test, 34 volunteers have been selected, 26 male and 8 female, with age ranged from 22 to 30 (mean 24.5, standard deviation 1.9). The number of participants has been chosen according to the most influential articles on the topic of sample size [16, 17].

As abovementioned all the subject involved in the comparative test were from an engineering background in order to assure the involvement in the study of users that have, as prerequisite knowledge, abilities in the use of CAD systems and preparation and interpretation of technical drawings and then they have the skills and the knowledge to comprehend if the developed AR tool could be a valid alternative to the traditional instruments for the detection of design variations. In fact, all of them were familiar with 2D engineering drawings and expert in the use of 3D modelling and CAD systems. On the contrary, none of the participants had any previous experience of augmented reality devices for the visualization of 3D models superimposed on a real scenario. This lack of knowledge and practical experience is functional for the user study because it reflects a characteristic of the user profile. In fact, as abovementioned in Sect. 1, the targeted users are workers that could adopt the proposed AR tool directly at the workplace.

Figure 4 shows the instruments and metrics adopted for the independent groups design test. The 34 subjects have been in fact separated in three homogeneous (age, gender and visual acuity) groups in order to assign to each group a specific instrument, i.e., a technical drawing, a CAD system and the proposed AR tool.

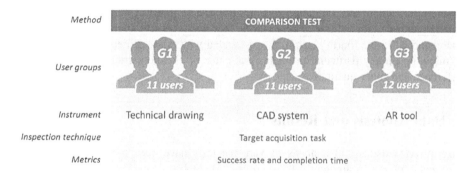

Fig. 4. Comparison test plan.

Each specific instrument has been then evaluated in terms of its effectiveness and efficiency on the basis of users' results while performing a target acquisition task. This evaluation has been carried out according to the ISO 9241-11:2018 [18] standard. In particular, the effectiveness is the accuracy and completeness with which users achieve specified goals while the efficiency is related to the resources expended in relation to the accuracy and completeness with which users achieve goals. In this study, the metrics adopted to measure effectiveness and efficiency are respectively: success rate and task completion time.

About the task, each participant was asked to detect the design variances between the physical and the virtual prototype (Fig. 1) through the assigned instrument (Fig. 5).

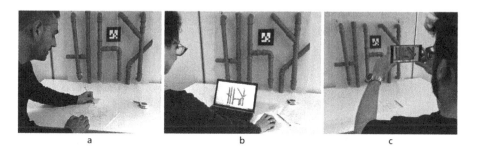

Fig. 5. Participants while using technical drawing (a), CAD system (b) and the AR tool (c).

The comparison test consisted of four sessions. In the first session, the participants began the test filling out a background questionnaire. In the second session, a tutor provided the participant with the assigned instrument and only for the AR tool a very brief demonstration (usually lasting about 30 s) about its usage has been provided. Then the tutor assigned the task to the participants. In particular, the participants were informed to adopt the assigned technical drawing or digital data as reference in relation to which detect the design variations made on the physical prototype. No mention was made about the number of discrepancies. In the third session, each participant carried out the task without any limitation in time. In order to perform the required task, the participants had at their disposal a measuring tape and a pencil to note down the design discrepancies. In the fourth session, the test ended with post-test debriefing interviews to follow up on any particular problems that came up for the participants, and collect preference and other qualitative data.

5   Data Analysis and Results

Descriptive statistics and analysis of variance tests have been used for statistically significant differences among the three groups. All analyses have been conducted using the statistical packages Microsoft Excel and IBM SPSS. The statistical significance level has been set at $p > 0.05$.

As abovementioned in the previous section, two separate statistical models have been adopted to analyse the effect of the instrument in the identification of design discrepancy task on time and success rate.

The following bar chart (Fig. 6) shows the success rate of the number of design variations identified by participants by means of the three different instruments. Error bars represent 95% confidence interval. The chart presents similar value of the average success rate, specifically 74.54% with SD of 9.3 for the group that performed the test with technical drawings, 74.54% with SD of 16.9 for the group with the CAD system, and 72.50% with SD of 6.2 for the AR tool.

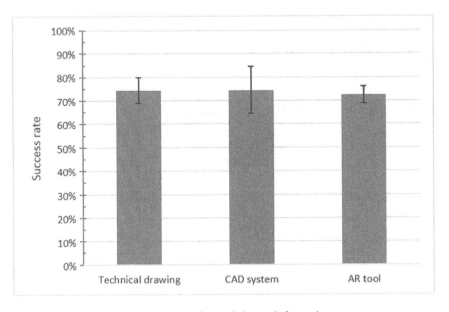

Fig. 6. Success rate of completion task for each group.

These data have been compared by means of an ANOVA test. The Levene's test for the homogeneity of the variances presents a value of 0.05, so it has been possible to confirm the null hypothesis. The ANOVA results show that $F(2,31) = 0.12$, $p = 0.88$ that means that there is not a significant effect of the tool on the success rate.

The following bar chart shows the average time of each group to complete the task, error bars represent 95% confidence interval. In particular, the participants with the AR tool completed the test in an average time of 1 min and 43 s, and standard deviation SD of 0.43 min. Differently, the participants with technical drawings and CAD models completed the test respectively in 12.50 min (SD 1.73) and 10.89 min (SD 2.20) (Fig. 7).

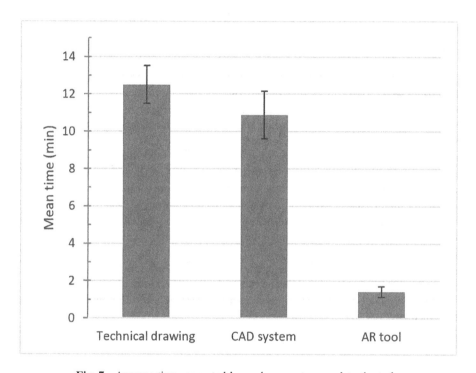

Fig. 7. Average time requested by each group to complete the task.

The completion time was analyzed using one-way ANOVA test which results confirm a significant effect of the tool on the time required by users to identify design variations with $F(2,31) = 165.46$ and $p < 0.05$. However, the Levene's test has come out as highly significant with a significance value that is < 0.05. So the null hypothesis is violated, that means that the variances of the groups are significantly different. This is quite evident in the abovementioned descriptive statistics results in which the standard deviation of the G3 group is 3 times less than the G1 group, and 5 times less than the G2 group. Since the variances are significantly different alternative statistics have been adopted. In particular, the Brown-Forsythe and Welch versions of the F-ratio have been investigated. Both these tests are still highly significant in fact their results show a significance value < 0,05. In particular, the Brown-Forsythe F-ratio presents a value of $F(2,20.11) = 158.52$ and $p < 0.001$, and the Welch F-ratio is $F(2,14.85) = 289.07$ and $p < 0.001$. As a consequence, it is possible to attest that there is a significant effect of the adopted instrument on the time required to identify the design discrepancies. A Games-Howell post-hoc test has been carried out to find out where the differences between the groups lie. The results confirmed that the group that uses the AR tool presents significantly different results from the other two groups ($p < 0.001$). On the contrary, there is not a statistically significant difference between the G1 and G2 groups ($p = 0.159$). In particular, the group that makes use of the AR tool presents a reduction respectively of 87.3% and 85.5% in time if compared to the group with the technical drawing and CAD model. Similar results are obtained when the median score is adopted as comparison tool [19], also in this case the AR tool brings down the completion time by 88.8% and 84.9% respectively of the G1 and G2 groups.

In conclusion, the comparison test produced interesting results about the adoption of AR technology for the detection of design discrepancies. In fact, even if all the participants were novice in the use of the AR tool they performed relatively well and reached similar results in terms of effectiveness of the groups that used technical drawings and CAD systems. Regarding the efficiency, considering the mean time to complete the task scored with the technical drawing and CAD system as reference, since these are the instruments at the moment commonly used to perform this kind of operations, it is surprisingly evident that the group with the AR tool performed better than the other two groups. In fact, the detection of design variations is significantly faster using the proposed AR tool compared to technical drawings and CAD systems. This outcome suggests that the use of AR minimizes the cognitive load of divided attention induced in attending to both the physical prototype and the related design data. In fact, divided attention tasks demand high mental workload [20] and increase the difficulty rather than simplify the operations [21]. As a consequence, this leads to a reduction in the efficiency of the user' processing capacity. Furthermore, for users with poor divided attention, any interference may alter the task that they are doing simultaneously.

Then the AR tool has proved to be an interesting solution that simplifies comparison of real objects and three-dimensional models and, consequently, can support users for the detection of design discrepancies. Its main advantage is related more to the efficiency than to the effectiveness when compared to the paper- and CAD-based solutions. It is, therefore, a valid solution, which is worth continuing to work on and to develop additional features and make it a complete user-friendly instrument able to support workers at the workplace within a smart factory context.

6   Conclusions

The paper has proposed a novel tool, based on AR technologies, that has been specifically developed to support workers for the detection of design variations between the 3D models prepared by the designers and the physical prototypes, as they have been made and assembled at the workplace.

The paper has focused on an exploratory and comparison test that is of fundamental importance because it lays the ground for the following development stages. In particular, the test aimed to compare the effectiveness and efficiency of three different instruments by means users can detect design variances. The test has been carried out within a controlled laboratory environment that is not affected by the many issues that usually arise into an industrial environment. Then this exploratory test represents a preliminary study of the proposed AR tool, carried out in an environment with ideal conditions with a specific typology of users, that will be further subjected to an iterative cycle of lab tests to expose usability deficiencies and field experimentations under real conditions of use.

What emerges from the experimentation is that the proposed AR tool provides similar results with the other instruments in term of effectiveness and an evident benefit in term of its efficiency when compared to other instruments traditionally adopted for the detection of design discrepancies. In conclusion, it is possible to validate the assumption that the proposed AR tool could be efficiently adopted as an alternative to technical drawing and CAD systems to easily support users for detecting design discrepancies that occur between the 3D models and the correspondent physical prototypes.

Furthermore, differently from the other two instruments taken into consideration for the comparison test, the proposed AR tool has been conceived as a handheld device in order to be used by workers at the workplace. In addition to these main features, with a view to the Industry 4.0 context, the AR tool could be integrated with the product data management (PDM) system to automatically store and share the data acquired by the other professional figures involved in the development and production processes.

References

1. Nee, A.Y., Ong, S.K., Chryssolouris, G., Mourtzis, D.: Augmented reality applications in design and manufacturing. CIRP Ann. **61**(2), 657–679 (2012). https://doi.org/10.1016/j.cirp.2012.05.010
2. Palmarini, R., Erkoyuncu, J.A., Roy, R., Torabmostaedi, H.: A systematic review of augmented reality applications in maintenance. Robot. Comput.-Integr. Manuf. **49**, 215–228 (2018). https://doi.org/10.1016/j.rcim.2017.06.002
3. Wang, X., Ong, S.K., Nee, A.Y.: A comprehensive survey of augmented reality assembly research. Adv. Manuf. **4**(1), 1–22 (2016). https://doi.org/10.1007/s40436-015-0131-4
4. Fraga-Lamas, P., Fernández-Caramés, T.M., Blanco-Novoa, Ó., Vilar-Montesinos, M.A.: A review on industrial augmented reality systems for the industry 4.0 shipyard. IEEE Access **6**, 13358–13375 (2018). https://doi.org/10.1109/access.2018.2808326
5. Ong, S.-K., Zhang, J., Shen, Y., Nee, A.Y.C.: Augmented Reality in Product Development and Manufacturing. In: Furht, B. (ed.) Handbook of Augmented Reality, pp. 651–669. Springer, New York (2011). https://doi.org/10.1007/978-1-4614-0064-6_30
6. Park, M., Schmidt, L., Schlick, C., Luczak, H.: Design and evaluation of an augmented reality welding helmet. Hum. Factors Ergon. Manuf. Serv. Ind. **17**(4), 317–330 (2007). https://doi.org/10.1002/hfm.20077

7. Regenbrecht, H., Baratoff, G., Wilke, W.: Augmented reality projects in the automotive and aerospace industries. IEEE Comput. Graph. Appl. **25**(6), 48–56 (2005). https://doi.org/10.1109/MCG.2005.124

8. Uva, A.E., Gattullo, M., Manghisi, V.M., Spagnulo, D., Cascella, G.L., Fiorentino, M.: Evaluating the effectiveness of spatial augmented reality in smart manufacturing: a solution for manual working stations. Int. J. Adv. Manuf. Technol. **94**(1–4), 509–521 (2018). https://doi.org/10.1007/s00170-017-0846-4

9. Zhou, J., Lee, I., Thomas, B., Menassa, R., Farrant, A., Sansome, A.: Applying spatial augmented reality to facilitate in-situ support for automotive spot-welding inspection. In: Proceedings of the 10th International Conference on Virtual Reality Continuum and Its Applications in Industry, pp. 195–200. ACM (2011). https://doi.org/10.1145/2087756.2087784

10. Stutzman, B., Nilsen, D., Broderick, T., Neubert, J.: MARTI: mobile augmented reality tool for industry. In: 2009 WRI World Congress on Computer Science and Information Engineering, vol. 5, pp. 425–429. IEEE (2009). https://doi.org/10.1109/csie.2009.930

11. Olbrich, M., Wuest, H., Riess, P., Bockholt, U.: Augmented reality pipe layout planning in the shipbuilding industry. In: 2011 10th IEEE International Symposium on Mixed and Augmented Reality, pp. 269–270. IEEE (2011). https://doi.org/10.1109/ismar.2011.6143896

12. Project Tango. https://developers.google.com/tango

13. Lenovo Phab 2 Pro. https://www3.lenovo.com/ee/et/tango

14. ARCore. https://developers.google.com/ar

15. 3DUnity. https://unity3d.com/

16. Lewis, J.R.: Sample sizes for usability studies: additional considerations. Hum. Factors **36**, 368–378 (1994). https://doi.org/10.1177/001872089403600215

17. Virzi, R.A.: Refining the test phase of usability evaluation: how many subjects is enough. Hum. Factors **34**(1992), 457–468 (1994). https://doi.org/10.1177/001872089203400407

18. Internal Organization for Standardization. 9241–11:2018: Ergonomics of human-system interaction — Part 11: Usability: Definitions and concepts (2018)

19. Rubin, J., Chisnell, D.: Handbook of Usability Testing: How to Plan. Design and Conduct Effective Tests. Wiley, Hoboken (2008)

20. Wickens, C.D., Hollands, J.G., Banbury, S., Parasuraman, R.: Engineering Psychology & Human Performance. Psychology Press, Boston (2015)

21. Backs, R.W., Seljos, K.A.: Metabolic and cardiorespiratory measures of mental effort: the effects of level of difficulty in a working memory task. Int. J. Psychophysiol. **16**(1), 57–68 (1994). https://doi.org/10.1016/0167-8760(94)90042-6

A New Loose-Coupling Method for Vision-Inertial Systems Based on Retro-Correction and Inconsistency Treatment

Marwene Kechiche[1](✉), Ioan-Alexandru Ivan[1], Patrick Baert[1],
Rolnd Fortunier[2], and Rosario Toscano[1]

[1] École Nationale d'Ingenieurs Saint-Étienne (ENISE),
58 rue Jean Parot, Saint-Étienne, France
marwene.kechiche@enise.fr
[2] École Nationale Suppérieure de Mécanique et d'Aérotechnique,
Téléport 2, 1 Avenue Clment Ader, 86360 Chasseneuil-du-Poitou, France

Abstract. Real time pose estimation of a mobile rigid-body in an unknown environment and without adding constraints (markers, antenna, ultrasound, Radio Frequency IDentification (RFID)...) is a crucial issue for Augmented Reality (AR) applications. One of the most advanced indoor/outdoor pose estimator is the Simultaneous Localization and Mapping algorithm (SLAM) based on monocular or binocular images. The complexity of this algorithm and his processing time present the main drawbacks of this type of pose estimator. It is difficult to use SLAM on mobile or embedded devices in real time applications, because they suffer from low computational resources. On the other hand Inertial Measurement Units (IMU) allow indoor/outdoor localization without important time processing but require signal integration which generates drifts over longer periods of time. In this paper we propose a new method for coupling SLAM with IMU. In this approach we take into account the SLAM processing time in order to avoid incoherences of timestamps in fused SLAM and IMU poses. To this end we propose a retro-correction method that synchronizes all poses with a general clock timestamping all events. In addition the quality of the poses is improved with detection and treatment of inconsistency. For this purpose an Adaptive Complementary Filter (ACF) was developed. Finally simulated and experimental results validate the efficiency of the proposed method.

1 Introduction

In the virtual and Augmented Reality (AR) applications the rendered image is computed according to the camera position, which is localized in the virtual world according to a reference in the real world. For an acceptable and stable rendering, a fluid tracking and a fluid processing must be performed at a minimum rate of 25 frames per second. Since most of the applications are

© Springer Nature Switzerland AG 2019
L. T. De Paolis and P. Bourdot (Eds.): AVR 2019, LNCS 11614, pp. 111–125, 2019.
https://doi.org/10.1007/978-3-030-25999-0_10

executed on low computation power embedded devices like mobile phones or single-board computers, it is difficult to conjugate the accuracy with the real time processing. This paper describes a method aimed at producing an accurate, stable and real time image rendering in AR applications. Among all the tracking techniques presented in the literature, this paper concentrated on those applicable to general AR applications, i.e. with indoor and outdoor situations, and without any added constraints in the environment like markers, antennas, cameras, ultrasound... In these conditions, the localization techniques are essentially based on vision and inertial data. Vision data coming from monocular or binocular images are mostly processed by a Simultaneous Localization And Mapping (SLAM) algorithm, whereas inertial data obtained from an Inertial Measurement Unit (IMU) are implemented into a time integration software. According to the SLAM algorithm, the pose of an operator in any environment is estimated without prior learning, preparation, or addition of markers like radio transmitters, RFID, etc. There are many types of SLAM algorithms in the literature, like for example RDSLAM [16], which detects the appearances and the change of structure of the scene, and which proposes a variant of RANdom SAmpling Consensus (RANSAC) [5] to support the localization in the dynamic scenes. Other algorithms are ORB-SLAM [12], LSD-SLAM [3], and SlidAR [13]. The common drawback of these techniques is the CPU time consumption. In particular, for real time applications, parallel processing of localization and mapping is performed [7], but this approach cannot be easily implemented on small and portable devices used in AR applications. Concerning the IMU, two basic sensors produce data in its coordinate system: an accelerometer for the movement of the attached rigid body and a gyroscope for the angular velocities. Although many other sensors can be present in a single IMU, like magnetometer, altimeter, or temperature sensors, this paper will concentrate on gyroscope and accelerometer. 3D localization is theoretically possible by using these two sensors. The orientation and the position are obtained by a single time integration of the angular velocity delivered by the gyroscope and a double integration of the acceleration measured by the accelerometer, respectively. Orientation can be also estimated by the accelerometer by observing the gravity direction and assuming no motion. Since the acquisition frequency of an IMU is relatively high (classical values range from 50 to 1000 Hz), position and orientation can be estimated in real time. However, the quality of these estimations is very poor, because the data is noisy and sensitive to external factors, the time integration causes drifts.

In this paper, we focus on a particular data fusion localization method called visual inertial technique. In this technique, a SLAM produces precise data using the vision algorithm at a relatively low frequency, while the time integration of high frequency data coming from an IMU can be used to calculate accurate poses. The main principle of this data fusion localization technique is to use IMU data for the calculation of the pose, this data being periodically corrected by SLAM informations.

The paper is structured as follows: SLAM-IMU coupling techniques, proposed method, and examples. After giving a general description of the SLAM-IMU coupling techniques, the proposed method is detailed. It includes a retro-correction technique that synchronizes all poses on a general time clock, and it applies a new adaptive complementary filter (ACF) to correct poses by detecting and treating inconsistency. Finally, a simpled worked example is used to illustrate this method, and the results obtained are discussed.

2   SLAM-IMU Coupling Techniques

Visual inertial localization techniques consist of coupling SLAM with an IMU in order to reduce the latency of the visual localization on low power computers. According to the literature, coupling can be performed via a tightly [8,10] or loosely [2,15] approach. The former consists of a standalone visual-based pose estimation module [3,7,12] and a separate IMU propagation module, whereas the latter proposes an individual visual-based pose estimation used to correct the IMU propagation [17]. For example, Konolige et al. [9] integrated IMU measurements as independent inclinometer and relative yaw measurements into an optimization framework using stereo vision measurements. However, few attention is paid on jitter and latency for AR applications. Tomazic et al. [1] have proposed a fusion approach combining visual odometry and an inertial navigation on mobile devices, but this method is inclined to drift due to the lack of optimization at the back-end. Kim et al. [6] have proposed an inertial and landmark-based integrated navigation method for poor vision environments. With the help of the inertial sensor, the system can provide reliable navigation when the number of landmarks is not sufficient for visual navigation. However, the dependency on the landmarks limits its adaptability. Li et al. [11] have proposed a novel system for pose estimation using visual and inertial data, and only three-axis accelerometer and colored marker are used for a 6-DoF motion tracking. Nevertheless, the pose calculation process is carried out on the server side for real-time performance. As a conclusion, although these sensor fusion methods can perform a robust 6-DoF motion tracking, limited attention is paid on real-time and smooth 6-DoF tracking. This may result in jitter and latency, making these method not suitable for mobile AR applications [4].

In this paper the coupling technique is based on a loosely approach. The vision SLAM algorithm is thus considered as a black box, which provides poses to correct the IMU estimation. To the authors knowledge, the existing coupling techniques between SLAM and IMU does not take into account the SLAM latency. This is a serious drawback for real-time applications executed on embedded devices. Then two main questions arise:

1. what is the origin of a given SLAM pose?
2. what are the timestamps of the SLAM poses and are they coherent with the IMU corrected poses?

Finally, in most of the developed methods using loosely coupling the inconsistent-pose problem is not treated [17]. In our case, for augmented reality applications these detections must be treated. A complementary filter and a low-pass filter are used to this end.

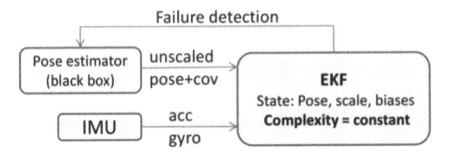

Fig. 1. Classic loosely-coupling technique schema with treatment of unscaled poses and detection of inconsistency, from [17] (EKF- Extended Kalman Filter)

3   Proposed Method

Figure 1 shows that SLAM estimator is considered as a black box and work on standalone mode. It gives unscaled poses and covariance matrix. The IMU gives an acceleration and angular velocities using the accelerometer and gyroscope. Then an Extended Kalman Filter (EKF) was developed to treat these data with detection of inconsistency. The inconsistency considered as failure detection and ignored without any treatment.

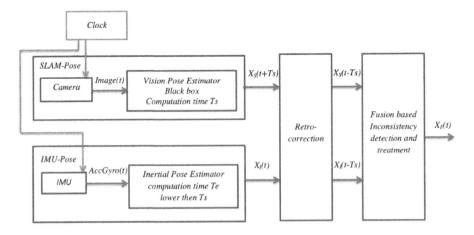

Fig. 2. The proposed approach including a general clock timestamping events with retro correction module and fusion module

The structure of the proposed method is depicted in Fig. 2. The pose of a considered mobile rigid-body object evolving in an unknown environment, is estimated with two timestamped "Smart Sensors" (SS): the SLAM-POSE and the IMU-POSE. The SLAM pose gives an estimation of the pose using acquired images by the embedded camera in the mobile rigid-body. The IMU-POSE gives an estimation of pose from accelerations and angular velocities measured by IMU. The IMU is attached to the camera which itself part of the mobile rigid-body. Due to the important acquisition rate difference between those SS, the estimated poses are corrected via a retro-correction algorithm. Then the pose is followed by fusion using an Adaptive Complementary Filter (ACF) with inconsistency detection and treatment. The last two blocks in the Fig. 2 incorporate the contribution on this paper to fix the problem of fusing the information obtained from different timestamped sources. These two blocks will be detailed in the next section.

Table 1. Comparison between the performances of computer and UdooX86

Criteria	DELL PC	UdooX86
CPU	Intel Core i7-7820HQ CPU 2,90 GHz x 8	Intel Pentium N3710 2.56 Ghz x 4
GPU	**Quadro M620/PCIe/SSE2**	**Intel HD Graphics 405 Up to 700 MHz 16 execution units**
RAM	32 GB	8 GB

The used SLAM on this work is developed by another partner [14]. This SLAM is considered as a black box, which cannot be modified. The given pose by the SLAM is considered as "reliable Pose". We note that contrary to [17] the used SLAM provides scaled Poses. Two computers were used, a PC and a single-board computer. A comparison between the performance of two devices is shown in Table 1. The computing frequencies of SLAM measured on those two different devices Table 1 are summarized in Table 2.

Table 2. SLAM frequencies on both computer and UdooX86

Frequency	DELL PC	UdooX86
Min	10 Hz	3 Hz
Max	33 Hz	16 Hz
Mean	18 Hz	8 Hz

For the IMU-Pose estimator, in contrast to SLAM, the estimation was totally developed for the coupling. To compute rotations and translations we use integration and double integration of the gyroscope and accelerometer data.

$$P(t) = \iint (A(t) - g(t))dt^2 \tag{1}$$

$$R(t) = \int \omega(t)dt \tag{2}$$

We used first-order Euler to integrate gyroscope data and Euler second order to integrate accelerometer data.

$$R_I(t) = R_I(t-1) + \omega(t) * Te \tag{3}$$

$$V_I(t) = V_I(t) + A(t) * Te \tag{4}$$

$$P_I(t) = P_I(t-1) + V_I(t) * Te + \frac{1}{2} * A(t) * Te^2 \tag{5}$$

3.1   SLAM-Pose and IMU-Pose

A performance comparison between Udoo and the PC are presented in the Table 1.

When the SLAM based vision is executed on a low computational system the resulting processing time cannot be ignored. Indeed let $X(t)$ be the actual pose of a mobile rigid-body at the time instant t, which is related to the acquired image at the same instant. Due to the computation time Ts, of the SLAM, the estimated pose is available at the time instant $t + Ts$. The delay Ts must be taken into account for efficient localization in AR applications. We note also that Ts is variable. The quality of SLAM depends directly of the quality of acquired images. More the image is textured, the better are the results.

More formally we have:

$$X_S(t + Ts) = F_S(Image(t), Ts) \tag{6}$$

where $Image(t)$ is the acquired image at the time instant t, Fs presents the SLAM algorithm that estimate the pose from $Image(t)$, Ts is the SLAM computation time, and $Xs(t + Ts)$ represents the resulting estimated pose of the actual pose of the mobile body $X(t)$. Note That we can rewrite the relation (6) into the following compact form:

$$X_S = F_S(Image(t - Ts)) \tag{7}$$

which shows clearly that the available SLAM pose at the time instant t corresponds to the estimation of the pose of the rigid-body at the time instant $t - Ts$.

Due to its high rate and fast processing time an IMU based pose estimator has been developed and combined with the SLAM-POSE estimator. Let Xi(t) the IMU-POSE estimated at time instant t, we have:

$$X_I(t) = F_I(GyroAcc(t)) \tag{8}$$

where $GyroAcc(t)$ are 3D gyroscopic and acceleration measurements of the mobile rigid-body. Fi represents the IMU pose algorithm that estimates the pose from $GyroAcc(t)$. In contrast to Fs, the computational time of Fi is smaller than 1 ms. It can estimate the pose of the mobile rigid-body in real-time. The main advantage of IMU based pose estimation is its fast computation time, but the drawback is the inevitable drifts due the time integration. Thus we can obtain a fast and reliable pose estimator by a correct coupling of Fs with Fi. This correction has been obtained by an original retro-correction algorithm followed by an original inconsistency detection and treatment.

3.2 Retro Correction

Loose-coupling runs two algorithms on standalone mode. The two types of poses (SLAM-Pose and IMU-Pose) are not ordered. The addition of a general clock allows to observe dated events. Using the general clock, the sources and the origin of each pose are identified. For the SLAM detection it is important to know the origin of the given pose. All poses for SLAM are dated by the timestamp of the origin image. Origin images are dated after acquisition. Acquisition and propagation times are unknown thus, ignored. Inertial poses are also dated and the date of each pose correspond to the acquisition instant. To order these data two buffers are created:

1. buffer for SLAM poses: A circular buffer with size 5 type FIFO;
2. buffer for IMU poses: A circular buffer with size 120 type FIFO

The size of the first buffer is chosen to have a correct historic if we want to perform analyses or compute predictions based on past data.

The size of the second buffer is chosen based to the mean frequency of SLAM on UdooX86 shown in Table 2 ($952/8 = 119$).

We have called this method retro-correction because we have two sensors: a high rate one (IMU) and slow rate one (SLAM). Looking to the timestamped data which arrives with computation delays, the future of SLAM is the past of IMU. Therefore the used SLAM poses to correct IMU poses are in the past and that is why we called this method retro-correction.

The retro-correction algorithm is:

1. Once an available timestamped IMU-Pose arrives, it is pushed back in the IMU circular buffer (mode free run).
2. Once an available timestamped SLAM-Pose arrives it is pushed back in the SLAM circular buffer.
3. Then the following operations occur:

- Search for the corresponding timestamped IMU-Pose.
- Send the two (SLAM-Pose and the corresponding IMU-Pose) poses to the Fusion Module.
- Send the last timestamped IMU-Pose to the Fusion module.

3.3 Inconsistency Detection and Treatment

Once the data ordered, an Adaptive Complementary Filter (ACF) is used to correct the poses. The ACF was chosen due to his fast processing time, which is better than EKF or KF (Kalman Filters). In this paper the rotations are solely treated for two reasons:

- In indoor AR applications the translations amplitudes are less significant than the rotations.
- The treatment of the translations (positions) is the same as for the rotations.

Adaptive Complementary Filter. The inputs of ACF filter are:

* SLAM-Pose
* IMU-Pose (corresponding to the SLAM-Pose)
* LastIMU-Pose (lastIMU-Pose available on the buffer sent by the retro-correction module)

$$X_S = (PX_{SLAM}, PY_{SLAM}, PZ_{SLAM}, Rx_{SLAM}, Ry_{SLAM}, Rz_{SLAM}) \quad (9)$$

$$X_I = (PX_{IMU}, PY_{IMU}, PZ_{IMU}, Rx_{IMU}, Ry_{IMU}, Rz_{IMU}) \quad (10)$$

$$Rx_{fusuion} = \alpha_x * Rx_{SLAM} + (1 - \alpha_x) * Rx_{IMU} \quad (11)$$

$$Ry_{fusion} = \alpha_y * Ry_{SLAM} + (1 - \alpha_y) * Ry_{IMU} \quad (12)$$

$$Rz_{fusion} = \alpha_z * Rz_{SLAM} + (1 - \alpha_z) * Rz_{IMU} \quad (13)$$

where:

$$\alpha_x = \frac{Rx_{SLAM}}{Rx_{IMU}} \quad (14)$$

$$\alpha_y = \frac{Ry_{SLAM}}{Ry_{IMU}} \quad (15)$$

$$\alpha_z = \frac{Rz_{SLAM}}{Rz_{IMU}} \quad (16)$$

Detection and Correction. After alignment, the poses can be fused. When fusing, a quality analysis is performed. The given SLAM pose can be incoherent with the corresponding computed IMU pose. Hence the quality of the pose must be analyzed. In [17] the detection of inconsistency is done by thresholding and no correction was proposed (direct rejection). In contrast, our approach not only detect the inconsistency but also corrects it. For the inconsistency detection we threshold the ratio between the IMU and SLAM poses. The following threshold limits are chosen: $\alpha \notin [0.8, 1.2]$, outside of which the inconsistency is detected. If within the range we get the last-SLAM (consistent) timestamped and we look for the corresponding IMU-Pose, and afterwards we accumulate all rotations to the last IMU-Pose rotation.

$$\Delta R = R_{lastImuPose}^{-1} * R_c \tag{17}$$

We note that R_c is the corresponding timestamped IMU-Pose to generate the last coherent timestamped SLAM-Pose

$$R_{fusion} = \Delta R * R_{fusion} \tag{18}$$

4 Simple Worked Examples

Two simple worked examples are used to illustrate the proposed coupling method. The first example is based on data generated by a theoretical path, whereas the second is exposed with experimental data obtained during a real test. In both examples, the effect of synchronizing all poses with a general timestamping clock is outlined. In the second example, the effect inconsistency treatment is also emphasized.

4.1 Simulated Data

A single axis pure sinusoidal rotation was applied to a rigid body. The time period is 0.8333 s, and the amplitude is 1.2 radians. A small random white noise was added to the original path to simulate data obtained from SLAM. For the IMU gyro simulation the original signal was derived to generate the angular velocities, then a bias and a white random noise was added to the angular velocities in order to simulate IMU behavior.

Figure 3 shows the reference sinusoidal path, together with the path obtained with a non timestamped SLAM algorithm applied to noisy data. It can be seen in this figure that a temporal shift due to the processing time (simulated) is present, whereas in Fig. 4 the SLAM poses are much closer to the reference. This figure is simply obtained by timestamping the events, and the agreement is very good. However, when applied on low power computer, real time estimation is not possible this way because the SLAM frequency becomes too small.

In Figs. 5 and 6, the complete fusion algorithm is applied on data simulating IMU and SLAM. It can be seen in these figures that, when the retro-correction method proposed in this paper is applied, the temporal shift is reduced, and

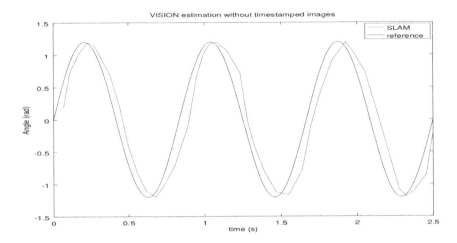

Fig. 3. Reference path and non timestamped SLAM path

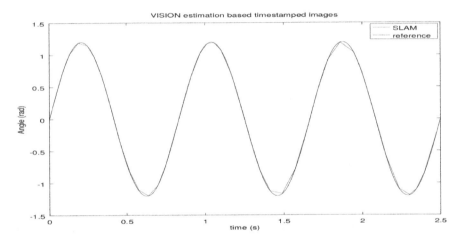

Fig. 4. Reference path and timestamped SLAM path (offline)

the reference and reconstructed paths are very similar. The differences observed in Fig. 6 are mainly due to the ACF treatment. Thus the proposed method is validated on these simulated data. The next step is to validate it by using experimental data.

4.2 Experimental Data

Experimental data were generated by using the workbench depicted in Fig. 7. A stereo camera is mounted on an actuated vertical axis rotating from 0 to π radians with an angular velocity of 90 °/s. The experimental data was generated

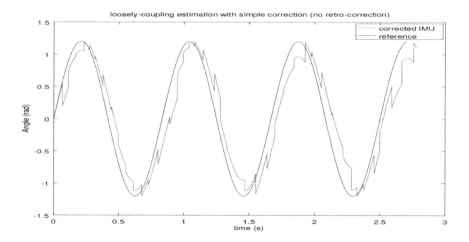

Fig. 5. Reference path and reconstructed path without retro-correction

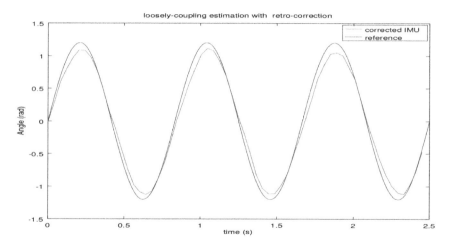

Fig. 6. Reference and reconstructed path with retro-correction (Real-time)

by the camera, for SLAM algorithm, and by an embedded IMU (LSM6DS0) mounted between the two camera chips.

Figure 8 shows the reference and the recalculated paths without any correction. Although the fused data in this figure presents the SLAM behavior with integration of IMU data, it can be seen that the poses are mainly controlled by the SLAM algorithm and the agreement is very poor.

In Fig. 9 the data were timestamped according to the proposed method. It can be seen in this figure that the estimated pose is closer to the reference than in Fig. 8, with a relatively good agreement. However, the peak at the top of the reference curve is not reproduced with a good accuracy. This is mainly due to the SLAM latency. Finally, it can be observed in Fig. 9 that the calculated

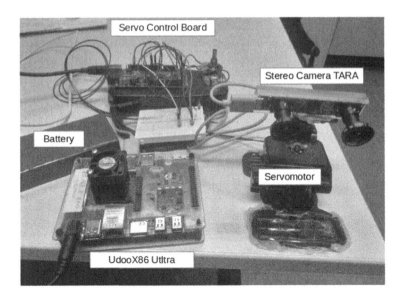

Fig. 7. Test support

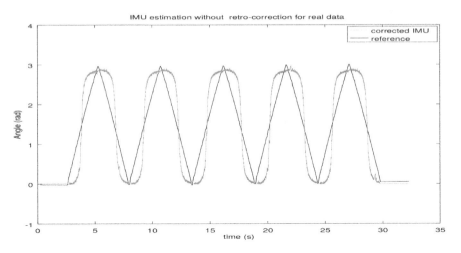

Fig. 8. Actual and recalculated paths with data fusion from IMU and SLAM, without any correction

curve (inred) presents two spikes per period, which are inconsistent poses. Thus Fig. 10 gives the calculated poses when the ACF is applied for inconsistency detection and treatment. It can be seen in this figure that the calculated path is in relatively good agreement with the reference one.

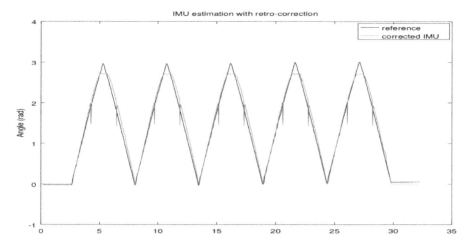

Fig. 9. Actual and recalculated paths with data fusion from IMU and SLAM, with only retro-correction

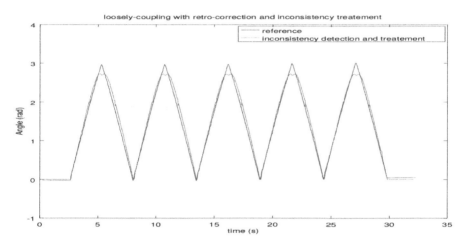

Fig. 10. Actual and recalculated paths with data fusion from IMU and SLAM, with retro-correction and inconsistency treatment

5 Concluding Remarks

In this paper a localization method that fuses SLAM and IMU data was presented. This method is based on a loose-coupling technique that runs two threads (SLAM and IMU). The use of a classic approach, i.e. without taking into account the time processing, produces computed poses that are shifted from the reference path. The retro-correction method proposed in this paper decreases this shift, and can be applied on low power devices. This method is based on the insertion of a master time clock that synchronizes all events, while the low frequency

SLAM poses correct the high speed but progressively drifted path obtained by integrating the IMU data. The proposed method is currently being improved by introducing a Kalman Filter to further reduce the IMU-Pose drift between two SLAM poses.

Acknowledgments. The authors wish to thank Thales, Institut Pascal, SFI, Nexter training, VapRail, and Studio bouquets. This work was done in context of FUI REVE5D project.

References

1. Fusion of visual odometry and inertial navigation system on a smartphone. Comput. Ind. **74**(C), 119–134 (2015). https://doi.org/10.1016/j.compind.2015.05.003
2. Bleser, G.: Towards visual-inertial slam for mobile augmented reality (2009)
3. Engel, J., Schöps, T., Cremers, D.: LSD-SLAM: large-scale direct monocular SLAM. In: Fleet, D., Pajdla, T., Schiele, B., Tuytelaars, T. (eds.) ECCV 2014. LNCS, vol. 8690, pp. 834–849. Springer, Cham (2014). https://doi.org/10.1007/978-3-319-10605-2_54
4. Fang, W., Zheng, L., Deng, H., Zhang, H.: Real-time motion tracking for mobile augmented/virtual reality using adaptive visual-inertial fusion. Sensors **17**(5) (2017). https://doi.org/10.3390/s17051037. http://www.mdpi.com/1424-8220/17/5/1037
5. Fischler, M.A., Bolles, R.C.: Random sample consensus: a paradigm for model fitting with applications to image analysis and automated cartography. Commun. ACM **24**(6), 381–395 (1981). https://doi.org/10.1145/358669.358692
6. Kim, Y., Hwang, D.H.: Vision/INS integrated navigation system for poor vision navigation environments. Sensors **16**(10) (2016). https://doi.org/10.3390/s16101672. URL http://www.mdpi.com/1424-8220/16/10/1672
7. Klein, G., Murray, D.: Parallel tracking and mapping for small AR workspaces. In: 2007 6th IEEE and ACM International Symposium on Mixed and Augmented Reality, pp. 225–234 (2007). https://doi.org/10.1109/ISMAR.2007.4538852
8. Kong, X., Wu, W., Zhang, L., Wang, Y.: Tightly-coupled stereo visual-inertial navigation using point and line features. Sensors **15**(6), 12816–12833 (2015). https://doi.org/10.3390/s150612816. http://www.mdpi.com/1424-8220/15/6/12816
9. Konolige, K., Agrawal, M., Solà, J.: Large-scale visual odometry for rough terrain. In: Kaneko, M., Nakamura, Y. (eds.) Robotics Research. STAR, vol. 66, pp. 201–212. Springer, Heidelberg (2011). https://doi.org/10.1007/978-3-642-14743-2_18
10. Leutenegger, S., Lynen, S., Bosse, M., Siegwart, R., Furgale, P.: Keyframe-based visual-inertial odometry using nonlinear optimization. Int. J. Robot. Res. **34** (2014). https://doi.org/10.1177/0278364914554813
11. Li, J., Besada, J.A., Bernardos, A.M., Tarrío, P., Casar, J.R.: A novel system for object pose estimation using fused vision and inertial data. Inf. Fusion **33**, 15–28 (2017). https://doi.org/10.1016/j.inffus.2016.04.006. http://www.sciencedirect.com/science/article/pii/S1566253516300239
12. Li, W., Nee, A.Y.C., Ong, S.K.: A state-of-the-art review of augmented reality in engineering analysis and simulation. Multimodal Technol. Interact. **1**(3) (2017). https://doi.org/10.3390/mti1030017. http://www.mdpi.com/2414-4088/1/3/17

13. Polvi, J., Taketomi, T., Yamamoto, G., Dey, A., Sandor, C., Kato, H.: SlidAR: a 3D positioning method for slam-based handheld augmented reality. Comput. Graph. **55**, 33–43 (2016). https://doi.org/10.1016/j.cag.2015.10.013. http://www.sciencedirect.com/science/article/pii/S0097849315001806
14. Ramadasan, D., Chevaldonné, M., Chateau, T.: MCSLAM: a multiple constrained SLAM. In: Proceedings of the British Machine Vision Conference 2015, BMVC 2015, Swansea, UK, 7–10 September 2015
15. Sirtkaya, S., Seymen, B., Alatan, A.A.: Loosely coupled Kalman filtering for fusion of visual odometry and inertial navigation. In: Proceedings of the 16th International Conference on Information Fusion, pp. 219–226 (2013)
16. Tan, W., Liu, H., Dong, Z., Zhang, G., Bao, H.: Robust monocular slam in dynamic environments. In: 2013 IEEE International Symposium on Mixed and Augmented Reality (ISMAR), pp. 209–218 (2013). https://doi.org/10.1109/ISMAR.2013.6671781
17. Weiss, S., Siegwart, R.: Real-time metric state estimation for modular vision-inertial systems. In: 2011 IEEE International Conference on Robotics and Automation, pp. 4531–4537 (2011). https://doi.org/10.1109/ICRA.2011.5979982

Ultra Wideband Tracking Potential for Augmented Reality Environments

Arnis Cirulis[(✉)]

Faculty of Engineering, Vidzeme University of Applied Sciences,
Valmiera, Latvia
arnis.cirulis@va.lv

Abstract. The main objective of this research is to implement a solution for precise visualization of 3D virtual elements where distances from the environment's participant to the object are up to ∼200 m. The primary results of this research should ensure a dynamic and animated three-dimensional (3D) computer model depiction in augmented reality (AR) mode without the use of fiducial or image-based markers. Such technological innovation will offer ample cases of augmented reality use in various industries and have economic ramifications. Current outdoor AR solutions lack precision, stability, operational range and multiple 3D object depiction in one scenario, dynamic properties to allow a participant to move freely in an environment without the loss of immersion.

Keywords: Internet of Things · Position simulation ·
Location based augmented reality · Outdoor augmented reality ·
Ultra-wideband tracking

1 Introduction

The evolution of the Internet in the last decades has taken several leaps, but the most fundamental influence on our life stems from the Internet of Things (IoT) phenomena. A recent IHS Markit report chronicles this quick-moving arena and gauges its explosive growth. The installed base for Internet-connected devices reached an estimated 12.1 billion in 2013, a number that is expected to more than quadruple to nearly 50 billion by 2025 [1]. The use of micro sensors and networking technologies connects the everyday physical world to the Internet. This is not only about connecting things, but also about providing services and opportunities to improve our life. 95% of corporate executives surveyed by *The Economist* say they plan to launch an IoT business within 3 years. According to PRWeb.com, 87% of manufacturers surveyed have not yet taken advantage of IoT to transform their facilities, but among the 13% of manufacturers who did implement IoT solutions, there were reports of increased efficiency, fewer product defects and a noted higher customer satisfaction. In addition, the automotive industry affects everyone's life and as Parks Associates claims, 89% of the new cars sold worldwide will have embedded connectivity by 2024 [2].

These are only some of the facts, but, nevertheless, the global tendencies and direction are quite clear. In the whole context of IoT technologies, when life is assigned

L. T. De Paolis and P. Bourdot (Eds.): AVR 2019, LNCS 11614, pp. 126–136, 2019.
https://doi.org/10.1007/978-3-030-25999-0_11

to the world of physical things, it is important to provide a natural interface between it and IoT elements, through the use of visualization technologies. In communication where people, things, processes and data are involved, feedback is crucial. The lower the communication barrier between humans and IoT elements, the more precise and correct the decisions that will be made. Undeniably, nowadays M2P (machine-to-people), M2M (machine-to-machine) and P2P (people-to-people) provide successful execution of these processes [3], but there is still plenty of space for innovation and improvements to complete and develop the next level. Respectively, the core idea of the research is to use real time augmented reality (AR) during the communication process. The information flow between IoT parties is bi-directional, so the more naturally information will be perceived by human senses, the more precise will be the reaction and a decreased response time. It has crucial meaning in many areas and industries (medicine, emergency services, military, logistics, Smart cities, manufacturing - Industry 4.0 etc.). Nowadays information perception via computer monitors and smartphone screens is quite typical, though rapid development of augmented reality technologies should be taken into account, such as future possibilities with devices like Magic Leap, Microsoft HoloLens 2, Meta 2, Daqri and others. These devices just recently appeared in the US market or will appear in the immediate future [4–6]. In the annual Internet of Things World Forum (IoTW) held in Dubai [7], it was already announced that besides services such as connected parking, connected lighting and waste management, alongside other vertical industries, more important is the opportunity to visualize these solutions for attendees. Also, Horizon2020 financed projects by the European Commission have a focus area on IoT technologies, emphasizing IoT wearables; user interfaces and object virtualization for an IoT ecosystem [8].

Standard visualization provides us with visual data depiction on screens of various devices, but augmented reality (AR) provides the depiction of virtual objects (text, images, graphics, video, 3D models) on top of the real world. The AR concept dates back to the Fifties of the previous century in cinematography, and modern definitions already appeared twenty years ago [9], but real use cases were developed in the last decade, when computing performance, graphical resolution and sensor (gyroscopes, magnetometers, accelerometers etc.) precision notably increased.

CCS Insight Global predicts that the dedicated augmented reality device market is expected to reach 5 million devices (AR smart glasses and mixed reality glasses) by 2021 and that augmented reality technology is already widely used in the education field for advanced learning and for teaching technologies [10]. Augmented reality and virtual reality technology will be used to contribute to projects with smart innovations in the future due to its great fascination and potential. At the moment several AR mobile applications (Vuforia, HP Reveal, Augment, Wikitude, Blippar etc.) [11] which mostly use fiducial or image-based markers for virtual object positioning are available. Future AR solutions mostly will integrate marker less solutions. For now, the growth in the marker less AR area is fostered by the smartphone industry and widely available sensors, such as GPS positioning, compass, video camera and connections to the Internet. Unfortunately, GPS positioning is insufficient in solutions where high precision is demanded and indoor capability provided. That is why nowadays marker less solutions are mostly oriented towards entertainment (PokemonGo) [12], advertising activities and training needs.

Without precise virtual object positioning, an important step in AR technology development will be switching from smartphone displays to head-mounted displays (HMD), thereby providing an essential increase in the environment's immersion level (HoloLens, Magic Leap) [4, 6]. At the end of 2018, the U.S. Army awarded Microsoft a 480 million USD contract to supply the military branch with as many as 100,000 HoloLens augmented reality headsets for training and combat purposes [13]. This fact not only makes the U.S. Army the most important HoloLens consumer, but this deal highlights the readiness of AR technologies for serious applications.

2 Problem Area and Global Practice in Object and Avatar Positioning

The previously stated facts prove the undoubted increase in AR significance in the future. However, there are still factors that limit the use of AR in the greater scope of areas and industries. To specify the research's goal and the use of AR technologies in the context of IoT, several problem areas arise for which solutions are being devised. The following functionality should be achieved:

1. Indoor and outdoor AR, regardless of weather conditions and lighting. However, nowadays marker-based solutions are quite precise, but they have a fundamental disadvantage in relation to immersion level, interactivity and mobility. Furthermore, in outdoor conditions it is inconvenient to use them. There are some AR solutions available for marker less outdoor use (SightSpace, PokemonGo) [12, 14] through the calculation of GPS coordinates, but in reality, inaccuracy is too high, preventing the participant from moving freely in the augmented environment. Basically, a target 3D model is statically positioned, and internal sensors' data are used for further projection calculations [15].
2. Virtual 3D model positioning at distances greater than 5 m. An additional disadvantage of marker-based AR systems is the short range of the operating distance where AR libraries should recognize the marker in video frames and place a virtual object on top of it. The latest solutions replace marker-based close-range solutions by spatial mapping and the use of depth cameras. However, phones with depth cameras like the Asus ZenfoneAR and the Lenovo Phab2 are quite expensive and the latest libraries like Google ARCore and Apple ARKit recognize the environment and the distance to the physical objects based on camera (not depth camera) by identifying interesting points, called features, and tracking how those points move over time. With a combination of the movement of these points and the readings from the phone's inertial sensors, the library determines both the position and orientation of the phone as it moves through space [16]. A surrounding 3D model is constructed in real time, thereby allowing the virtual object precise placement at distances from 10 cm to 5 m [17]. This approach offers high precision, participant mobility and a high immersion level in a typical living room or office conditions (indoor, close range).
3. The provision of high precision and stability of depicted 3D models. The main disadvantage of GPS-based AR solutions is insufficient precision in a virtual object

coordinate system, resulting in a disturbing effect, where the virtual object is jumping and its position is unstable, especially if it is observed with 3D models. Some algorithms are used for stabilization, such as the Kalman filter, but it does not solve issues if a participant is in motion and observing a virtual object, because displacement occurs [18]. Quite a lot of research has been done in the field of line and blob detection algorithms to recognize real world objects by analysing video frames [19, 20] but in real-time solutions this is impracticable because of changing weather and lighting conditions, as well as the fact that participant movement is not suggested.

4. Real time depiction of 3D models set in AR mode, including animations and photorealistic rendering. At the present moment there are no AR platforms which could offer simultaneous independent 3D model depiction. Mostly nowadays AR solutions provide depiction of single 3D or grouped 3D models (Vuforia, Augment, Wikitude, Blippar etc.) in marker-based systems. By avoiding this limitation, more intelligent environments could be developed, which are necessary for bi-directional communication among human and IoT elements in the form of visualized objects.

3   The Importance of Position Tracking for Multiple Entities in a Dynamic Environment

In virtual and augmented reality (VR/AR) environments, the collaboration among different parties can be implemented at various levels where each has its own challenges. There are meaningful differences in visualization and position calculation aspects. If the environment participant's position is static and also the position of the virtual object is static, visualization can be achieved quite easily, even if the viewing angle is changing. But if the environment's participant is in motion and also the virtual object or objects are moving, the complexity increases. Generally, sixteen modes can be estimated (see Fig. 1), where the most challenging is to provide an environment with several participants moving around and also several moving virtual objects.

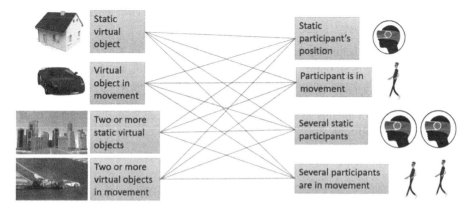

Fig. 1. General collaboration modes in VR/AR systems

For separate modes a sub-mode can also be estimated with additional requirements, especially in a VR environment. It depends on the implemented scenario. A significant difference is in the visualization of several participants' avatars for one participant or for all the participants, accordingly only one person is wearing VR HMD at a time and he sees other persons' 3D avatars, or everyone in the environment is wearing VR HMDs and see each other's avatars. In the first case it is useful implementation as well, because a participant in a VR environment can still easily communicate with real persons surrounding him and avoid collisions when he is moving. The other case should be implemented using multiplayer mode, where the synchronization process and scene depiction on several VR HMDs should be implemented via a data transmission network.

To depict virtual objects in the correct size, position and rotation, calculations for each frame should be done according to the flowchart in Fig. 2.

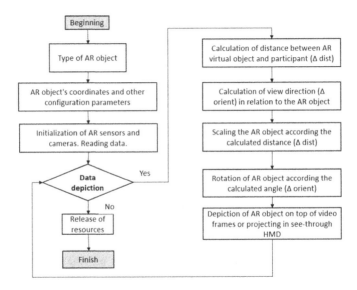

Fig. 2. General flowchart of a virtual object in an AR environment

These are the general steps which were implemented in the City 3D-AR project [18] based on GPS positioning, unfortunately, due to a lack of precision, this project was left only as a test platform. Nowadays for these calculations we can rely on Unity or Unreal engines' libraries and provide basic positioning in more natural way, nevertheless, issues with the precision of multiple dynamic objects in multiuser large-scale environment are still topical.

4 The Necessity for Precision Improvement and Flexibility in Large Scale AR Environments

As there are various modes of collaboration among parties in the VR/AR environment, it is important to test these modes in real 3D scenarios, without the use of a real positioning system, e.g. different technologies offered under real-time locating systems (RTLS). Therefore, a simulator was developed, which generates coordinates for static and dynamic objects in space (see Fig. 3).

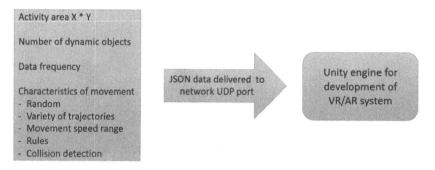

Fig. 3. Coordinate generator for the simulation of objects' positions

It is possible to specify the measurements of area, number of agents, frequency of JSON data objects generated (see Fig. 4). As advancements in the simulator, different types of movement can be implemented instead of random positions, e.g. a variety of trajectories, range of movement speed, avatar movement rules and collision detection. After the JSON data objects are formatted, they are delivered over the network to a specified IP address and UDP port number.

```
"id": 1,
"address": "0x85568CFE61EE",
"datastreams": [
  {
    "id": "posX",
    "current_value": "7,7",
    "at": "2018-10-13 18:57:26:249"
  },
  {
    "id": "posY",
    "current_value": "1,09",
    "at": "2018-10-13 18:57:26:249"
  },
  {
    "id": "posZ",
    "current_value": "106,32",
    "at": "2018-10-13 18:57:26:358"
  },
  {
    "id": "clr",
    "current_value": "2,26",
    "at": "2018-10-13 18:57:26:358"
  },

  {
    "id": "numberOfAnchors",
    "current_value": "6",
    "at": "2018-10-13 18:57:26:358"
  },
  {
    "id": "acc",
    "current_value": "26;188;34",
    "at": "2018-10-13 18:57:26:358"
  },
  {
    "id": "gyro",
    "current_value": "2;18;3",
    "at": "2018-10-13 18:57:26:358"
  },
  {
    "id": "mag",
    "current_value": "119,08;192,81;18,36",
    "at": "2018-10-13 18:57:26:479"
  }
]
}
```

Fig. 4. Generated JSON data object

The next phase involves Unity scripts to correctly use data objects for 3D models and cameras in the VR/AR environment (see Fig. 5). By changing simulation parameters, the number of agents and complexity of 3D models, it is possible to evaluate the performance of the environment. Evaluation can be done in a qualitative way, by visual investigation during runtime or various quantitative values can be acquired by the use of a Unity Profiler. General values include CPU usage for frame processing during rendering and script processing.

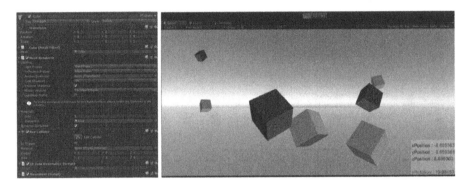

Fig. 5. Unity JSON deserialize and transformation scripts for 3D visualization

In this case measurements were made from 1 to 10 agents, generating JSON data objects at random intervals ranging from 50 ms to 500 ms (see Fig. 6).

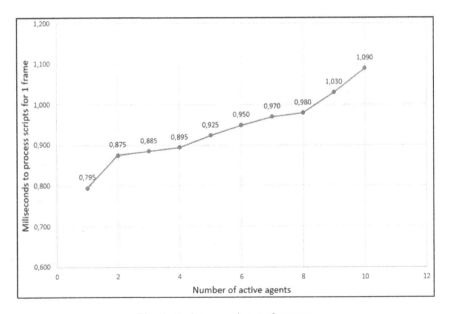

Fig. 6. Script execution performance

This is because usually raw data from positioning systems are processed to remove noise, e.g. the Kalman filter, and intervals could vary for one session. To also evaluate the influence of network latency and jitter, data objects were sent over 1 Gbps Ethernet and Wi-Fi IEEE802.11ac standards, but there was not significant difference in performance, because latency in both cases is 1 ms or less in perfect conditions with two connections. However, situations can be very different, if other Wi-Fi standards and frequencies are used, because of higher latency and high jitter (from 1 ms to 200 ms), especially if there are several Wi-Fi connections. Specific tests were executed on a VR compatible MSI GT75VR Titan laptop with 7th Gen. i7 processor and GeForce GTX 980M SL graphical adapter. For each test 30,000 frames were captured, and an average value was calculated, getting results from 0.795 ms with one agent to 1.090 ms with 10 moving agents. Rendering time was even less with lower variance. In specific Unity composition there are not noticeable performance issues, but this slight tendency must be considered if more complicated environments are developed, meaning not only higher quality animated 3D models, but also more serious collaboration logic and gaming mechanics implemented through scripting.

Thanks to a data generator for movement simulation, it is possible to save a lot of financial resources and time and specify the general requirements, to choose the most appropriate positioning system for a VR/AR system.

5   UWB Tracking as the Solution for AR and VR Systems

To prepare a real testbed, ultra-wideband (UWB) positioning technology was chosen. Last year's results with high precision has been achieved in UWB tracking, meaning that this technology could be suitable not only for sport tracking and logistics, but also for multiple user collaborative VR/AR environments. The 9 × 12 m room was setup to test precision and also validate the accuracy of the coordinate generator (see Fig. 7).

Ultra-wideband uses short-range radio communication, in contrast to Bluetooth Low Energy and Wi-Fi, the position determination is not based on the measurement of signal strength (Receive Signal Strength Indicator, RSSI), but on a runtime method Time of Flight (ToF) or time-difference-of-arrival (TDoA). The light propagation time between an object and several anchors is measured. At least three receivers are required for the exact localization of an object using trilateration. There must also be a direct line of sight between the receiver and the transmitter [21]. UWB utilizes a train of impulses rather than a modulated sine wave to transmit information. This unique characteristic makes it perfect for precise ranging applications. Since the pulse occupies such a wide frequency band (3–7 GHz according to the IEEE 802.15.4a standard), its rising edge is very steep, and this allows the receiver to very accurately measure the arrival time of the signal. The pulses themselves are very narrow, typically no more than two nanoseconds [22]. UWB positioning systems offer 5–30 cm accuracy in both indoors and outdoors and enable to get both 2D and 3D data [23]. Companies like Sewio, Eliko, Insoft, Decawave offer setups for the implementation of UWB wireless real-time locating systems.

Similar to a data generator, UWB systems provide data delivery over a network via UDP port in the form of lightweight data-interchange JSON data objects.

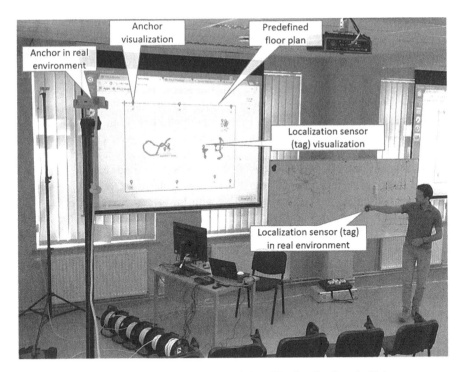

Fig. 7. UWB RTLS testbed for real-time 3D visualizations in Unity

6   Conclusions

This work focused on the large scale indoor and outdoor positioning necessary for use in augmented and virtual reality systems. Such systems are becoming more and more relevant in the context of smart cities and also Industry 4.0 where more natural involvement of society and professionals can be achieved. To determine and evaluate the requirements in 3D environments, a data generator was developed to simulate coordinates and different parameters related to nowadays positioning technologies (frequency, speed, precision, etc.). This study allowed the testing of the potential of UWB tracking before a real system is bought and set up. Now at the VR lab. at Vidzeme University of Applied Sciences in collaboration with EchoSports UWB tracking system is set up for further tests and real-time experiments with high quality 3D models for various surroundings and avatars, allowing to evaluate a large scale free-room concept, when physical movement is not limited and is performed in a natural way, without the use of controllers. These experiments will allow us to find out more information related to the mitigation of cybersickness, use cases and performance in augmented reality, as well as the potential of multi-user collaboration locally and remotely, by connecting remote free-rooms via 5G technology.

Acknowledgements. This work is a post doctorate research project funded by ERAF, project number: 1.1.1.2/VIAA/1/16/105. Project name: Dynamic 3D visualization of the Internet of

Things (IoT) elements in outdoor augmented reality (AR) modes. Research activities take place at the Faculty of Engineering at the Vidzeme University of Applied Sciences, and specifically, in the virtual reality technologies laboratory. The project relates to Latvia's Smart Specialization Strategy (RIS3). Specifically, the project aims to contribute to the number 4 priority and number 5 specialization "Modern Information and Communication Technologies".

References

1. Morelli, B.: The Internet of things explodes. IHS Markit Technology Blog, 17 January 2014
2. Cisco Jasper Infographics, The Internet of Things: It's not about things, it's about service. https://www.jasper.com/infographics/internet-things-its-not-about-things-its-about-service
3. Vermesan, O., Friess, P.: Digitising the Industry - Internet of Things Connecting the Physical, Digital and Virtual Worlds. The River Publishers Series in Communications (2016)
4. Ewalt, D.M.: Inside Magic Leap, The Secretive $4.5 Billion Startup Changing Computing Forever. Issue of Forbes, 29 November 2016
5. Matney, L.: Google launches Tango AR smartphone system. TechCrunch, 1 November 2016
6. Fitzsimmons, M.: Hands on: Microsoft HoloLens review. Techradar, 3 November 2016
7. Rich, S.: Internet of Things World Forum Showcases Smart City Technologies in Dubai at 3rd Annual Conference. Cisco Press Relations, 12 June 2015
8. ICT Proposers' Day 2016, IoT integration and platforms, Internet of Things Networking session, Bratislava, 26 September 2016
9. Azuma, R.: Location-based mixed and augmented reality storytelling. In: Barfield, W. (ed.) Fundamentals of Wearable Computers and Augmented Reality, 2nd edn, pp. 259–276. CRC Press, Boca Raton, August 2015
10. Rohan: Augmented Reality and Virtual Reality Market worth $1.06 Billion by 2018. MarketsandMarkets Research, 13 March 2014
11. Wood, L.: Global Mobile Augmented Reality Market 2016–2020 with Augmented Pixels, Aurasma, Blippar, Catchoom, DAQRI, Metaio & Wikitude Dominating, Research and Markets, 22 February 2016
12. Tabbitt, S.: The Pokémon GO phenomenon. Cisco's Technology News Site, 9 August 2016
13. Joshua, B.: Microsoft Wins $480 Million Army Battlefield Contract, Bloomberg L.P., November 2018. https://www.bloomberg.com/news/articles/2018-11-28/microsoft-wins-480-million-army-battlefield-contract
14. Errin: SightSpace Free-D Mobile App for SketchUp Is Here! The Limitless Computing Blog, 30 October 2012
15. Schuster, H.: Android: compass implementation – calculating the azimuth. DeviantDev J. (2015)
16. ARCore overview, Google Developers, February 2019. https://developers.google.com/ar/discover/
17. Ashley, J.: Understanding HoloLens Spatial Mapping and Hologram Ranges. The Imaginative Universal, 23 March 2016
18. Cirulis, A., Brigmanis, K.B.: 3D outdoor augmented reality for architecture and urban planning. Procedia Comput. Sci. **25**, 71–79 (2013)
19. Comport, A.I., Marchand, É., Chaumette, F.: A real-time tracker for markerless augmented reality. In: Proceedings of the 2nd IEEE/ACM International Symposium on Mixed and Augmented Reality. IEEE Computer Society (2003)

20. Reitmayr, G., Drummond, T.: Going out: robust model-based tracking for outdoor augmented reality. In: Proceedings of the 5th IEEE and ACM International Symposium on Mixed and Augmented Reality. IEEE Computer Society (2006)
21. Julia, L.: UWB: Two Localization Techniques in Comparison, infsoft GmbH, August 2018. https://www.infsoft.com/blog-en/articleid/305/uwb-two-localization-techniques-in-comparison
22. UWB Technology, Sewio Networks, s.r.o. (2018). https://www.sewio.net/uwb-technology/
23. Ultra Wideband Positioning, Eliko Tehnoloogia - Sensing the Future (2018). https://www.eliko.ee/services/ultra-wideband-positioning/

HoloHome: An Augmented Reality Framework to Manage the Smart Home

Atieh Mahroo$^{(\boxtimes)}$ ⓘ, Luca Greci, and Marco Sacco

Institute of Intelligent Industrial Technologies and Systems for Advanced Manufacturing (STIIMA), National Research Council of Italy (CNR), 23900 Lecco, Italy
{atieh.mahroo,luca.greci,marco.sacco}@stiima.cnr.it

Abstract. This paper introduces the HoloHome, an Augmented Reality framework which aims to provide new means of interaction with the Smart Home and its components. HoloHome integrates multiple technological paradigms encompassing Internet of Things, Ambient Assisted Living, and Augmented Reality to offer the ultimate tailored comfort experience for the Smart Home's inhabitants including frail elderlies and people with mild cognitive disabilities. The main purpose of the HoloHome is to provide a Mixed Reality environment implemented on the Microsoft HoloLens, to allow the user interaction with the Smart Home devices and appliances through the Augmented objects. HoloHome tackles the issue of real world locations of the objects and alignment of the virtual objects on the real objects within the spatial environment, using Vuforia image processing engine. It also defines the modality of the interconnection between the Mixed Reality application and the distributed network of smart devices through the communication channel of WiFi-enabled microcontrollers. Two use cases depict the HoloHome interaction methods and functionalities in two typical scenarios: regulating the domestic devices – such as turning on/off light – through the Augmented Reality framework, and providing visual hints and clues to help the users with mild cognitive impairments to find the objects they need easily.

Keywords: Smart home · Augmented Reality · Mixed Reality · Microsoft HoloLens · Vuforia

1 Introduction

The evolution of computers has emerged a new generation of users whose expectations, in terms of computer interaction, have been raised. This new generation of users is not only seeking for the information, but they also would like to be able to create, collaborate, and interact with the digital world. Augmented Reality (AR) [1] is one of the human-computer interaction technologies that has shown great potential in the recent decade. The goal of AR is to provide users the enhanced version of the real world by overlaying virtual computer-generated objects to the real physical environment. It aims to seamlessly blend the virtual objects with the real world environment in order to bring a new improved experience for the users [2]. The term "Mixed Reality" (MR) [3] and

© Springer Nature Switzerland AG 2019
L. T. De Paolis and P. Bourdot (Eds.): AVR 2019, LNCS 11614, pp. 137–145, 2019.
https://doi.org/10.1007/978-3-030-25999-0_12

AR found to be used interchangeably – despite some minor distinctions – however, MR is designed to combine both features of the Virtual Reality (VR) [4] and AR.

In recent years, AR has faced an increasing potential and acceptance in different domains including the activities that need continuous and smart assistance. The superimposed virtual objects are able to guide the users to find the location of the real objects within the domestic environment, and provide them a complete step by step 3D animated instruction – with visual and voice feedbacks – on how to complete a task in regards to each of the related object. This could bring strong advantage in the field of Smart Homes [5] and Ambient Assisted Living (AAL) [6], in which inhabitants' constant support in performing the Activities of Daily Living (ADLs) [7] is required. The paradigm of smart home encompasses multiple methods and techniques to represent all the house appliances in a connected network of smart devices in which they are able to transmit and exchange data and information.

This work describes an AR framework, called "HoloHome", within the smart house in order to investigate a new means of interaction for the inhabitants to manage and regulate the domestic environment and to perform their daily activities in a more convenient and independent way to fulfill the requirements of an AAL environment. HoloHome is part of the Italian project "Future Home for Future Communities (FHfFC)" [8], which aims to develop the "house of the future" by integrating multiple technological paradigms to promote inhabitants' comfort, safety, and independence while providing them continuous support in performing different ADLs. The "house of the future" has been implemented inside the STIIMA's Living Lab in Lecco and is currently under development phase.

Although some ADLs within the house might seem to be simple and intuitive for most of us, but some people like elderlies and people with cognitive impairments face a lot of challenges to perform regular daily tasks [9]. These tasks could vary from operating a kitchen appliance, opening and closing the window, or a simple act of turning on/off the light.

In the context of AAL, it is of pivotal importance to enable Ambient Intelligence (AmI) [10] to equip the domestic environment with Context-Aware (CA) [11] devices which would guarantee smart services for the inhabitants. The aim of these services is to help people with special needs to live more independently, in a safe and healthy environment. In this regard, providing a continuous interactive guide and support for the users to complete their daily activities is crucial.

The remainder of this paper is organized as follows: Sect. 2 highlights some of the notable works in the field of smart homes and AmI; Sect. 3 delves into the detailed architecture of the HoloHome framework; Sect. 4 depicts two use cases in which HoloHome features are illustrated; and finally the Conclusion summarizes the main outcomes of this paper in addition to the future works.

2   Related Works

Although there have been many studies concerning the AR in general, little study has been conducted in regards to the exploitation of AR in the field of smart homes. AR was first been reviewed by Azuma in 1997 [1] alongside with describing the characteristics of AR systems and its limitations. Ever since that date, there has been vast amount of research on AR within different fields. With release of Microsoft HoloLens smart glasses in 2016, many researchers started exploiting HoloLens in several contexts. Evans et al. [12] described a thorough evaluation on the HoloLens functionalities on the guided assembly instruction. Recently, Kučera et al. [13] depicted the modality of the connection between the AR applications and the smart devices through microcontrollers. There exists variety of research works related to the Smart Homes and simulation of the domestic environment; however, little study has been conducted in regards to the managing the smart home through AR. In the field of Smart Home, Mahroo et al. [14] investigated the possibility of deploying a Smart Home and enabling a CA system, exploiting the protocol of Internet of Things (IoT) and semantic web technologies. In another study [15] smart home and architecture of connected ubiquitous devices for increasing the customized comfort metrics has been discussed.

3   The Smart Home Architecture of FHfFC

This section describes the architecture of the HoloHome, which enables the users to interact with the smart home. As mentioned in Sect. 1, the possibility to interact with the home appliances and to receive visual feedback from them would bring a huge benefit in the context of AAL environment. In this regard, this work proposes an AR application installed on the Microsoft Hololens smart glasses that is capable of interacting with real objects within the physical environment. HoloLens is able to keep track of the spatial environment around the Hololens in order to understand the real world position of each object. Spatial mapping provides a mesh representation of the real objects surfaces within the environment, allowing developers to implement more realistic AR applications [12]. Knowing the spatial mapping of the domestic environment, the HoloHome is able to seamlessly align the virtual objects to real world objects.

Although the proposed AR application must be flexible enough to adjust to different positioning of the real objects within the environment, it also needs to be rigid enough to have the virtual objects always anchored to the real appliances as well.

HoloHome must be capable of positioning each hologram – virtual object made by Hololens – on the corresponding real-world object with the exact same positioning and precise scale alignment to create the ultimate realistic feelings for the users. In this regard, HoloHome exploits the Vuforia's image processing library which allows HoloLens to track a target image from the front-facing camera's coordinate system. In this way, the HoloHome is able to find the accurate location of each appliance within the real world environment and instantiate the virtual version of that object in a way that both real and virtual objects align on each other perfectly.

Moreover, the HoloHome leverages the ability to control and regulate the home appliances exploiting IoT system to bring the new way of interaction between the users and the smart home platform (Fig. 1).

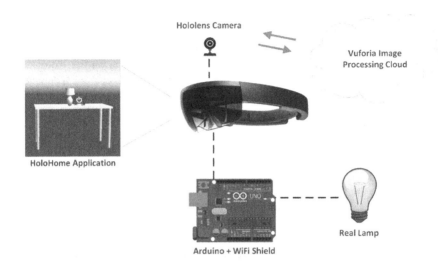

Fig. 1. Conceptual model of different technologies involved in HoloHome & their interactions.

In the following subsections, the technical details of the development phase of AR framework are discussed.

3.1 Physical Infrastructure

In this work, the "house of the future" is simulated within the Living Lab of Lecco as part of the national project FHfFC. The lab environment comprises kitchen major appliances, while their functionalities are simulated via AR. The kitchen environment consists of a refrigerator, a washing machine, a dishwasher, a sink, a pantry, few cabinets, and a desk. In addition to the home appliances that are needed in this project lab, the physical infrastructure of this smart house also encompasses a network of interconnected and interrelated smart devices – sensors and actuators – to enable the transmission and exchange of data over the internet exploiting IoT. In order to allow the inhabitants to interact with the "house of the future", a pair of AR glasses is needed to be used in this environment. One of the best AR devices on the market is the Microsoft HoloLens which is chosen to use in this project. HoloLens has a pair of translucent screens for its eye-pieces that allows the injection of the holograms into the user's line of sight without completely blocking the user off the world.

HoloLens uses an accelerometer (to measure the speed of the head while moving), a gyroscope (to measure the tilt and orientation of the head), and a magnetometer (to function as a compass) to provide an ideal AR experience. HoloLens is also equipped with depth sensing camera and a standard camera to collect data about the surrounding area in order to determine where surrounding physical objects are located and to draw a digital picture of the surroundings.

3.2 AR Application on HoloLens

The HoloHome is implemented with Unity 3d software [16] for the Universal Windows Platform (UWP). The application exploits the Microsoft Mixed Reality Toolkit to handle the AR environment, and Vuforia SDK [17] to augment the holograms on the target images registered in the Vuforia portal (Vuforia is described further in the Sect. 3.3). Although it is important to be able to move the real object and re-instantiate the corresponding virtual object alongside with that – as it may happen in real life scenarios to redecorate the house and change the place of home appliances – it is also not convenient to locate the objects with Vuforia each time the user launches the application. As a result, HoloHome has been developed in such way to save the real-world location of the objects – after image target is being detected by Vuforia – and load the previously anchored positions of the objects next time the user launches the application. In order to meet both requirements simultaneously, HoloHome adopted a particular approach which foresees two different options at the very first scene when the application is launched; asking the user if she/he would like to locate the home appliances via Vuforia target image tracker, or she/he would like to restore the location of the objects as it was anchored and saved in the previous session. If the user chooses the option to turn on the Vuforia and locate the objects, HoloHome application would start with no virtual object, waiting for the image targets to be detected by HoloLens front camera to locate the virtual objects on the image target. However, if the user chooses the other option to turn off the Vuforia and restore the last object positions as in the last session, HoloHome would turn off the HoloLens front camera – which makes the Hololens camera free for other uses – and load the latest locations of the objects anchored the most recent time HoloHome had run (Fig. 2).

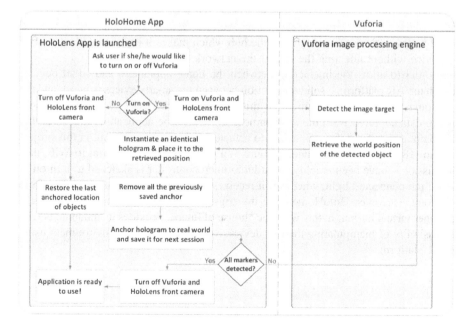

Fig. 2. The overall architecture of HoloHome and its interaction with Vuforia engine.

3.3   Vuforia Image Processing Engine

Vuforia is the most widely used platform to develop AR applications with robust tracking and performance. Vuforia engine is a marker-based AR system in which the markers are the images or 3D objects registered with the application – via the Vuforia library – that act as information triggers in your application.

When HoloLens camera detects any of the registered markers within the real environment (while running HoloHome), this prompts the display of virtual objects over the real world position of the marker in the camera view. Marker-based tracking can use a variety of different marker types, including QR codes, physical reflective markers, Image Targets, and 2D tags. In this work, we use Image Target which is the simplest and most common type of markers. There exist few images uploaded on the Vuforia image processing portal, in which each of them is associated with a specific kitchen appliance. When the user launches HoloHome via the HoloLens, she/he can look for the image target in the real environment allowing the Hololens camera to recognize the marker and augment the virtual object associated with that particular image.

Another advantage that Vuforia would bring to the application is the fact that any real-world object's position becomes dependent on the tracked image target. Not only does this approach improve the hologram stability, but also it allows the user to relocate the real object and be able to easily re-instantiate the hologram as well.

3.4   Arduino Interaction with AR Platform

In order to develop the idea of the smart home and regulating the heterogeneous devices within the domestic environment through an automatic remote system, the need for deploying an IoT network is crucial. The "house of future" forms a network of WiFi-enabled devices – preferably in a star typology where there is one central hub and several nodes or devices connected to the hub which makes it easier to add or remove any device without affecting the rest of the network.

In order to allow the interaction between the home appliances and smart devices through the AR platform, a solid connection between the smart devices network and the HoloHome application is required. In this network, each device is capable of both sending data captured from the environment to HoloHome application, and receiving the prompt action from the application to actuate the device. As a result, in this project Arduino [18] microcontroller and Arduino WiFi shield [19] – for Serial to WiFi data transmission – have been employed. Arduino microcontroller is sketched to turn on or turn off the connected light, whenever it receives the related prompt data coming from HoloHome. However, HoloHome sends this trigger to the Arduino when the user clicks the proper virtual button. In this way, the "house of future" provides its inhabitants with the possibility of manipulating the real devices within the domestic environment using the AR platform.

4   Use Case Scenario

The HoloHome is designed and implemented to provide continuous help and support for the smart house inhabitants for performing ADLs. Combining various technological paradigms, HoloHome enhances user interaction with the smart home platform. Considering the complexity of the system, it is important to define some use cases to illustrate the interaction between the dwellers and the smart house AR platform; to demonstrate how HoloHome boosts the user experience while interacting with their smart house; and finally to demonstrate how HoloHome help the dwellers in performing their ADLs in a more convenient and independent way. In the following use cases, a middle-aged male user is considered, who is 58 years old and suffers from mild executive function deficit [20], which causes him difficulties in performing sequential and executive tasks.

4.1   User Turning on/off the Light

The first use case depicts how the inhabitant of the "house of future" is able to interact with the house devices in general and with the lights in particular. HoloHome is able to create a virtual version of the home appliances and overlay the virtual object over the real object. The HoloHome application then provides the dwellers, visual step by step instruction and feedback over each appliance object. Each time a dweller launches the HoloHome application and has the virtual objects located in place – either by reading the image targets or by restoring the previous locations –, HoloHome is ready to provide help and support regarding each activity he is gazing at. When the dweller looks around the room wearing the HoloLens, each time he stops moving his head and looks at each specific device – his eye gaze collides with the hologram –, the virtual images, animations, sound, or any necessary information appears on that particular object to help him regulate the device.

4.2   User Looking for the Specific Object with the Help of HoloHome

The second use case demonstrates how the HoloHome supports the dweller to find different objects he needs. This can be done by providing him with visual hints to find the place of the ingredients he needs for his meal recipe, or the particular appliance and/or its functionality that he is looking for. Considering the fact that people with mild cognitive impairments tend to suffer from amnesia and face difficulties remembering the order of execution sequences, HoloHome offers features and functionalities to provide some hints and clues in a form of hologram graphics, to guide the user in finding the anticipated object faster and easier. Assume the user would like to turn on the washing machine with specific settings. He has already located the washing machine using Vuforia so the HoloHome knows the exact location of the washing machine, its components, and the control panel. When he looks at the washing machine, a virtual panel appears on the washing machine which aligns on the exact same position as the real washing machine's control panel and its buttons. Each button or setting will be lit with different colors in a sequential order to avoid the user's confusion by facing too many buttons at the same time. HoloHome also provides the explanation in the form of virtual text or voice command on each setting. When he sets the first setting by clicking on the virtual button, the first setting's hologram disappears and the second setting that needs to be done would be lit with another color.

5 Conclusion and Future Works

This work introduces HoloHome, an AR application deployed on the HoloLens to manage and regulate the "house of the future". Exploiting the protocol of IoT, AAL, and AR technologies, this framework aims to provide continuous guide and support for the inhabitants to complete their ADLs. The AR framework investigates the possibility of aligning virtual objects over the real objects to offer specific services for each domestic devices in AR form. Moreover, HoloHome is able to communicate with real devices and regulate them through WiFi-enabled Arduino microcontrollers.

Future works foresee the deployment of the domestic appliances within the "house of the future", and connecting all the appliances to the HoloHome application. It is also envisioned to validate the HoloHome framework in the near future to evaluate the usability of the application amongst diverse group of people.

References

1. Azuma, R.T.: A survey of augmented reality. Presence: Teleoperators Virtual Environ. **6**(4), 355–385 (1997)
2. Hettiarachchi, A., Wigdor, D.: Annexing reality: enabling opportunistic use of everyday objects as tangible proxies in augmented reality. In: Proceedings of the 2016 CHI Conference on Human Factors in Computing Systems, pp. 1957–1967 (2016)
3. Billinghurst, M., Kato, H.: Collaborative mixed reality. In: Proceedings of the First International Symposium on Mixed Reality, pp. 261–284 (1999)
4. Steuer, J.: Defining virtual reality: dimensions determining telepresence. J. Commun. **42**(4), 73–93 (1992)
5. Harper, R.: Inside the Smart Home. Springer, Berlin (2006)
6. Wichert, R., Eberhardt, B.: Ambient Assisted Living. Springer, Heidelberg (2012)
7. Katz, S.: Assessing self-maintenance: activities of daily living, mobility, and instrumental activities of daily living. J. Am. Geriatr. Soc. **31**(12), 721–727 (1983)
8. Future home for future communities. http://www.fhffc.it
9. Stuss, D.T.: Functions of the frontal lobes: relation to executive functions. J. Int. Neuropsychol. Soc. **17**(5), 759–765 (2011)
10. Aarts, E., Wichert, R.: Ambient intelligence. In: Bullinger, H.J. (ed.) Technology Guide, pp. 244–249. Springer, Berlin (2009). https://doi.org/10.1007/978-3-540-88546-7_47
11. Schilit, B., Adams, N., Want, R.: Context-aware computing applications. In: WMCSA, pp. 85–90 (1899)
12. Evans, G., Miller, J., Pena, M.I., MacAllister, A., Winer, E.: Evaluating the Microsoft HoloLens through an augmented reality assembly application. In: Degraded Environments: Sensing, Processing, and Display 2017, vol. 10197, p. 101970V (2017)
13. Kucera, E., Haffner, O., Kozák, Š.: Connection between 3D engine unity and microcontroller arduino: a virtual smart house. In: 2018 Cybernetics & Informatics (K&I), pp. 1–8 (2018)

14. Mahroo, A., Spoladore, D., Caldarola, E.G., Modoni, G.E., Sacco, M.: Enabling the smart home through a semantic-based context-aware system. In: 2018 IEEE International Conference on Pervasive Computing and Communications Workshops (PerCom Workshops), pp. 543–548 (2018)

15. Spoladore, D., Arlati, S., Sacco, M.: Semantic and Virtual Reality-enhanced configuration of domestic environments: the smart home simulator. Mobile Inf. Syst. **2017**, 1–15 (2017)

16. "Unity 3D" software. https://unity3d.com

17. "Vuforia" SDK. https://www.vuforia.com

18. "Arduino" microcontroller. https://www.arduino.cc

19. "Arduino" WiFi shield. https://www.arduino.cc/en/Guide/ArduinoWiFiShield

20. Traykov, L., et al.: Executive functions deficit in mild cognitive impairment. Cogn. Behav. Neurol. **20**(4), 219–224 (2007)

Training Assistant for Automotive Engineering Through Augmented Reality

Fernando R. Pusda$^{(\boxtimes)}$, Francisco F. Valencia$^{(\boxtimes)}$,
Víctor H. Andaluz$^{(\boxtimes)}$, and Víctor D. Zambrano$^{(\boxtimes)}$

Universidad de las Fuerzas Armadas ESPE, Sangolquí, Ecuador
{frpusda, ffvalencia, vhandaluz1, vdzambrano}@espe.edu.ec

Abstract. This article proposes the development of an augmented reality application for mobile devices with Android OS focused on the visualization and interaction of the user with the components, technical characteristics, location and the processes of disassembly and assembly of engine in a vehicle; facilitating the learning process referring to this automotive system. The application was developed in the graphic engine 3D Unity and the use of Vuforia for the recognition of objects in 3D, being a technological tool that allows vouching the learning in the internal combustion engine, guiding the user and changing the paradigms of the use of physical manuals with the new technological advances as the augmented reality.

Keywords: Augmented Reality · Internal combustion engine · Unity · Vuforia

1 Introduction

Industrial processes have evolved fastly, and it is very important to train personnel with specific steps and explicit instructions that guarantee personal safety and the correct use of machines to obtain better performance in production, leading industries to create environments that simulate specific events in purpose of user's capacitation in areas such as firefighter training, mine safety, construction and civil engineering, etc., [1, 2]. Inadequate training and insufficient work experience have been identified as one of the main causes of occupational accidents involving the use of modern machinery and equipment. Many of these occupational accidents could be avoided if operators are adequately trained in the work area. New technological advances have been used widely in different types of industries and even in the training of workers, projecting presentations or explanatory videos [3].

In the automotive industry, technological advances have also been developed, both in processes that the operator must perform and in the training of personnel to improve their quality and efficiency [4]. This sector is strongly influenced by the new tools created such as maintenance assistants, intelligent manuals, assembly guides, and others; reinforcing theoretical knowledge and complementing it with practice through the manipulation and control of equipment to carry out industrial processes [5, 6]. Assembly process planning uses 3D CAD models to manipulate components in order to improve quality levels in terms of accuracy and reduce production time on workstations [7]. Innovate with 3D visualization technologies in manual tasks, use of tools and equipment

L. T. De Paolis and P. Bourdot (Eds.): AVR 2019, LNCS 11614, pp. 146–160, 2019.
https://doi.org/10.1007/978-3-030-25999-0_13

in workplaces like automobile assembly lines, increases operator welfare and performance in the automotive field by improving general safety conditions [8].

On recent years, the automotive industry has begun to apply Virtual Reality and Augmented Reality, with the aim, allowing designers and engineers to visualize and modify new prototypes of vehicles and their systems, significantly reducing time and development costs, increasing productivity, efficiency of these processes and resulting in an improvement in the quality of new products [6]. The use of Virtual Reality can be seen in vehicle manufacturers, integrating these concepts in the processes of Induction of New Products (NPI) in the design phase, for 3D digitalization, providing an ideal environment for the analysis of possible modifications on the base design [6, 9]. Augmented Reality has been a great help in the simulation of properties and physical phenomena to which vehicles are subjected, obtaining results quickly and accurately [10]. Although this has benefited overall performance in several industrial applications, only in recent years, that AR based assembly systems have been developed that focus on the process of dynamic and interactive instruction by superimposing digital information and components on a real object [11, 12].

This application seeks breaking paradigms with respect to traditional capacitation and trainings, strengthening and improving the learning process in the different work environments, *e.g.*, automobile assembly, automotive systems manufacturing and repair workshops, among others. Given the complexity of the assembly processes and the identification of vehicle components, the proposed application focuses on providing assistance to the user to visualize detailed information of the internal combustion engine of the vehicle, using AR that gives a more dynamic and interactive learning and offers a virtual training support, which can be implemented in the automotive industry to increase productivity on industrial processes. This work divides on three parts:

I. *Components description,* details each of the parts that make up the engine of the vehicle, using AR to train the user in the recognition of automotive parts;

II. *Virtual assembly and disassembly,* it allows to visualize onto the mobile device, the components of the automotive systems, specifically the combustion engine of the vehicle and by means of an intelligent guide that train the operator to interact in the processes of assembly and disassembly of the motive joint;

III. *Function simulation,* gives a clear idea of the work carried out by different parts of the automotive systems present in the vehicle from 3D simulations shown on the screen of the mobile device, being possible to identify the importance of each mechanism and its precision in the cycles performed by the internal combustion engine.

This article is divided in 6 Sections, including the Introduction. Section 2 describes the architecture of the proposed system; Sect. 3 describes the unmarked tracking system. Section 4 shows the assembly/disassembly and functioning animation, then Sect. 5 presents the analysis of the results and finally, the conclusions are detailed in Sect. 6.

2 System Architecture

The automotive industry has evolved over the years in the area of production, manufacturing and maintenance of vehicles, therefore training for new workers is important in these modern technologies in order to improve the experience and increase the safety of operators in industrial processes, using training tools didactic and interactive simulate real situations in a safe way in the virtual environment.

Hence, this article is oriented to the development of an augmented reality application for training and assistance in the assembly processes of the components of a vehicle on Automotive Engineering, like a visual guide for the operator that provides the relevant information of virtual objects creating an innovative experience.

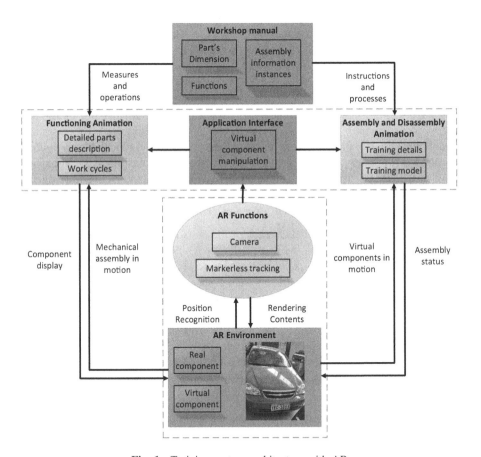

Fig. 1. Training system architecture with AR

The Fig. 1 shows training system's architecture; for its development the first thing is to obtain the essential information from the vehicle work manual, *i.e.*, the dimensions of the components, the instructions, the assembly and disassembly processes and the operations performed by each part within the automotive system in the work cycle. The application interface allows the user to select between two training areas through the mobile device screen, which are System Functioning and Engine Assembly. In the functioning simulation area, it allows the user to see the cycles that are performed in each cylinder of the vehicle engine, *e.g.*, admission, compression, explosion and escape, such as the interaction one by one, the components involved in the operation of the same, also allows to see the detailed description of each the elements that allow the internal combustion engine. In the assembly and disassembly simulation area it is observed step by step this process in an orderly and explanatory way the disassembly of an internal combustion engine, allowing the user to learn the process in a more didactic way, fast and similar to reality.

The Augmented Reality (AR) environment is based on of real and virtual components that are displayed on the mobile device and depending on the option selected in the interface described above will change the movements of digital models, through the functions provided by AR such as the device camera and the recognition pattern focused on the recognition system without marks, the application shows the main menu for the user to access the training.

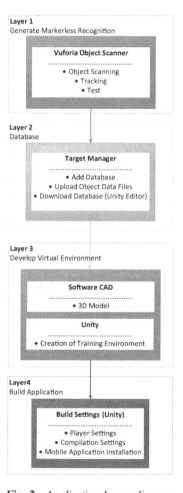

Fig. 2. Application layers diagram

The Fig. 2 describes and shows the process for the creation of the application from its initial phase, it is constituted by four general stages and each one defines specific tasks for the development and execution of the training assistant, these are: *(i) Marcking less recognition,* the 3D pattern of the object is determined, because the training assistant to be developed proposes the use of a markless tracking system by obtaining characteristic points; *(ii) Database,* allows to save the objects or characteristics to be used through the online portal of Vuforia developers, in the Targets Manager section are added files of 3D objects (*.OD) with compatible characteristics for use in the development of the application; *(iii) Develop virtual environment,* refers to the development of the virtual components of the automotive system to be used in the application, in which the 3D elements of the parts and selected parts of the object of study are created through CAD software [13], and also to the animations and tactile controls in Unity; *(iv) Build Application,* the presentation parameters such as application orientation, textures, sizes and menu options are configured; also define Android compilation compatible settings by selecting the minimum level of application programming interface (API), as well as application version configuration and Package Name. The created application is transferred to the mobile device for installation and use of the virtual assistant.

3   Markerless Tracking System

This system is developed using the Vuforia Object Scanner tool, which enables the use of the mobile device's camera as a scanner to generate a 3D pattern; to begin with, a default objective called "Object Scanning Target" (OST) is selected that defines the position and orientation of the object provided by the Vuforia page. The characteristic region of the target fulfills two functions; it precisely identifies the position of the physical object, and defines the limits of the space of the selected object according to its dimensions.

Fig. 3. Scanning environment origin

To initiate the scanning session of the object of study and because of its dimensions, an OST printout was used at a 30:1 magnification scale so that the virtual axes generated from the origin contain the entire vehicle area, as shown in Fig. 3. The scanning application allows the capture and recognition of vehicle characteristic points. It is recommended to perform this process with the object of study well aligned with the axes of the OST, in an environment with balanced lighting, placing the object on a light gray background and avoid containing shadows caused by object on a light gray background and avoid containing shadows caused by other objects or overexposure brightness.

By moving the camera around the vehicle, a grey virtual dome is displayed and as the characteristic points are captured, the color of the coverage area changes to green. At the Fig. 4 the captured characteristic points and the properties of the created marker are shown and the general data of the file is displayed on the summary screen.

Fig. 4. Creation of the pattern 3D

The target manager database can be accessed through the Vuforia developer portal, to configure the object data obtained from the scanning process, creating a license in the License Manager section, being used in Unity for the development of the application.

4 Assembly/Disassembly and Functioning Animation

3D models are created with the technical information collected from the vehicle, automotive systems and mechanisms, through a CAD software to be exported to the graphic development engine Unity, with the program 3ds Max as shown in Fig. 5, creating compatible files (*.FBX) to use them in the virtual environment.

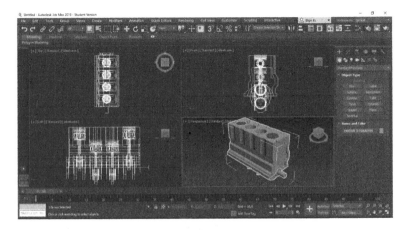

Fig. 5. Internal combustion engine in 3ds Max

These models allow to establish animation states based on the flowchart of Fig. 6, the initial process is to create a new animation clip and select the components to be used. In this new clip are recorded all the changes required by the scene and distributed through the timeline, taking into account the number of frames per second. A script is developed that allows the control of the states of animation and the game objects.

Unity uses the animation panel to create the motion clips of the components and specifies the transformation properties of each piece in mechanical systems; this allows each clip to be integrated into the Animator panel, which uses a graphic system of nodes to represent the animation states and each of them is connected by means of transitions, as shown in the Fig. 7.

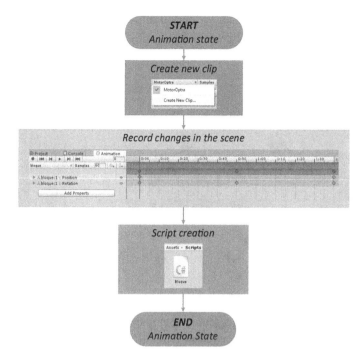

Fig. 6. Animation flowchart

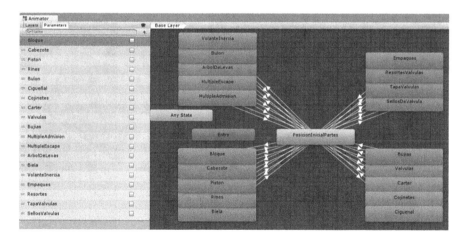

Fig. 7. Animation states connected with transitions

Transitions must achieve conditions to generate the environment of interaction with the user by creating parameters, that can be integers, floating, boolean or triggers, with scripts control the parameter's variations, generating changes into the state of the animation. In Fig. 8, buttons and scripts are used in Canvas to change scenes, *i.e.*, the procedures for assembly, functioning and technical data are programmed to show the environment of each option by touch controls.

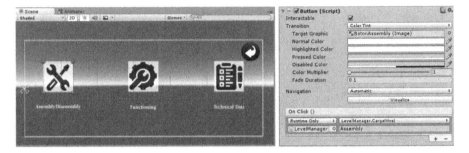

Fig. 8. Creation Touch Controls

5   Analysis and Results

This section shows the augmented reality performance application focused on user training and assistance, parts recognition automotive system assemblies and their operation. The application can be implemented on industries using this tool to improve the training process and use virtual data to allow innovative assistance.

Fig. 9. Application language selection

At the beginning of the training assistant, the application first must be installed on the mobile device, the main screen allows you to select the language that is shown in Fig. 9. Vehicle recognition is performed by the smartphone camera, allowing user interaction through application functions, *e.g.*, assembly, disassembly, parts recognition and simulation of automotive functioning system.

Figure 10 shows the design of a **template AR** that allows training assistant to be used with 2D recognition, to provide knowledge on the assembly and disassembly of an internal combustion engine to the general public without having to be present in the laboratory.

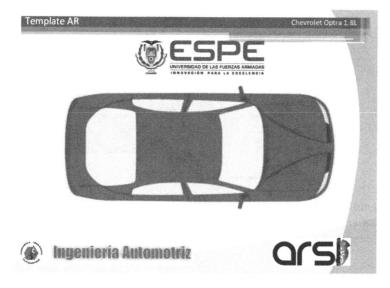

Fig. 10. Template 2D

When focused with the camera of the mobile device, the training assistant works normally generate the 3D model of the vehicle and their internal combustion engine parts, as can be seen in Fig. 11.

Fig. 11. Training assistant on the 2D template

Alternatively, using the assistant in the lab after selecting the language in the main menu should focus on the study vehicle and show the message "Please focus the vehicle" as seen in the Fig. 12, until the mobile device camera recognizes the default vehicle; shown on the main menu, leading to the study module selection.

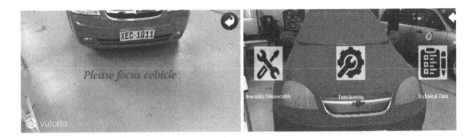

Fig. 12. Vehicle recognition

Recognition of parts, this option allows the user to visualize each part of the combustion engine of the vehicle, the same to the detailed technical information of each component as shown in Fig. 13.

Fig. 13. Parts recognition environment

In Fig. 14 the application allows to visualize the description of component using of a text box.

Fig. 14. Description of the selected component

Virtual Assembly of Fig. 15, assists the user with the procedure of internal combustion assembling and disassembling through the animations and steps described, which are displayed on the device screen.

Fig. 15. Assembly scene

The **simulation of operation** is shown in Fig. 16 and guided by animations that the user understands in a didactic way the work performed by the mechanical assembly inside the engine.

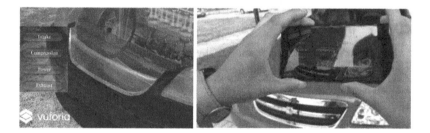

Fig. 16. Engine operation and Duty Cycles

On the vehicle's **technical data** set is shown three drop-down menus that indicates the engine specifications, fuel need and dimensions, giving the specifications that identify the automobile as shown on the Fig. 17.

Finally, a usability test is conducted with two groups of people belonging to automotive engineering: students and teachers, to establish the interest of the training assistant through Augmented Reality [14]. For the first group, it considers ten automotive engineering students of initial levels by the little knowledge in the subject of internal combustion engines, and the assistant is a training guide to reinforce the

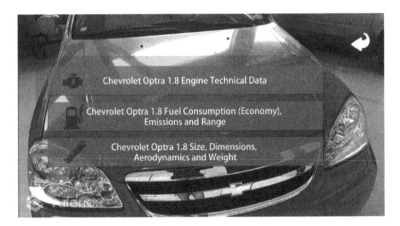

Fig. 17. Vehicle's technical data

learners' abilities and the second group included ten teachers of the Universidad de las Fuerzas Armadas ESPE, who made use of the training assistant to evaluate the interaction and familiarization with the application. Table 1 shows the questions posed to both groups, for the valuation of each item the scale from 1 to 5 is used, with 1 being the minimum acceptance and 5 the maximum.

Table 1. Usability evaluation tests

Q1. Is it difficult to get used to the application of augmented reality?
Q2. Do you need specialized support to use this application?
Q3. Do the different training assistant options work properly?
Q4. Do you need to have technical knowledge prior to using the application?
Q5. Is the interaction with the training assistant simple and intuitive?
Q6. Can the virtual training system be implemented in the area of professional education?
Q7. How difficult is the handling of automotive system components in a virtual environment?
Q8. Is induction necessary before using the training assistant?
Q9. Do you find inconsistencies in this training assistant?
Q10. How often would you use the application?

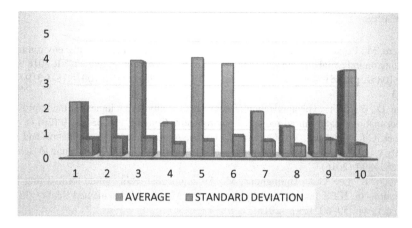

Fig. 18. Results of the surveys

The results obtained from the questions posed shown in Fig. 18 indicate a high acceptance rate of students and teachers who checked the training assistant.

6 Conclusion

The development of applications for training in the automotive industry improves the professional formation processes through the use of augmented reality providing benefits that contribute to recognize and manipulate the elements that compose the automotive systems, guiding the user in the different industrial processes for repair work or assembly of engines, reducing time and resources, improving learning in a safe environment. This work shows acceptation, its accessibility and the development of the Augmented Reality training assistant that provides relevant information and virtual simulations of the processes of assembly, disassembly, recognition and operation of the engine parts in the vehicle, using the markerless tracking system through the screen of a mobile device.

Acknowledgements. The authors would like to thanks to the Corporación Ecuatoriana para el Desarrollo de la Investigación y Academia–CEDIA for the financing given to research, development, and innovation, through the CEPRA projects, especially the project CEPRA-XI-2017-06; Control Coordinado Multi-operador aplicado a un robot Manipulador Aéreo; also to Universidad de las Fuerzas Armadas ESPE, Universidad Técnica de Ambato, Escuela Superior Politécnica de Chimborazo, Universidad Nacional de Chimborazo, and Grupo de Investigación ARSI, for the support to develop this work.

References

1. Rosero, M., Pogo, R., Pruna, E., Andaluz, Víctor H., Escobar, I.: Immersive environment for training on industrial emergencies. In: De Paolis, L.T., Bourdot, P. (eds.) AVR 2018. LNCS, vol. 10851, pp. 451–466. Springer, Cham (2018). https://doi.org/10.1007/978-3-319-95282-6_33
2. Tatić, D., Bojan, T.: The application of augmented reality technologies for the improvement of occupational safety in an industrial environment. Comput. Ind. **85**, 1–10 (2017)
3. Turner, C.J., Hutabarat, W., Oyekan, J., Tiwari, A.: Discrete event simulation and virtual reality use in industry: new opportunities and future trends. IEEE Trans. Hum.-Mach. Syst. **46**, 882–894 (2016)
4. Sarupuri, B., Lee, G.A., Billinghurst, M.: An augmented reality guide for assisting forklift operation. In: IEEE International Symposium on Mixed and Augmented Reality (ISMAR-Adjunct), pp. 59–60. IEEE (2016)
5. Lima, J.P., et al.: Markerless tracking system for augmented reality in the automotive industry. Expert Syst. Appl. **82**, 100–114 (2017)
6. Ortiz, J.S., et al.: Virtual training for industrial automation processes through pneumatic controls. In: De Paolis, L.T., Bourdot, P. (eds.) AVR 2018. LNCS, vol. 10851, pp. 516–532. Springer, Cham (2018). https://doi.org/10.1007/978-3-319-95282-6_37
7. Chen, C.J., Hong, J., Wang, S.F.: Automated positioning of 3D virtual scene in AR-based assembly and disassembly guiding system. Int. J. Adv. Manuf. Technol. **76**, 753–764 (2015)
8. Laudante, E., Caputo, F.: Design and digital manufacturing: an ergonomic approach for industry 4.0. In: Systems & Design: Beyond Processes and Thinking, pp. 922–934 (2016)
9. Ortiz, J.S., et al.: Teaching-learning process through VR applied to automotive engineering. In: Proceedings of the 2017 9th International Conference on Education Technology and Computers, pp. 36–40 (2017)
10. Zhang, J., Sung, Y.T., Hou, H.T., Chang, K.E.: The development and evaluation of an augmented reality-based armillary sphere for astronomical observation instruction. Comput. Educ. **73**, 178–188 (2014)
11. Hořejší, P.: Augmented reality system for virtual training of parts assembly. Procedia Eng. **100**, 699–706 (2015)
12. Jetter, J., Eimecke, J., Rese, A.: Augmented reality tools for industrial applications: what are potential key performance indicators and who benefits? Comput. Hum. Behav. **87**, 18–33 (2018)
13. Quevedo, W.X., et al.: Virtual reality system for training in automotive mechanics. In: De Paolis, L.T., Bourdot, P., Mongelli, A. (eds.) AVR 2017. LNCS, vol. 10324, pp. 185–198. Springer, Cham (2017). https://doi.org/10.1007/978-3-319-60922-5_14
14. Andaluz, V.H., et al.: Multi-user industrial training and education environment. In: De Paolis, L.T., Bourdot, P. (eds.) AVR 2018. LNCS, vol. 10851, pp. 533–546. Springer, Cham (2018). https://doi.org/10.1007/978-3-319-95282-6_38

Microsoft HoloLens Evaluation Under Monochromatic RGB Light Conditions

Marián Hudák, Štefan Korečko$^{(\boxtimes)}$, and Branislav Sobota

Department of Computers and Informatics,
Faculty of Electrical Engineering and Informatics,
Technical University of Košice, Letná 9, 041 20 Košice, Slovakia
{marian.hudak.2,stefan.korecko,branislav.sobota}@tuke.sk

Abstract. Mixed reality technologies provide more natural interaction with virtual objects integrated in physical environments. Considering global lighting conditions, the spatial mapping has limitations, which manifest when scanning under limited light intensity. This paper evaluates the impact of red, green and blue monochromatic lighting sources on the spatial map scanning and distance measurement process of the Microsoft HoloLens holographic computer and head-mounted display. The paper also compares luminous power values, measured by MS HoloLens, with the ones obtained from a professional luminous flux meter.

Keywords: Microsoft HoloLens · Monochromatic light source ·
Spatial sensing · Lighting conditions

1 Introduction

Mixed reality extends a physically accessible environment with three-dimensional virtual objects [3]. Thanks to Microsoft (MS) HoloLens [11], the mixed reality is more accessible for common use and virtual collaboration [12]. Its built-in depth camera technology allows to scan surrounding environments for physically available objects. The spatial depth detection can be performed by precise real-time mapping using an RGB-D camera [1].

The scan sensitivity of the surrounding environment is affected by global illumination [10]. Therefore, the quality of image detection and spatial mapping depends on the intensity of the light source placed in the physical space [13]. In the case of previous research works [7,9,14], MS HoloLens has been evaluated in environments with achromatic (white) light sources, which provide ideal conditions for the detection and scanning [5].

However, with the advent of RGB lighting, it is important to consider using mixed reality equipment under other than achromatic lightning conditions. The light of a specific color may be used for various purposes such as entertainment or research. With respect to the latter purpose, we plan to conduct experiments

© Springer Nature Switzerland AG 2019
L. T. De Paolis and P. Bourdot (Eds.): AVR 2019, LNCS 11614, pp. 161–169, 2019.
https://doi.org/10.1007/978-3-030-25999-0_14

focused on the human cognition, which we currently perform in a CAVE environment [8], in a more flexible collaborative mixed reality one. Before constructing such an environment, it is necessary to evaluate the performance of the corresponding devices under unusual lighting conditions, provided by RGB light sources.

In this paper we report on a set of experiments where the spatial scanning performance of MS HoloLens has been evaluated under the red, green and blue monochromatic illumination. As it has been observed [6] that the performance of depth cameras in a mixed reality setting decreases significantly when the global illumination is dimmed, we decided to emulate different light intensities by using one, two or three independent light sources. We focused on the luminous flux metering accuracy, maximum scanning distance and physical object detection performance of MS HoloLens. The rest of the paper starts with a short review of related researches in Sect. 2. Section 3 describes the object of the experiments, MS HoloLens, and Sect. 4 the corresponding setup. Section 5 presents the procedures and results of the individual experiments and Sect. 6 summarises the most important findings and outlines plans for more comprehensive experiments.

2   Related Work

The most related work is the study [9], where a series of experiments to quantitatively evaluate MS HoloLens performance has been carried out. These experiments focused on the accuracy and stability of MS HoloLens head posture estimation, its capability to reconstruct a real environment, spatial mapping and speech recognition. While the range of these experiments is greater than of the ones reported here, they all have been conducted under normal ambient indoor lighting with 25 W power output. Another difference is the measurement area size, which is 5 m × 8 m in [9] and 3 m × 3 m in our experiment. There is one methodological similarity between [9] and this work, namely a use of a reference measuring device for comparison: The work [9] uses OptiTrack motion tracking system to evaluate the head localization process of MS HoloLens while we use a professional luminous flux meter to evaluate similar measurements performed by MS HoloLens. The authors of [9] later focused on an evaluation of MS HoloLens from the user perspective and developed an appropriate quality-of-experience model [14]. This evaluation has been, again, performed under normal indoor lightning conditions.

The work [7] evaluates MS HoloLens with a specific goal in mind, namely to improve the spatial perception for humans with visual impairments by means of the depth mapping. It uses MS HoloLens as a tool for distance-based vision, which can show the user a colored depth scan in a real-time mixed visualization. Each recognized object is colorized in high-contrast with respect to its distance from the user. As a part of [7], the performance of MS HoloLens spatial mapping and object recognition was evaluated under a typical indoor lighting with a finding that the recognition is reliable up to the distance of 3 m from the object. The measurement area was a room with the size 5.3 m × 3.6 m.

3   MS HoloLens

MS HoloLens is a mixed reality device, combining a head-mounted stereoscopic see-through display with a set of input devices and a processing computer running MS Windows 10 operating system. The input devices are used to sense the surrounding environment, including user actions. They consist of an inertial measuring unit to determine the user position, four microphones for sound capturing, a two megapixel RGB camera for photo and video capturing, an ambient light sensor, a depth camera and a set of four gray-scale cameras. The experiments described in this paper focus on the last three devices.

The depth camera and the gray-scale cameras are used for the spatial mapping, i.e. a reconstruction of the surrounding environment in a form of a 3D model. The depth camera is also used for the collision detection between the real and virtual objects in a mixed environment and has the resolution of 1024×1024 pixels and the $120° \times 120°$ field of view. It utilizes an active infra-red illumination for more accurate measurements. The depth is estimated using the time-of-flight (TOF) approach with the speed of 1 to 5 frames per second (FPS) for far-depth sensing and up to 30 FPS for near-depth sensing [2].

4   Experimental Setup

In our experiments we tested MS HoloLens performance using three RGB light sources and a measurement area with the size of $3\,\mathrm{m} \times 3\,\mathrm{m}$. Our goal was to test the depth camera together with the gray-scale cameras and find out how they perform under monochromatic red, green and blue illumination. To be able to capture the data directly from MS HoloLens sensors in real time we used a special developer mode called *HoloLens Research mode*[1]. Two experiments have been carried out. The first experiment evaluated the luminous flux metering accuracy of MS HoloLens by comparing it to a professional flux meter. The second one focused on the MS HoloLens performance under a limited lighting, where it measured the maximum spatial mapping distance and the physical object detection speed.

The experiments have been conducted within a cubical indoor space of $3\,\mathrm{m} \times 3\,\mathrm{m} \times 3\,\mathrm{m}$. The exact arrangement of the space can be seen in Fig. 1, where L1, L2 and L3 are the light sources, PU is the position of HoloLens and P1, P2 and P3 are specific points, used in the experiments. For each of them, 3D Cartesian coordinates are given in meters in the format [x;y;z]. The orientation of the axes is given in the bottom left corner of Fig. 1.

The cubical space has one wall only. It is the rear wall, formed by the front side of a rear projection screen. The white area of the projection screen is $2.1\,\mathrm{m}$ high and thus covers 70% of the rear wall. The length between the white area and the floor or ceiling is $0.45\,\mathrm{m}$. The three identical light sources have been placed immediately in front of the screen, so the homogeneous matte surface of the screen provides a diffuse area for the lights. The light sources are *LED PAR*

---

[1] https://docs.microsoft.com/en-us/windows/mixed-reality/research-mode.

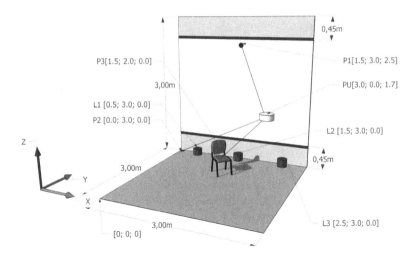

Fig. 1. Arrangement of the cubical indoor space used for the experiments

56 Floor 36x1W Black RGB spotlights, manufactured by Ignition. As the name suggests, one source contains 36 RGB LEDs, each with 1 W power output. The pulse width modulation of the light sources is 500 Hz, which is more than 16 times higher than the FPS rate of HoloLens cameras. The chair (P3) was used in the second experiment only.

Three separate light sources were utilized to be able to adjust the light intensity. While it is possible to dim the lights, we used them on full power, because the color reproduction is negatively affected otherwise. This resulted in 12 different light configurations, shown in Fig. 2. To clearly see the configurations, the individual photos in Fig. 2 have been shot with different negative exposure compensation. Therefore, they do not represent the actual illumination of the area as perceived by the user. From the user point of view, the whole space has been as illuminated by a diffused light. The overlapping borders between the light sources were only slightly visible on the projection screen. To eliminate the effect of external light sources, all experiments have been conducted after dark with darkened windows and all the lights on the surrounding hallways turned off.

5   Experiment Procedures and Results

Each experiment utilized the setup in a slightly different way. The experiment procedure descriptions in this section use the designations from Fig. 1 to identify the position of MS HoloLens and other important locations.

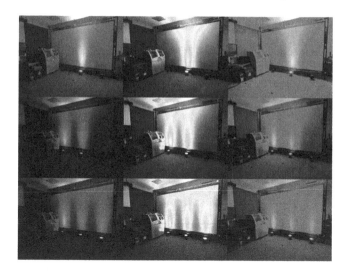

Fig. 2. Experimental setup under all light configurations used

5.1   Luminous Flux Measurement Accuracy

The objective of the first experiment was to compare the light power metering of MS HoloLens with a professional digital luminous flux meter under various RGB lighting conditions. The flux meter used was TES-133, manufactured by TES Electrical Electronic Corp.

Procedure. Both the MS HoloLens and TES-133 were placed at point PU and pointed to the point P1 on the projection screen (Fig. 1). PU was 1.7 m above the ground to emulate the position of MS HoloLens on the head of a person. The position of P1 was 2.5 m above the light L2, where the outputs produced by individual lights merged into one. This merging can be also observed in Fig. 2.

Nine sets of nine measurements were performed. The first three sets used only the red parts of the LEDs of the lights, the sets *4* to *6* used only the green ones and the last three used only the blue ones. During the measurements of the sets *1*, *4* and *7* the only light turned on was L2. The sets *2*, *5* and *8* used L1 and L2 and the sets *3*, *6* and *9* used all three lights. Some measurements were also carried out with the lights L2 and L3 but the results were similar to L1 and L2, so it has been decided to continue with the latter configuration only. During all the measurements the lights that were turned on used all the LED components of the corresponding color at 100% intensity.

Results. The final results of the experiment are shown in Table 1. The luminous flux values in the 4th and 5th row are averages of all the measurements in the given set. Both TES-133 and MS HoloLens consider the green light as the most intense and the red light as the least intense. For each set s, $1 \leq s \leq 9$, the difference is computed in percent as the value δ_s:

Table 1. Comparison of luminous flux measured by MS HoloLens and TES-133

Set of measurements		1	2	3	4	5	6	7	8	9
Lights	color	R	R	R	G	G	G	B	B	B
setup	count	1	2	3	1	2	3	1	2	3
	TES-133 (lm)	24	37	51	39	54	**67**	36	49	62
Results	MS HoloLens (lm)	**23**	32	46	31	48	61	28	41	55
	Difference (%)	10.79	14.15	15.02	20.40	11.72	9.12	22.93	16.46	11.28

$$\delta_s = \frac{\sum_{i=1}^{9}\left(\frac{|\Phi_i^T - \Phi_i^H|}{\Phi_i^T} \times 100\right)}{9}. \tag{1}$$

In (1), Φ_i^T is the luminous flux value from the i-th measurement of the set measured by TES-133 and Φ_i^H is the flux value measured by MS HoloLens. We consider Φ_i^T the accepted value and Φ_i^H the observed value.

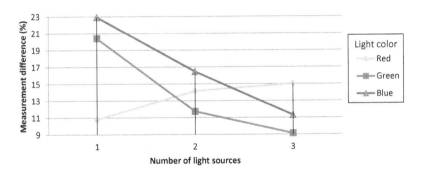

Fig. 3. Relation between the number of lights turned on and the difference between the luminous flux metering of TES-133 and MS HoloLens

As Fig. 3 shows, in the case of the blue and green lights the difference is decreasing with the increasing light intensity, i.e. the number of lights turned on. In the case of the red lights there is an opposite trend. The measurement sets with the red lights were also the only ones containing cases where the flux value measured by MS HoloLens was higher that the one measured by TES-133. There were three such cases in the set 1, one in the set 2 and two in the set 3.

5.2   Performance Under Low Light

In the second experiment we focused on the MS HoloLens performance under low light. Only one light source, namely L2, was turned on. We were interested in the influence of the light color on the maximum distance, detectable by the

120° spatial sensing system of MS HoloLens and the spatial mapping speed of an object, positioned 3 m far from MS HoloLens.

Procedure. To measure the maximum detectable distance, we placed MS HoloLens at point PU and pointed it to the point P2 (Fig. 1). The chair at P3 was removed, so there were no objects between PU and P2. The distance from PU to P2 was 4.57 m. Three sets of 30 measurements were carried out. The first set used all the red, the second one all the green and the third one all the blue LED components of L2 at 100% intensity.

For the spatial mapping speed evaluation we placed a conference room chair at the point P3 (Fig. 1). The chair (Fig. 5 left) has got a black frame and dark blue seat and back cushions. MS HoloLens was placed at PU and pointed to the chair. The distance between MS HoloLens and the chair was 3 m. Three sets of measurements with the same light settings as in the maximum detectable distance case were conducted. There were 10 measurements in each set.

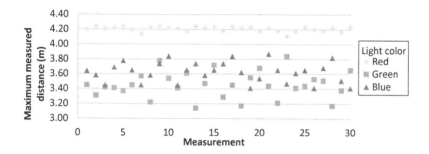

Fig. 4. Maximum detectable distance under low light (Color figure online)

Results. As Fig. 4 shows, the distance metering was most successful under the red light with the average of 4.21 m, minimum of 4.11 m and maximum of 4.24 m. However, even in this case, it didn't manage to reach P2 (4.57 m). The performance under the green and blue light was considerably worse with the averages of 3.45 m and 3.62 m. The variance of the values was higher, too. They ranged from 3.14 m to 3.84 m under the green light and from 3.41 m to 3.87 m under the blue light.

The spatial mapping speed evaluation results (Fig. 5 right) also show the best performance under the red light with the average detection time of 2.4 s. Again, the performance under the green and blue light was similar with the average times being 3.7 s and 3.5 s. However, in this case the variance was highest for the red light where the measured times ranged from 2.07 s to 2.94 s. For the green color the range was from 3.51 s to 3.95 s and for the blue one it was from 3.14 s to 3.74 s.

 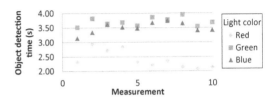

Fig. 5. The chair as detected and seen by MS HoloLens under the red light (left) and the object detection time measurement results (right) (Color figure online)

6 Conclusion

While several experiments evaluating the performance of the MS HoloLens holographic computer have been conducted and documented, none of them considered specific, monochromatic light conditions. The uniqueness of the experiments presented here dwells in the evaluation under monochromatic red, green and blue light. They were performed inside a $3 \times 3 \times 3$ m cubical indoor space, which was considered sufficient according to the findings of the previous studies [9] and [7].

The results obtained show that MS HoloLens performs better under the red light. The performance under the green and blue light was worse than under the red one but similar to each other. There were also other variations. Considering the first experiment, the difference between the luminous flux value measured by MS HoloLens and the TES-133 professional meter decreased with the increasing light intensity under the green and blue light, while under the red light it increased. In the case of the maximum detectable distance evaluation there was less variance in the values obtained under the red light than under the green and blue light. On the contrary, the variance in the spatial mapping speed evaluation was higher in the red set and lower in the green and blue set. The different results for the red light may be caused by the use of the infra-red illumination in the MS HoloLens depth camera.

To confirm the findings presented here, we plan to repeat the experiments with considerably more measurements. We also intent to perform similar evaluation in a larger indoor space with an ambient RGB illumination. Regarding the planned utilization of MS HoloLens, the results obtained find the device suitable for the human condition-related experimentation and also for an ongoing research [4] focusing on ambient user interfaces.

Acknowledgements. This work has been supported by the APVV grant no. APVV-16-0202: "Enhancing cognition and motor rehabilitation using mixed reality".

References

1. Aguilar, W.G., Rodríguez, G.A., Álvarez, L., Sandoval, S., Quisaguano, F., Limaico, A.: Real-time 3D modeling with a RGB-D camera and on-board processing. In: De Paolis, L.T., Bourdot, P., Mongelli, A. (eds.) AVR 2017. LNCS, vol. 10325, pp. 410–419. Springer, Cham (2017). https://doi.org/10.1007/978-3-319-60928-7_35
2. Anandapadmanaban, E., Tannady, J., Norheim, J., Newman, D., Hoffman, J.: Holo-SEXTANT: an augmented reality planetary EVA navigation interface. In: 48th International Conference on Environmental Systems (2018)
3. Brown, L.: Holographic micro-simulations to enhance aviation training with mixed reality. Technical report, Western Michigan College of Aviation (2018)
4. Galko, L., Porubän, J.: Tools used in ambient user interfaces. Acta Electrotechnica et informatica **16**(3), 32–40 (2016)
5. Ghosh, A., Ranjan, R., Nirala, A., Yadav, H.: Design and analysis of processing parameters of hololenses for wavelength selective light filters. Optik-Int. J. Light Electron Opt. **125**(9), 2191–2194 (2014)
6. Jacobs, K., Loscos, C.: Classification of illumination methods for mixed reality. In: Computer Graphics Forum, vol. 25, pp. 29–51. Wiley Online Library (2006)
7. Kinateder, M., Gualtieri, J., Dunn, M.J., Jarosz, W., Yang, X.D., Cooper, E.A.: Using an augmented reality device as a distance-based vision aid-promise and limitations. Optom. Vis. Sci. **95**(9), 727 (2018)
8. Korečko, Š., et al.: Assessment and training of visuospatial cognitive functions in virtual reality: proposal and perspective. In: Proceedings of 9th IEEE International Conference on Cognitive Infocommunications (CogInfoCom), Budapest, pp. 39–43 (2018)
9. Liu, Y., Dong, H., Zhang, L., El Saddik, A.: Technical evaluation of hololens for multimedia: a first look. IEEE MultiMedia **25**(4), 8–18 (2018)
10. Mandl, D., et al.: Learning lightprobes for mixed reality illumination. In: 2017 IEEE International Symposium on Mixed and Augmented Reality (ISMAR), pp. 82–89. IEEE (2017)
11. Microsoft: Microsoft hololens homepage (2019). https://www.microsoft.com/en-us/hololens
12. Noor, A.K.: The hololens revolution. Mech. Eng. Mag. Select Articles **138**(10), 30–35 (2016)
13. Rohmer, K., Jendersie, J., Grosch, T.: Natural environment illumination: coherent interactive augmented reality for mobile and non-mobile devices. IEEE Trans. Vis. Comput. Graph. **23**(11), 2474–2484 (2017)
14. Zhang, L., Dong, H., El Saddik, A.: Towards a QoE model to evaluate holographic augmented reality devices: a hololens case study. IEEE MultiMedia (2018)

Towards the Development of a Quasi-Orthoscopic Hybrid Video/Optical See-Through HMD for Manual Tasks

Fabrizio Cutolo[1,2]([✉]), Nadia Cattari[2], Umberto Fontana[1,2], and Vincenzo Ferrari[1,2]

[1] Information Engineering Department, University of Pisa, Pisa, Italy
fabrizio.cutolo@endocas.unipi.it
[2] EndoCAS Center, Department of Translational Research and New Technologies in Medicine and Surgery, University of Pisa, Pisa, Italy

Abstract. Augmented Reality (AR) based on head-mounted displays (HMDs) represent the most ergonomic and efficient solution for aiding complex manual tasks performed under direct vision and it is considered an enabling technology of the fourth industrial revolution (Industry 4.0). This work intends to give a harmonized and consistent overview of the workflow carried out so far in the framework of the European Project VOSTARS towards the development of a novel AR HMD specifically designed to aiding complex manual tasks within arm's reach. First results and future developments are being presented.

Keywords: Augmented Reality · Head-mounted displays · Orthoscopy · Visual perception · Industry 4.0

1 Introduction

Augmented Reality (AR) allows merging natural perception of the real environment with computer-generated information and it is considered an enabling technology of the fourth industrial revolution (Industry 4.0); it is indeed aimed at aiding complex tasks as a tool in the manufacturing industry such as in assembly tasks, maintenance and repair, training, quality control, and commissioning [1–4]. In recent years, AR has already proven to be a promising technology also in the education field [5–7] and in the healthcare industry, as demonstrated by the increasing number of publications in the medical training field [8, 9], surgical field [10–15] and rehabilitation [16].

In the ideal AR system, especially if designed for aiding manual tasks in the peripersonal space, there should not be any perceivable difference between the user's natural experience of the world and his/her augmented experience through the AR interface [17]. For this goal, the conditions to be satisfied are twofold: accurate spatial registration between real world scene and computer-generated content and ergonomic interaction with the AR scene [18]. According to the words of one of the pioneer researchers in the field of AR [19], "*The basic goal of an AR system is to enhance the user's perception of and interaction with the real world through supplementing the real world with 3D virtual objects that appear to coexist in the same space as the real world*".

© Springer Nature Switzerland AG 2019
L. T. De Paolis and P. Bourdot (Eds.): AVR 2019, LNCS 11614, pp. 170–178, 2019.
https://doi.org/10.1007/978-3-030-25999-0_15

For this reason, AR systems based on head-mounted displays (HMDs), are intrinsically the most ergonomic and efficient solutions for aiding manual tasks, thanks to their ability to preserve a natural and egocentric viewpoint thus allowing the user to freely interact with the augmented scene with his/her own hands [20, 21]. Since 2016, the authors have been coordinating the European project VOSTARS (Video and Optical See-Through Augmented Reality Surgical Systems, Project ID: 731974) [22], whose aim is to develop and validate an innovative and immersive HMD to act as surgical navigator. The aim of the project is to bring 'surgical navigation' directly in front of the surgeon's eyes and so to aid him/her during complex manipulative tasks. More generally the ambition is to overcome the drawbacks of state-of-the-art AR HMDs and so massively revolutionize the paradigms through which visual AR is delivered. In line with this, this paper provides a detailed insight into the achievements of the early stage of development of the ongoing European project. The paper explicitly refers to some previously published works where early results were individually unveiled. Therefore, it intends to give a harmonized and consistent overview of the workflow carried out so far towards the development of a new AR visor specifically devoted to aiding complex manual tasks within arm's reach.

2 Materials and Methods

In AR HMDs, the see-through capability can be accomplished either through the video see-through (VST) mechanism or through the optical see-through (OST) mechanism. In binocular VST HMDs, the view of the real world is captured by a pair of stereo cameras rigidly anchored to the visor with an anthropometric interaxial distance (Fig. 1). The stereo views of the world are presented onto the binocular micro displays of the HMD after being digitally combined with the virtual content. AR image registration is here easily achieved but the real view of the world is mediated by the

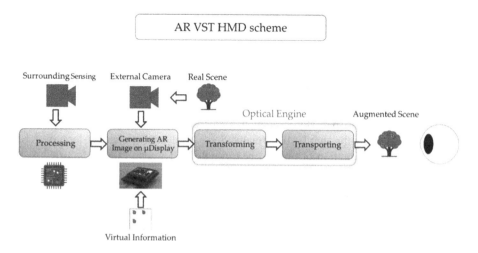

Fig. 1. Schematic representation of the Video See-Through Augmented Reality paradigm

cameras. By contrast, in OST HMDs, the user's direct view of the world is mostly preserved albeit at expenses of a reduced AR image registration. Here, the direct view of the real world is optically merged, through a beam combiner, with the computer-generated content (Fig. 2). This aspect confers a clear advantage over VST solutions, particularly when used to interact with objects at close distances, since it allows the user to maintain almost unaltered his/her own natural visual experience. The computer-generated content is projected onto a microdisplay (i.e., image source) and then optically redirected to the user's retina with a bit of delay.

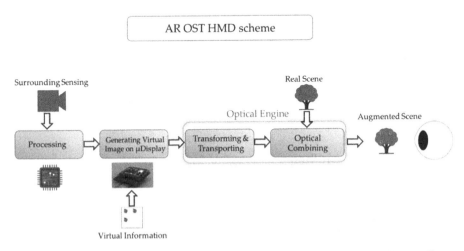

Fig. 2. Schematic representation of the Optical See-Through Augmented Reality paradigm

Our goal is to design and validate a hybrid OST/VST HMD able to provide the benefits of both see-through paradigms and to ensure a perceptually consistent AR experience to the user at close distances. To do so, we are presenting, in a coherent fashion, the technical solutions implemented so far towards the realization of a novel AR HMD devoted to aid complex manipulative tasks. Each technical solution was tested on a different embodiment of hybrid OST/VST.

This section is structured as follows: Sect. 2.1 reports on a novel solution that allows the development of stereoscopic AR HMDs able to provide both the see-through mechanisms (optical see-through and video see-through); Sect. 2.2 describes a software solution that is able to partially recover natural stereo fusion of the scene in quasi-orthoscopic video see-through HMDs; finally, Sect. 2.3 describes a closed-loop method for the automatic calibration of optical see-through HMDs with infinity focus.

2.1 Hybrid Video/Optical See-Through HMD

In 2017, we presented a novel approach for the development of an HMD able to provide both the see-through modalities [23]. The proposed solution is based on the use of a pair of liquid-crystal (LC) shutters that can be electronically controlled so to modify the transparency of a generic see-through display. This allows us to switch from the augmented and camera-mediated view, with the optical shutter on closed-state (i.e., VST mode), to the see-through (i.e., OST mode) view, with the optical shutter on open-state. Under OST mode, only the computer-generated elements are rendered onto the microdisplay whereas under VST mode the real 2D scene and virtual content are digitally blended and then rendered onto the display. A first custom-made prototype implementing such mechanism was realized by assembling the two optical shutters on a commercial waveguide-based OST display opportunely modified for housing also the stereo cameras needed for the VST modality (Fig. 3).

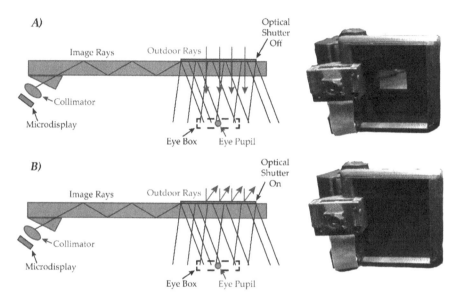

Fig. 3. Switch between Optical and Video See-Through modality by means of the optical shutter. On the top (A), the shutter in open-state, on the bottom (B) the shutter in closed-state

The proposed solution is therefore able to combine the benefits of both the AR modalities since it allows switching to the most suitable modality during use and without relying on any moving part (i.e. as with a physical shutter).

2.2 Perspective Preserving Solution for Quasi-Orthoscopic Binocular Video See-Through HMDs

From a human-factor standpoint, non-rigorously stereo-orthoscopic VST HMDs, as the one presented in the previous section, raise issues related to the user's interaction with the augmented content. In such systems, depth perception through stereopsis is adversely affected by sources of spatial perception errors. The issues affecting depth perception are mostly bound to the non-orthoscopic placement of the stereo cameras on the HMD.

In a study published in 2018 we presented a software solution aimed at mitigating the effect of the eye-to-camera parallax [24]. The idea is to partially resolve eye-camera parallax by opportunely warping the camera images through a perspective-preserving homographic transformation that accounts for: the intrinsic parameters of the cameras (K_C), the intrinsic parameters of the see-through displays (K_D), the geometry of the binocular VST HMD ($R_C^D|\vec{t}_C^D$), and the distance ($d^{C \to \pi}$) and orientation ($\vec{n}^T$) of the reference plane with respect to the user (Fig. 4) [25, 26]. The plane-induced homography is:

$$H_C^D = \left(K_D \left(R_C^D + \frac{\vec{t}_C^D \cdot \vec{n}^T}{d^{C \to \pi}} \right) K_C^{-1} \right) \tag{1}$$

While the pixel-to-pixel relation between camera points and display points is:

$$\lambda x_D = H_C^D \left(R_C^D, t_C^D, K_C, K_D \right) x_C \tag{2}$$

By means of this plane-induced homography applied over the camera frames, the perceived distortion of space around the pre-defined reference plane can be dramatically reduced and therefore the user is able to interact with the augmented scene with reduced discomfort.

In Fig. 5 a new custom-made prototype of the hybrid OST/VST HMD used for testing the proposed solution is shown. As in the first prototype, the visor is based on a reworked version of a commercial binocular OST HMD (DK-33 by LUMUS). The

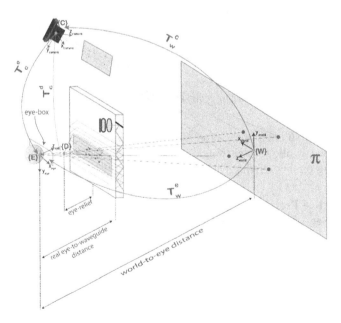

Fig. 4. Rigid transformation involved in the evaluation of the perspective preserving homography induced by a reference plane.

binocular HMD features a quasi-orthoscopic placement of the external cameras with an anthropometric interaxial distance.

Fig. 5. Embodiment of the hybrid Video/Optical See-Through HMD used for assessing the perspective preserving solution that mitigates the parallax due to a non-orthoscopic setting of the.

2.3   Closed – Loop Calibration for Optical See-Through HMDs with Infinity Focus

To achieve an accurate alignment between the user's view of the real 3D world and the computer-generated 2D graphics projected onto the see-through display, in any OST HMD a dedicated calibration routine is needed to estimate the projective parameters of the combined eye-display pinhole model. State-of-the-art OST calibration methods models the eye-NED system as an off-axis pinhole camera model, and therefore include the contribution of the eyes' positions also into the modelling of the intrinsic matrix of the eye-NED. In a study recently presented at ISMAR 2018 [27], we presented a method for robustly calibrating OST NEDs that explicitly ignores this assumption and that is specifically suited for OST displays having the imaging plane at infinity as the one used in our second prototype of hybrid OST/VST HMD.

Our assumption is that, being the imaging plane at infinity, the eye-NED can be modelled as an on-axis pinhole camera and therefore the eye-to-display parallax only affects the extrinsic component of the homographic transformation between eye and display in its centered position (i.e., at the center of the display eye-box). The overall transformation matrix that properly maps the world points onto the image points of the display is:

$$P_{def} = H_E^D \cdot K_D \cdot T_W^E = \lambda K_{eq} \cdot [R_{eq}|t_{eq}] \tag{3}$$

Where $K_D \cdot T_W^E$ is the projection matrix associated to the display. In order to determine the correct points onto the display (with no parallax), a perspective preserving homography (H_C^D) is to be estimated to correlate the two different pinhole models, in our case the eye/display model in its actual position and the eye/display model in its centered position. This transformation compensates the eye to display parallax. By opportunely decomposing the resulting P_{def}, we obtain an equivalent intrinsic matrix and an equivalent extrinsic matrix. These two matrices are then used for modelling the parameters of the virtual camera correctly. By using K_{eq} and $R_{eq}|t_{eq}$, in the rendering engine, the OST AR system can project pixels onto the display so to be properly aligned with their corresponding 3D point in the world.

The calibration procedure was tested on the same prototype presented in Sect. 2.2, here used under OST modality. The promising results of the calibration are shown in Fig. 6.

Fig. 6. Example of OST view after calibration.

3 Discussion and Future Works

In this work, we have unveiled an overview of the workflow carried out so far towards the development of a novel AR visor specifically devoted to aiding complex manual tasks within arm's reach. Overall, the achievements of the development phase carried out so far are:

- Implementation of a hybrid HMD which can work both under OST and VST modality and switch between each modality upon request. The switching mechanism is made possible by means of a pair of opto-electronic LC-shutters placed ahead of the optical combiners of the OST HMDs.
- A quasi-orthoscopic setting of the visor under VST modality, with a reduced eye-to-camera parallax and an anthropometric interaxial distance that reduces distortions in depth perception.
- A software solution that further mitigates the effect of the eye-to-camera parallax by means of a perspective-preserving image warping applied to the camera frames under VST modality. This transformation allows the recovering of a correct

perception of the relative depths in the peripersonal space around a pre-defined reference plane.

- A closed-loop calibration routine designed for an OST HMD with infinity focus. The calibration routine yields accurate results in terms of real-to-virtual alignment also under OST modality.

As future developments, a third hybrid visor is currently under development that will comprise all the above described key features:

- A pair of OST near-eye-displays with the associated optical engines.
- A pair of opto-electronic shutters.
- A pair of external cameras needed for capturing real-world views under VST modality.
- A dedicated software framework that manages the OST/VST switching and all the needed image transformations needed for mitigating the perceptual issues related to the eye-to-camera parallax and for ensuring a perceptually consistent AR experience to the user at close distances.

This visor will undergo specific usability tests in order to assess the overall efficacy of all the proposed solutions.

Acknowledgments. Funded BY THE HORIZON2020 Project VOSTARS, Project ID: 731974. Call: ICT-29-2016 - Photonics KET 2016.

References

1. Ong, S.K., Yuan, M.L., Nee, A.Y.C.: Augmented reality applications in manufacturing: a survey. Int. J. Prod. Res. **46**, 2707–2742 (2008)
2. Henderson, S., Feiner, S.: Exploring the benefits of augmented reality documentation for maintenance and repair. IEEE Trans. Vis. Comput. Graph. **17**, 1355–1368 (2011)
3. Nee, A.Y.C., Ong, S.K., Chryssolouris, G., Mourtzis, D.: Augmented reality applications in design and manufacturing. CIRP Ann.-Manuf. Technol. **61**, 657–679 (2012)
4. Wang, X., Ong, S.K., Nee, A.Y.C.: A comprehensive survey of augmented reality assembly research. Adv. Manuf. **4**, 1–22 (2016)
5. Bower, M., Howe, C., McCredie, N., Robinson, A., Grover, D.: Augmented reality in education - cases, places, and potentials. In: IEEE Annual International Conference (2013)
6. Radu, I.: Augmented reality in education: a meta-review and cross-media analysis. Pers. Ubiquit. Comput. **18**, 1533–1543 (2014)
7. Bacca, J., Baldiris, S., Fabregat, R., Graf, S.: Kinshuk: augmented reality trends in education: a systematic review of research and applications. Educ. Technol. Soc. **17**, 133–149 (2014)
8. Qu, M., et al.: Precise positioning of an intraoral distractor using augmented reality in patients with hemifacial microsomia. J. Cranio Maxill. Surg. **43**, 106–112 (2015)
9. Voinea, A., Moldoveanu, A., Moldoveanu, F.: Efficient learning technique in medical education based on virtual and augmented reality. In: ICERI Proceedings, pp. 8757–8764 (2016)
10. Meola, A., Cutolo, F., Carbone, M., Cagnazzo, F., Ferrari, M., Ferrari, V.: Augmented reality in neurosurgery: a systematic review. Neurosurg. Rev. **40**, 537–548 (2017)

11. Kersten-Oertel, M., Jannin, P., Collins, D.L.: The state of the art of visualization in mixed reality image guided surgery. Comput. Med. Imaging Graph. **37**, 98–112 (2013)
12. Fida, B., Cutolo, F., di Franco, G., Ferrari, M., Ferrari, V.: Augmented reality in open surgery. Updates Surg. **70**, 389–400 (2018)
13. Eckert, M., Volmerg, J.S., Friedrich, C.M.: Augmented reality in medicine: systematic and bibliographic review. JMIR mHealth uHealth **7**, e10967 (2019)
14. Ferreira Reis, A., Wirth, G.J., Iselin, C.E.: Augmented reality in urology: present and future. Rev. Med. Suisse **14**, 2154–2157 (2018)
15. Yoon, J.W., et al.: Augmented reality for the surgeon: Systematic review. Int. J. Med. Robot. Comput. Assist. Surg. MRCAS **14**, e1914 (2018)
16. Dunn, J., Yeo, E., Moghaddampour, P., Chau, B., Humbert, S.: Virtual and augmented reality in the treatment of phantom limb pain: a literature review. Neurorehabilitation **40**, 595–601 (2017)
17. Cutolo, F., et al.: A new head-mounted display-based augmented reality system in neurosurgical oncology: a study on phantom. Comput. Assist. Surg. **22**, 39–53 (2017)
18. Cutolo, F., Freschi, C., Mascioli, S., Parchi, P., Ferrari, M., Ferrari, V.: Robust and accurate algorithm for wearable stereoscopic augmented reality with three indistinguishable markers. Electronics **5**, 59 (2016)
19. Azuma, R., Baillot, Y., Behringer, R., Feiner, S., Julier, S., MacIntyre, B.: Recent advances in augmented reality. IEEE Comput. Graph. Appl. **21**, 34–47 (2001)
20. Cutolo, F., Badiali, G., Ferrari, V.: Human-PnP: ergonomic AR interaction paradigm for manual placement of rigid bodies. In: Linte, C.A., Yaniv, Z., Fallavollita, P. (eds.) AE-CAI 2015. LNCS, vol. 9365, pp. 50–60. Springer, Cham (2015). https://doi.org/10.1007/978-3-319-24601-7_6
21. Cutolo, F., Carbone, M., Parchi, P.D., Ferrari, V., Lisanti, M., Ferrari, M.: Application of a new wearable augmented reality video see-through display to aid percutaneous procedures in spine surgery. In: De Paolis, L.T., Mongelli, A. (eds.) AVR 2016. LNCS, vol. 9769, pp. 43–54. Springer, Cham (2016). https://doi.org/10.1007/978-3-319-40651-0_4
22. http://www.vostars.eu/
23. Cutolo, F., Fontana, U., Carbone, M., D'Amato, R., Ferrari, V.: Hybrid video/optical see-through HMD. In: Adjunct Proceedings of the 2017 IEEE International Symposium on Mixed and Augmented Reality (Ismar-Adjunct), pp. 52–57 (2017)
24. Cutolo, F., Fontana, U., Ferrari, V.: Perspective preserving solution for quasi-orthoscopic video see-through HMDs. Technologies **6**, 9 (2018)
25. Hartley, R., Zisserman, A.: Multiple View Geometry in Computer Vision. Cambridge University Press, Cambridge (2003)
26. Tomioka, M., Ikeda, S., Sato, K.: Approximated user-perspective rendering in tablet-based augmented reality. In: International Symposium on Mixed and Augmented Reality, pp. 21–28 (2013)
27. Fontana, U., Cutolo, F., Cattari, N., Ferrari, V.: Closed – loop calibration for optical see-through near eye display with infinity focus. In: Adjunct Proceedings of the 2018 IEEE International Symposium on Mixed and Augmented Reality (2018)

An Empirical Evaluation of the Performance of Real-Time Illumination Approaches: Realistic Scenes in Augmented Reality

A'aeshah Alhakamy[1,2]([envelope]) [iD] and Mihran Tuceryan[1]([envelope]) [iD]

[1] Indiana University - Purdue University (IUPUI), Indianapolis, IN 46202, USA
aalhakam@iupui.edu, {aalhakam,tuceryan}@iu.edu
[2] University of Tabuk, Tabuk, TA, Saudi Arabia

Abstract. Augmented, Virtual, and Mixed Reality (AR/VR/MR) systems have been developed in general, with many of these applications having accomplished significant results, rendering a virtual object in the appropriate illumination model of the real environment is still under investigation. The entertainment industry has presented an astounding outcome in several media form, albeit the rendering process has mostly been done offline. The physical scene contains the illumination information which can be sampled and then used to render the virtual objects in real-time for realistic scene. In this paper, we evaluate the accuracy of our previous and current developed systems that provide real-time dynamic illumination for coherent interactive augmented reality based on the virtual object's appearance in association with the real world and related criteria. The system achieves that through three simultaneous aspects. (1) The first is to estimate the incident light angle in the real environment using a live-feed 360° camera instrumented on an AR device. (2) The second is to simulate the reflected light using two routes: (a) global cube map construction and (b) local sampling. (3) The third is to define the shading properties for the virtual object to depict the correct lighting assets and suitable shadowing imitation. Finally, the performance efficiency is examined in both routes of the system to reduce the general cost. Also, The results are evaluated through shadow observation and user study.

Keywords: Direct illumination · Indirect illumination · Augmented reality · Image-based lighting · Incident light · Reflected light

1 Introduction

The illumination information extracted from the physical world can provide the means to realistically render augmented objects into the final scene. Acquiring an illumination model featuring accurate precision that would capture the

L. T. De Paolis and P. Bourdot (Eds.): AVR 2019, LNCS 11614, pp. 179–195, 2019.
https://doi.org/10.1007/978-3-030-25999-0_16

whole real environment can be challenging under reduced assumptions. A photo-realistic and dynamic augmented reality scene that integrates the illumination model requires addressing several aspects in each frame. In this paper, we evaluate two of our previous developed methods in details [23,24] where both extracted the light information from the real environment producing an illumination model that is applied onto the virtual objects to render a visually cohesive final scene. Both routes include three main aspects of the system but have different reflected light simulation method, however, they both have similar obligation to operate simultaneously in real-time. The system assumes a live-feed 360° camera is instrumented on any AR device (e.g., head mounted display, projection display, handheld mobile, or webcam camera).

The first (1) aspect of the system is to estimate the angle of incident light (i.e. direct illumination). The incident light is known as the light falling from the source onto the objects directly and is then depicted by the eyes of the observer. This estimation utilizes the 360° camera to capture the entire environment map of the real scene.

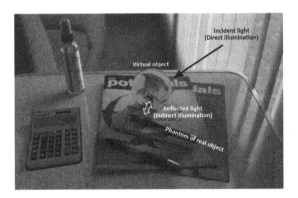

Fig. 1. Illustration of the incident light and the reflected light differences and their interactions with the real and virtual objects.

The second (2) aspect simulates the reflected light (i.e. indirect illumination) captured from the surroundings of the real world onto the virtual object. The reflected light is defined as the inter-light bouncing off the surfaces of one object into another (see Fig. 1). The indirect illumination has two separate sub-methods that are used and tested to reduce the performance cost. One of them is (a) global cube map textures captured from 360° camera view, while the other one (b) samples the local regions surrounding the virtual object form the main AR view. In the shading process these textures are added and updated as part of the image-based lighting (IBL) mode.

The third (3) aspect passes the estimated incident light and the simulated reflected light on the virtual objects through shading language and CG techniques. The environment light conditions and shadow effects from the preceding methods can be defined through different types of shaders as required. The study aims to evaluate and explore the efficient techniques to render a realistic and coherent AR scene. The comparison process of both routes in the system would provide us with pros and cons for each method.

2 Related Work

The realistic perception of a virtual object in augmented, mixed or virtual realities has been the interest of many previous research exploration in several techniques. While the major focus of this study involves augmented reality where the virtual objects were added to the physical scene, some of the techniques used can be applicable to other fields. Although prior work assumed certain restrictions for the system to succeed, they also addressed some limitations and challenges for future work. However, their work was an inspiration for the current system.

Estimation and Detection of the Incident Light. Debevec et al. [1] presented the first state-of-the-art methods that attempt to solve the problem of the interactive rendering of a virtual object in augmented reality scenarios. Karsch et al. [2] developed a method that capture and represent part, or subset of the incident light field (ILF) for a virtual object placed in the specific region of the scene during rendering process which then facilitated the development of post-capture refocusing, depth estimation and small viewpoint transformations applications.

One of the main techniques used is spatial sampling of the incident light at multiple points which can be obtained in preprocess using the Iterated Closest Point (ICP) algorithm [17]. Besides the precomputed radiance, some assume a pre-known geometry of the real scene with more information about the angle of incident light to acquire an effective Bidirectional Reflection Distribution Function (BRDF) [7].

Furthermore, Nowrouzezahrai et al. [12] assumed the possibility of aggregating the spatial variation of the scene lighting into directional distribution represented as an environment map of the incident light. Their factorization of the light seeks to compute the dominant light direction and color separately using spherical harmonic (SH) coefficients which is obtained by a projection onto the SH basis functions. As these functions are orthonormal, the dot product was the amount of incident light from all directions at a certain vertex [15]. Gruber et al. [6] also expressed the incident light field with SH to estimate the distance of the light field which represent every incident light ray on a hemisphere from many different observers of an arbitrary geometry.

The traditional image-based lighting approach usually utilized a light probe to capture one angle in space represented in the HDR image, Unger et al. [19] used a sequence of light probes to capture a path of incident light then rendered the objects with a corresponding light probe as if illuminated by that light. Similar techniques used either a perfect reflector mirror sphere or a fish-eye camera where both could exhibit specular reflections for higher resolution. The prefect reflector would improve additional extensions of the BRDF such as the Fresnel factor and the Schlick approximation [13,18]. In general, the development of capturing HDR images become more critical for realistic results by mining approximate illumination information directly from s video sequences to avoid physical measurements in the real scene [10].

Richter-Trummer et al. [14] used inverse rendering which factored the texture color into incident light and the diffuse albedo color through the radiance transfer method. For more material estimation Rohmer et al. [15] assumed a Lambertian environment for incident light integration over the surface of the upper hemisphere.

Stimulation of the Reflected Light. The reconstruction of the direct and indirect illumination using the radiosity algorithm for diffuse lighting on the virtual object in the real scene [4] is one of the methods to simulate the reflected light. Loscos et al. [7] also used the hierarchical radiosity to compute the indirect illumination employing a rough subdivision of the scene. The reflected light can be estimated from the photon map when the ray-tracing hit the diffuse surfaces. The rendered result composited from the real video image is directly exhibited on the output device.

The calculation of multiple bounces of indirect illumination were possible through Monte Carlo integration to evaluate the irradiances in cache records by shooting recursive rays into arbitrary directions over the surface point [8]. Metha et al. [11] used two Monte Carlo for sampling as path-tracing passes to estimate the per pixel illumination. They sampled the cosine hemisphere for indirect illumination on the diffuse surfaces on the real and virtual objects and sampled the Phong lobe for glossy surfaces on the virtual objects only. Other sampling techniques were applied on the shadow mapping to create Virtual Point Lights (VPLs) which then enables the calculation of the indirect illumination [9,15].

Global Cub Map. The cube maps have been supported as the environment map by most graphics cards to demonstrate proper global reflections with high-resolution on the virtual objects [7]. While each pixel on the unit sphere represented an individual direction, cube map were used to evaluate the SH coefficient [6]. Also, Rohmer et al. [15] rendered a low-resolution cube map from the position of the virtual object of the indirect illumination atlas as texture of the surrounding environment. Schwandt et al. [16] created a global environment map for all the virtual objects in parallel. Their calculations included both diffuse and glossy cube map for unique material properties.

Local Sampling. A 2D texture mapping was enabled by Agusanto et al. [3] which was stored in a frame buffer of an image to render the scene. A few HDR images or environment maps were used to capture the light information represented as 2D texture or 4D surface light fields then projected onto a geometric model [10]. Franke [5] formulated two shading systems to estimate the natural illumination and compute the reflected light on several materials of the virtual objects in real-time.

Utilization of 360° Panoramic Videos. The utilization of 360° video sequence to capture the environment map is not a novel notion. Rhee et al. [20] used a conventional low dynamic range (LDR) 360° videos to render an interactive mixed reality scene. Iorns et al. [21] developed a system to use a live streaming 360° video as an input for image-based lighting where real-time shadowing and reflection were investigated. Michiels et al. [22] used an appropriate 360° environment map linked to a car position for real-time lighting added to the rendering equation.

3   Method

This section describes the entire system briefly where rendering a visually coherent final scene is the ultimate goal. The system consists of our previous and current three main methods where the second method is branched into two sub-methods. For a complete overview of the entire system (see Fig. 2).

1. **Estimate Incident Light.** The live-feed of the physical environment is captured using a 360° camera with the support of parallel computations to sample the light area and estimate the angle of the real light source. Then, A virtual light in the virtual scene is created to imitate the incident light.
2. **Simulate Reflected Light.** The inter-bouncing light between the objects and the surrounding is captured using two methods. The resulted texture is rendered into the virtual object based on the material properties.
   - *Global Cube Map.* The panoramic HDR of the 360° camera is used to create the 6 faces as a 2D texture to construct a cube map for the image-based lighting mode which can be modified while defining the shading properties for each virtual object.
   - *Local Sampling.* The main camera of the AR view device is employed to sample the region below and surrounding each virtual object. The resulted texture is then used in IBL mode where the material property of each object can be reflected on the objects.
3. **Define Shading Properties.** the virtual object materials and characteristics are defined in this method as needed. The object's main texture, normal map, specular map, and diffuse elements are also addressed in this method in addition to many other properties that we are going to discuss in details later.

Rendering. The global illumination for the entire system is used to create more realistic appearance. The suited rendering path for real-time lights is the deferred shading which has the best lighting and shadow reliability. The forward rendering and legacy deferred are also used when trade-offs are necessary. These rendering paths support the per-pixel lighting including normal maps and light cookies. Additional rendering passes would not be required for reflection depth and normal buffers. They support semitransparent objects and anti-aliasing. The number of pixels illuminated influence the performance cost of a per-pixel light instead of the number of lights.

184   A. Alhakamy and M. Tuceryan

Tracking. Various interactive AR/VR experiences are developed using rotational and positional trackers. The device location is relative to the physical world and provides the device tracker with positional or rotational information. We employed the Vuforia engine in our system which supports the positional device tracker. This type of tracker is suitable when content is added in the environment for a robust 6 degrees-of-freedom target tracking. However, various scripts were developed to provide additional configurations for the different light conditions while the objects, camera or marker is moving.

Hardware Description. The system operates on a personal computer with Intel®, Core™ i7-3930k CPU @ 3.20 GHz 3201 MHz, six core(s), 64.0 GB RAM, and NVIDIA GeForce GTX 970 GPU. As input devices, two cameras are used: DSLR Nikon D7200 and live-feed RICOH THETA S 360°.

Additional Software. The camera live-feed requires a broadcasting software and system registry to display the input video through the 3D engine. It is recommended to check devices compatibility while transferring the code onto different machines.

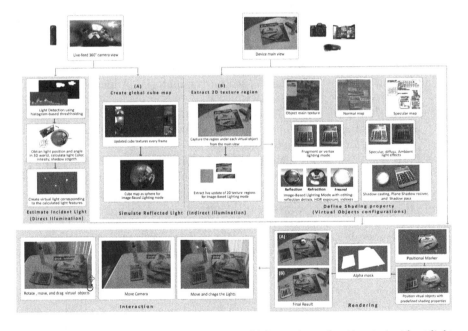

Fig. 2. A full overview of the entire system which consists of: estimate incident light, simulate reflected light, and define shading property followed by rendering and interaction.

3.1 Estimate Incident Light

The panoramic 360° view provides a complete environment map of the physical world where the illumination information can be estimated for an arbitrary number of lights. The 3D engine reads the camera feed as a 2D texture which is then converted to an OpenCV Mat object using parallel computation support, and finally converted to an OpenGL 3D vector as the 3D engine interacts with the data. This procedure is then reversed to represent the outcomes.

The estimation process samples the incident light based on the saturated bright areas of the image frame from the live-feed. The threshold is based on the histogram median of the entire frame. A suitable noise reduction through Gaussian blur is used for pure contours. Any additional noise blobs are removed by performing a series of erosion and dilation functions for specific structuring elements.

Each light is represented as a contour which is sorted based on the size to determine the main light followed by the second significant light and so on. The contour moments are used to specify the light location on the panoramic view.

The spherical projection is used to transform the screen coordinates (x, y) of the light location into the spherical coordinates (r, θ, ϕ) in the engine form for the light position in the Cartesian coordinates (x, y, z). The corresponding light position on the spherical environment map would be represented by a virtual light that has most of the features of the real light including angle, color, intensity, shadow strength and more.

The spherical coordinates are represented as follows, after normalization form Texel to sphere transformation.

$$(r, \theta, \phi) = \int_1^{lights} (\sqrt{x^2 + y^2}, \tan^{-1} \frac{y}{x}, \cos^{-1} \frac{z}{r}) \tag{1}$$

Where *lights* is the number of light detected and The measurement and dimensions of the panoramic 360° view affects the spherical projection as follows:

$$(dw, dh) = \int_1^{lights} (\frac{(360 \times 2 - 1)(x - 0)}{(w - 0) + 1}, \frac{(360 - 0)(x - 0)}{(w - 0) + 1}) \tag{2}$$

The virtual light location corresponding to the real light is represented as the resulting coordinate, if the light is placed onto the left side of the entire image frame.

$$(x, y, z) = \int_1^{lights} (\frac{\sin \phi \cos \theta}{\pi \times dh}, \frac{\cos \phi}{\pi \times dw}, \frac{\sin \phi \cos \theta}{\pi \times 90}) \tag{3}$$

However, if the real light is placed onto the right side of the entire view the location is represented as:

$$(x, y, z) = \int_1^{lights} (\frac{\sin \phi \cos \theta}{\pi \times (dh - 90)}, -\frac{\cos \phi}{\pi \times (dw - 180)}, \frac{\sin \phi \cos \theta}{\pi \times 90}) \tag{4}$$

Then, the light color is sampled based on the mean color of each contour. While the light intensity is influenced by the light area and the view dimensions, the shadow strength is applied based on the light intensity. The values are scaled to a suitable range for the 3D engine.

3.2   Simulate Reflected Light

In this section, two sub-methods are used to simulate the reflected light which embodies the surrounding environment for each virtual object in the scene.

Global Cube Map. This sub-method was discussed in our earlier work [24] but it is presenting briefly here for the context and further analysis. The cube map method also utilizes the panoramic 360° view to create 6 faces (left, back, right, front, top, bottom) which consist of 2D textures to construct a cubic HDR that can be assigned to the image-based lighting mode property. Information about the indirect illumination such as additional lights, real objects and surfaces are included in the 360° view which enrich the cube map surrounding the virtual object with details. The cube map is created from a live-feed every frame, therefore a coroutine concept is found to be useful to accelerate the rendering process. The unfolded cube is formatted from the appropriate six faces whose area is arranged typically in horizontal cross configuration. Thus, the cube faces would fit perfectly when the texture maps are folded.

Each face of the cube textures is saved as a separate image file then collected again in the arrangement. Each pixel in every face has a color that is calculated through the spherical projection then normalized for any further rotation and movements as follows:

$$(\theta, \phi) = \int_0^{points} (\tan^{-1}\frac{z}{x}, (\cos^{-1} z) + \frac{dir \times \pi}{180}) \quad (dx, dy) = \int_0^{points} (\frac{\theta}{\pi}, \frac{\phi}{\pi}) \quad (5)$$

The values should be maintained inside the height and the width of the created texture, to represent the final color as:

$$(px, py) = \int_0^{points} (dx \times w, py \times h) \qquad C = \left(\begin{array}{c} \dot{} \\ h - py - 1 \end{array} px \right) \quad (6)$$

The resulted cube map is assigned to the image-based lighting mode when the shading properties are defined. Many other features can be manipulated in order to find the perfect material for each virtual object.

The cube map constructed was purely developed for 3D engine C# which allow reading from a live streaming video in real-time. There are many cube map tools available but having a previous video recording or a panoramic static picture are required for these tools.

Local Sampling. The creation of cube maps in every frame raises the performance cost, so we sample the region surrounding each virtual object from the view of the main camera which depicts a close-wide illumination information. A similar but short version was covered in our previous work [23] but got developed and improved to fit the local sampling approach.

Therefore, A plane attached to the virtual object is added to sample the live-video texture of the AR main camera. A mesh filter is initialized to keep the vertices of the mesh updated at 100 frames per second. A culling mask is used to hide the undesirable part of the view inside the object layer.

The mesh renderer of the background plane gets the main texture and assigns it to the video texture. The background plane is dispatched every 10 frames per second. Each vertex in the final mesh is transformed from the 3D world to view port point. The region capturing in progress the Vuforia device orientation is modified based on the screen orientation. Also, the local texture resolution is transformed based on the local scale of the entire camera output texture.

A virtual camera renderer captures the target texture to set the image-based lighting mode with that texture to define the material property of the virtual object.

The advantage of this sub-method is to sample only the local region on a specific object instead of the whole environment. It provides a suitable outcome for diffuse and glass materials, however the cube map provides better result with the specular materials.

3.3   Define Shading Properties

As known the shader is a collective computation of the shading properties during rendering. For a realistic object, the proper level of light, darkness, color must be considered. The virtual object consists of vertices, UV information and normals as part of the material features. Some of these properties are predefined such as the object's main texture and normal map, while other can be manipulated at the run-time like image-based lighting textures or cube maps.

In this system, two types of shaders are used to fully define the entire shading properties based on how the virtual objects receive the lights.

Lit Surface. The objects that receive and reflect the lights in the virtual scene have this type of shader which provides three passes: first forward base for the main light, second forward add for any additional lights, and third for casting the shadow on the other surfaces.

Also, it contains various properties such as vertex/fragment lighting mode, Ambient light, Lambert diffuse, Blinn Phong specular, Fresnel, image-based Lighting mode, Ashikhmin, Shirley and Premoze BRDF anisotropy.

Lambert Diffuse. It is calculated from the normal direction (N), light direction (L), color (c), diffuse factor (f) and attenuation (a), using the famous formula:
$d = \sum_{i=0}^{n} c \times f \times a \times max(0, \overrightarrow{N} \cdot \overrightarrow{L})$.

Blinn Phong Specular. Some selective surfaces depict shining parts and lean towards representing some highlights. The calculation of the specular effect, required the mesh normal (N), light direction (L) with world space view direction (v), specular color (S_c), specular factor (S_f), attenuation (a) and specular power (S_p). At the beginning we must calculate the halfway vector direction (h) by normalizing the light direction (L) and world space view direction (v) represented in the model: $S = \sum_{i=0}^{n} S_c \times S_f \times a \times max(0, \overrightarrow{N} \cdot \overrightarrow{h})_p^S$.

Image-Based Lighting Reflection. The simulated reflected light in the form of a cube map or local 2d textures are assigned in this part where the required computation for each type is redefined based on the texture form. The texture will stay updated with live-feed video in real-time.

Image-Based Lighting Refraction. Same as the previous description except the received texture is bent to mimic the refraction effect. The Snell's law is used to represent the refraction model: $R = e \times i + [(e\cos_i \theta) - \sqrt{1 - e^2(1 - \cos_r^2 \theta)}]\overrightarrow{N}$ Where (i) is the velocity of light in vacuum and (r) is the velocity of light in the medium.

Image-Based Lighting Fresnel. It simulates the glass and water reflection/refraction where the viewer's point of view influences the normal vector. $Fresnel(F) = f + (1 - f)(1 - \overrightarrow{v} \cdot \overrightarrow{h})^5$.

Ashikhmin, Shirley and Premoze BRDF Anisotropy. A function that defines a certain pattern of how the light is going to hit a surface material such as silver or copper, The BRDF secular reflectively model used is:

$$S = \frac{\sqrt{(n_u+1)(n_v+1)}(N \cdot h)^{\frac{n_u(H.T)^2 + n_v(H.B)^2}{1 - (N.H)^2}}}{8\pi \times (v.H) \times max((N.L),(N.V))} \times F$$

where (n_u, n_v) refer to width/height of tangent map, while (N) is normal vector, (h) halfway vector, (T) tangent vector, (B) BiTangent vector.

Unlit Surface. The object that only casts shadow without receiving any light are assigned to this shader. The background of the virtual scene should be hidden but must reflect the shadow of other virtual objects. Therefore, an alpha mask is used to cut-out the main color of the background plane to produce a transparent material lookalike that receives the shadow but nothing else.

4   Result Evaluation

The developed system is examined through multiple scenes with different lighting conditions and various environment settings. For further analysis, the results are evaluated based on three categories:

4.1   Incident Light Evaluation

Shadow observation is the most visible clue that can be used to determine the accuracy of the incident light estimation in our experiment, in addition to the light color and intensity of the whole scene. The shadow cast from the real objects in comparison with the shadow cast from the virtual objects is a tangible evidence that the estimated light angle is valid.

For the real objects, the angle of the real light is calculate manually by measuring the real object length (t) in the physical scene (see Fig. 3) with the aspect ratio to the shadow length (s) in six scenarios to obtain the measured angle $\theta = \tan^{-1}\frac{t}{s}$. The measured angle θ then is compared to the estimated angle ϕ presented in the virtual light (y) axis where both angles provide a small margin of error.

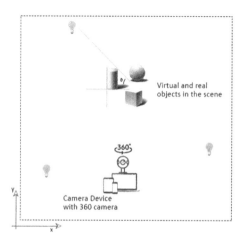

Fig. 3.   Illustration of the final scene setting including the physical scene and the augmented elements.

Table 1 shows that the difference between the measured angle θ and the estimated angle ϕ is a reasonable amount which proves that our estimation is nearly accurate.

The computed statistics of the errors combined are averaged using the root-mean-square error (RMSE):

$$RMSE = \sqrt{\frac{\sum_{i=1}^{6} Error_i^2}{6}} = 1.567 \tag{7}$$

Table 1. The error margin between the measured (θ) and estimated (ϕ) angles of the incident light in degrees.

Scene number	Shadow length (s)	Measured (θ)	Estimated (ϕ)	Error
1st	28	−167.04	−166.12	0.2
2nd	5.4	−50.19	−49.01	−1.18
3rd	15	27.47	26.24	1.23
4th	6	−127.56	−126.98	0.58
5th	18	203.42	203.01	0.41
6th	4	58.39	61.10	−2.71

4.2 Performance Evaluation

The reason for presenting two sub-methods for simulating the reflected light is to reduce the performance cost as mentioned above. In this section, we evaluate the performance through different scenes for both global cube map and local sampling.

Table 2. Performance evaluation for the global cube map against the local sampling with different number of objects

Scene no.	Global cube map			Local Sampling		
	1	2	3	1	2	3
FPS [1/s]	6	5	4	45	40	36
Update [ms]	173.4	177.3	183.2	35.9	37.8	40.2
Input [ms]	0.04	0.05	0.07	0.05	0.06	0.06
Surfaces [ms]	4.21	4.98	5.48	4.42	4.69	4.75
Camera render[ms]	0.66	0.68	0.84	1.08	1.40	1.69
Rendering [ms]	0.03	0.04	0.05	0.04	0.05	0.06

Table 2 shows that the number of virtual objects in each scene has limited influence on the performance cost. Although the local sampling reduces the performance cost significantly, the camera render is delayed by approximately 1 ms. In order to sample the required local area from the main view in certain regions related to the location of the virtual object, the performance cost has to be increased at a small fraction compare to the enhanced performance in the update function and other aspect of the system.

4.3   User-Feedback Evaluation

A user study is conducted as an online survey that contains several pictures and video of the results in different lighting conditions. The feedback data from 33 subjects who identified as college students shows that the incident light estimation is 51.2% more accurate than the average of faulty results. However, the angle of virtual objects shadow scored 20.4% higher than the estimation of the incident light. This observation is what led to the calculation of incident light angle in Sect. 4.1 above (see Fig. 4).

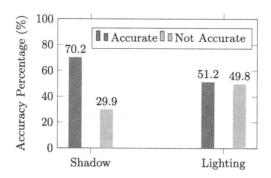

Fig. 4. User-feedback for the incident light estimation comparing to the shadow accuracy in general.

The results from the reflected light simulation method are not the same as shown (in Fig. 5) where the subjects are asked to evaluate which column presented more realistic results. The sampling of local regions had 66.67% more realistic results than the global cube map.

Furthermore, each object in the scene was rated based on how realistic they look compared to the other object whether it was real or virtual. The user ranked the object from 0 to 10 range as [(very virtual, 0), (virtual), (natural), (realistic), (very realistic, 10)].

Some users believed that some real objects were virtual by 1.5 points of the ranking which is a proof that the system works. While there is a slight misconception recognizing the virtual object from the real ones, the rating is also influenced by that misconception. Thus, the results are re-evaluated for the final feedback from the users based on the scene that provided the best realistic outcome according to the subjects' recommendations where all the virtual objects scores above the average (see Fig. 6).

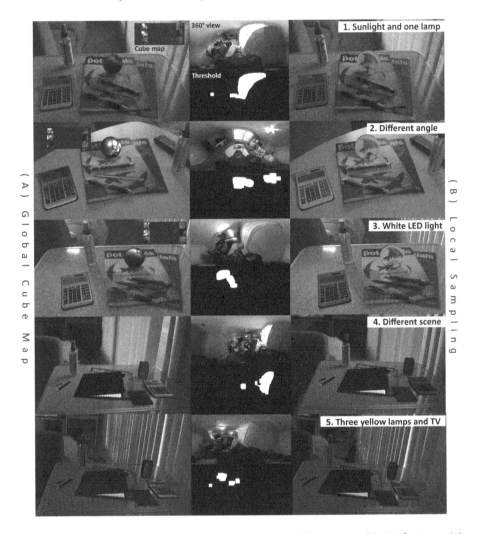

Fig. 5. System results where each environment condition has: (1) 360° view, (2) histogram-based thresholding, (3) cube map textures, (4) final scene of the global cube map creation result, and (5) final scene of the local sampling

The reflected light sub-methods influence the shading properties, and it seems that the global cube map creation provides more realistic results for the metal material objects where the local sampling has a more realistic outcome when the materials are glass or transparent. Finally, the lighting conditions are updated and changing in real-time accordingly, even when the object, camera or light are moving based their locations.

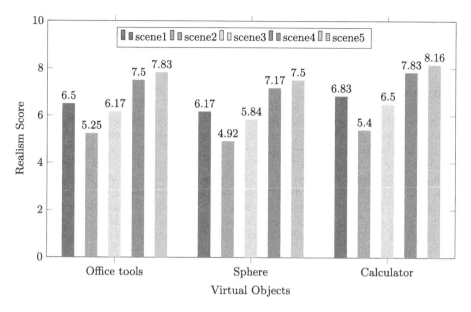

Fig. 6. Evaluating the virtual objects realism based on the scene environment and lighting condition.

5   Conclusion and Future Work

Light is an essential phenomenon to improve realism in augmented scenes, and we have been observing the light from different perspectives to estimate both incident and reflected light in the most practical methods. We are aware that our system can be improved in many ways, although current implementation generates plausible rendering. Estimating the incident light is the main goal to have the perfect light for any environment. The current method studies the light based on color brightness which is the most common method, however the physics-based lighting method is the focus for investigation in our future work.

Simulating the reflected light provides evidence from multiple evaluations that the global cube map is not an accurate method and reduces the overall performance, therefore, the local sampling can be explored further for a more realistic approach [25]. Also, a complete GPU implementation would provide a better performance cost in general. The experiment does not involve tracking development currently, but it can be improved in the future, as the tracking issue only appears in the video result where you can see the virtual object fluctuate sometimes.

It is reasonable to provide a step for hand gesture occlusion with the virtual elements, and real/virtual objects occlusion for more realistic view when the objects are moving in the final scene. Although the system solve the inter-reflection from the real object onto the virtual objects the reverse procedure where the virtual object is reflect onto the real object which required geometry registration and mapping is under development for future work. Some physics effects such as caustics which required shadow and photon mapping is also under investigation.

References

1. Debevec, P.: Rendering synthetic objects into real scenes: bridging traditional and image-based graphics with global illumination and high dynamic range photography. In: Proceedings of the 25th Annual Conference on Computer Graphics and Interactive Techniques, SIGGRAPH 1998, pp. 189–198. ACM, New York (1998)
2. Karsch, K., Hedau, V., Forsyth, D., Hoiem, D.: Rendering synthetic objects into legacy photographs. ACM Trans. Graph. (TOG) **30**(6), 157 (2011)
3. Agusanto, K., Li, L., Chuangui, Z., Sing, N.W.: Photorealistic rendering for augmented reality using environment illumination. Paper presented at the Second IEEE and ACM International Symposium on Mixed and Augmented Reality, 2003 Proceedings (2003)
4. Fournier, A., Gunawan, A.S., Romanzin, C.: Common illumination between real and computer generated scenes. Paper presented at the Graphics Interface (1993)
5. Franke, T.A.: Delta light propagation volumes for mixed reality. Paper presented at the 2013 IEEE International Symposium on Mixed and Augmented Reality (ISMAR) (2013)
6. Gruber, L., Richter-Trummer, T., Schmalstieg, D.: Real-time photometric registration from arbitrary geometry. Paper presented at the 2012 IEEE International Symposium on Mixed and Augmented Reality (ISMAR) (2012)
7. Jacobs, K., Loscos, C.: Classification of illumination methods for mixed reality. Paper presented at the Computer Graphics Forum (2006)
8. Kan, P.: High-quality real-time global illumination in augmented reality. Ph.D. thesis. Institute of Software Technology and Interactive Systems (2014)
9. Knecht, M., Traxler, C., Mattausch, O., Purgathofer, W., Wimmer, M.: Differential instant radiosity for mixed reality. Paper presented at the 9th IEEE International Symposium on Mixed and Augmented Reality (ISMAR) (2010)
10. Kronander, J., Banterle, F., Gardner, A., Miandji, E., Unger, J.: Photorealistic rendering of mixed reality scenes. Paper presented at the Computer Graphics Forum (2015)
11. Mehta, S. U., Kim, K., Pajak, D., Pulli, K., Kautz, J., Ramamoorthi, R.: Filtering environment illumination for interactive physically-based rendering in mixed reality. Paper presented at the Eurographics Symposium on Rendering (2015)
12. Nowrouzezahrai, D., Geiger, S., Mitchell, K., Sumner, R., Jarosz, W., Gross, M.: Light factorization for mixed-frequency shadows in augmented reality. Paper presented at the 10th IEEE International Symposium on Mixed and Augmented Reality (ISMAR) (2011)
13. Pessoa, S., Moura, G., Lima, J., Teichrieb, V., Kelner, J.: Photorealistic rendering for augmented reality: a global illumination and BRDF solution. Paper presented at the 2010 IEEE Virtual Reality Conference (VR) (2010)
14. Richter-Trummer, T., Kalkofen, D., Park, J., Schmalstieg, D.: Instant mixed reality lighting from casual scanning. Paper presented at the 2016 IEEE International Symposium on Mixed and Augmented Reality (ISMAR) (2016)
15. Rohmer, K., Büschel, W., Dachselt, R., Grosch, T.: Interactive near-field illumination for photorealistic augmented reality with varying materials on mobile devices. IEEE Trans. Vis. Comput. Graph. **21**(12), 1349–1362 (2015)

16. Schwandt, T., Broll, W.: A single camera image based approach for glossy reflections in mixed reality applications. Paper presented at the 2016 IEEE International Symposium on Mixed and Augmented Reality (ISMAR) (2016)
17. Sloan, P.-P., Kautz, J., Snyder, J.: Precomputed radiance transfer for real-time rendering in dynamic, low-frequency lighting environments. Paper presented at the ACM Transactions on Graphics (TOG) (2002)
18. Supan, P., Stuppacher, I., Haller, M.: Image based shadowing in real-time augmented reality. IJVR **5**(3), 1–7 (2006)
19. Unger, J., Gustavson, S., Ynnerman, A.: Spatially varying image based lighting by light probe sequences. Vis. Comput. **23**(7), 453–465 (2007)
20. Rhee, T., Petikam, L., Allen, B., Chalmers, A.: MR360: mixed reality rendering for 360 panoramic videos. IEEE Trans. Vis. Comput. Graph. **23**(4), 1379–1388 (2017)
21. Iorns, T., Rhee, T.: Real-time image based lighting for 360-degree panoramic video. In: Huang, F., Sugimoto, A. (eds.) PSIVT 2015. LNCS, vol. 9555, pp. 139–151. Springer, Cham (2016). https://doi.org/10.1007/978-3-319-30285-0_12
22. Michiels, N., Jorissen, L., Put, J., Bekaert, P.: Interactive augmented omnidirectional video with realistic lighting. In: De Paolis, L.T., Mongelli, A. (eds.) AVR 2014. LNCS, vol. 8853, pp. 247–263. Springer, Cham (2014). https://doi.org/10.1007/978-3-319-13969-2_19
23. Alhakamy, A., Tuceryan, M.: AR360: dynamic illumination for augmented reality with real-time interaction. In: 2019 IEEE 2nd International Conference on Information and Computer Technologies ICICT, pp. 170–175 (2019)
24. Alhakamy, A., Tuceryan, M.: CubeMap360: interactive global illumination for augmented reality in dynamic environment. In: IEEE SoutheastCon (2019, accepted and presented)
25. Alhakamy, A., Tuceryan, M.: Polarization-based illumination detection for coherent augmented reality scene rendering in dynamic environments. In: Proceedings of ACM Computer Graphics International (2019, accepted)

Cultural Heritage

Combining Image Targets and SLAM for AR-Based Cultural Heritage Fruition

Paolo Sernani[1]([⊠]), Renato Angeloni[2], Aldo Franco Dragoni[1], Ramona Quattrini[2], and Paolo Clini[2]

[1] Department of Information Engineering (DII),
Università Politecnica delle Marche, Via Brecce Bianche, 60131 Ancona, Italy
{p.sernani,a.f.dragoni}@univpm.it
[2] Department of Construction, Civil Engineering and Architecture (DICEA),
Università Politecnica delle Marche, Via Brecce Bianche, 60131 Ancona, Italy
r.angeloni@pm.univpm.it,
{r.quattrini,p.clini}@univpm.it

Abstract. Augmented Reality (AR) is one of the prominent technologies in Cultural Heritage (CH) exploitation. Taking advantage of commonly used tools as smartphones and tablets, digital contents have the potential to improve visitors' understanding and enjoyment of historical buildings and museums. In this regard, the early stage research described in this paper aims to develop an AR app combining image target-based AR and Simultaneous Localization And Mapping (SLAM). Leading visitors' attention, the app will enhance CH fruition turning it in an interactive learning experience. The presented case study, the "Studiolo" of the Duke in the "Palazzo Ducale" of Urbino, with its high concentration of depicted elements, is ideal to explain the advantages of combining image target and SLAM to achieve a stable and reliable AR. In addition to information superimposed to Points of Interest (POIs), SLAM can be used to anchor suggestions about different POIs into specific positions inside the "Studiolo", guiding the users' orientation during the visit.

Keywords: Cultural Heritage · AR · Immersive augmented reality · Image target recognition · SLAM

1 Introduction

Virtual, augmented and mixed realities are playing a key role in several industries completely redefining customer experiences [1]. In recent decades the spread in the use of virtual visualization systems in the field of Cultural Heritage (CH) aimed to improve user interaction with the artwork using highly interactive, physical-virtual connections [2]. Concerning AR, it has proved effective in increasing visitor satisfaction, attracting new target markets and contributing to a positive learning experience [3]. Mainly using very common devices, AR overlays information around the user enhancing content and telling hidden stories [4]. So, these possibilities make AR extremely valuable for CH with lots of represented subjects and hidden meanings as in [5].

© Springer Nature Switzerland AG 2019
L. T. De Paolis and P. Bourdot (Eds.): AVR 2019, LNCS 11614, pp. 199–207, 2019.
https://doi.org/10.1007/978-3-030-25999-0_17

Hence, the presented case study, the "Studiolo", is emblematic. Investigated in the CIVITAS framework, an interdisciplinary research project founded by the Polytechnic University of Marche aimed to develop and test new paradigms of digitization and fruition for CH in museal context, it has been chosen as test-bench in several application. The "Studiolo" is one of the most known attraction of the magnificent "Palazzo Ducale" in Urbino[1]. Moreover, it is a very closed space, about 3.5 by 3.5 meters, but tall more than 5 meters divided in two orders (Fig. 1). In the lower order of the "Studiolo", walls are covered by wood marqueteries representing objects alluding to the symbols of Arts and Virtues, the second presents the portraits of the "Uomini Illustri", great men, present and past, referring to ethical and intellectual thoughts nourishment for action [6]. So, due to this high concentration of elements and allegories in such a narrow space, without a thorough knowledge of Renaissance History, Literature, Philosophy and Policy it is hard to understand all the aspects of this unique masterpiece.

To facilitate it, this paper proposes the development of an AR App that guides the visitor in what becomes an interactive learning experience that reveals signifier's significance. The problems of major interest faced under this research concerned:

1. The acquisition of images of the Point of Interests (POIs) inside the "Studiolo", to have a rich dataset as reference for the AR app;
2. The design of the AR app, integrating image-target-based AR and Simultaneous Localization And Mapping (SLAM), in order to provide to the user specific and stable suggestions in addition to the information about the POIs.

To this end, Sect. 2 describes related works about the usage of AR in CH as well as the application of SLAM to AR. Section 3 presents the challenges faced for the POI acquisition inside the "Studiolo". Section 4 describe the proposed app, with the integration of image target-based AR and SLAM. Section 5 concludes the paper with a series of considerations on the limits of the proposed approach and suggestions for future improvements.

2   Related Works

Technological developments are changing the ways people experience reality creating an alternative immersive virtual environment or simply overlaying virtual elements to the physical one. These technologies are very effective to support the experience of CH by integrating tangible and intangible and defining new engaging ways to learn cultural contents. Focusing on AR, it has demonstrated capacity to enhance visit quality of CH sites often limited due to the difficulties in communicating the large amount of interesting meanings hidden in a work of art. Artist or client biography, artifact details about its historical period or usage and many other information are essential to understand and enjoy the visit; however, exceeding with knowledge might cause visitors loss of

---

[1] It is a famous Renaissance building, residence of one of the most influential personality in Italian Renaissance, the Duke of Urbino Federico Da Montefeltro.

focus as happen, particularly for young people, for visits leaded by human guide. As in [7] and [8], AR overcomes this issue implementing gamified visits, exploiting the game elements to encourage visitors to move towards specific POI or providing additional contents; AR in CH is even more effective when guided by scientific approaches to art perception [9]. AR also allows the visualization of no longer existing elements, destroyed by human action or natural disasters, or, on the contrary, to hide successive addition revealing the original appearance of the investigated item [10–13]. The approach to AR is also very user-friendly, since it takes advantage of everyday usage mobile devices as smartphones and tablets.

Fig. 1. 3D Pointcloud of the "Studiolo". The division in two orders and the high concentration of represented element clearly appears.

Concerning specific AR technologies presented in scientific literature, image target recognition is one of the most applied approaches: the image is a reference for computer graphics (2D and/or 3D) to be overlaid [14]. However, the image target recognition does not allow the user to move around the superimposed 3D models nor to maintain a realistic scale during user movements. In this regard, alternative techniques to anchor computer graphics artifact to real elements showed through a display are emerging. Specifically, SLAM is a mapping technique used in robotics to build a map of an unknown environment where a mobile robot should localize itself and operate autonomously [15, 16]. However, outside its original sector and scope (autonomous systems such as embedded robotics, unmanned aerial vehicles, or self-driving cars), it has been recently applied also to AR with the aim of improving the tracking of objects [17]. This means that the user can move around the virtual reconstructions offered by the 3D models as well as going towards them to be able to visualize some specific characteristics. One of the innovations of this work is to combine image targets and SLAM, to achieve a stable AR. In the proposed app, when the image target is acquired, i.e. the POI is recognized, the app automatically switches to SLAM mode to perform instant tracking, decoupling the virtual artifacts superimposed on the mobile device screen from the image targets and assigning them to a virtual ground plane. Therefore, image targets are used to recognize the POI and the SLAM to perform the object tracking and achieve a reliable immersive user-experience.

This work also focusses on POI acquisition that can be included in the larger context of CH digitization and it is strictly connected to the need of a heterogeneous documentation for studies, restoration, and valorization purposes. Multiple techniques as Terrestrial Laser Scanning (TLS) and Digital Photogrammetry (DP) can be applied [18]. As in [19] the integration of multi-source data acquired allows to obtain a "virtual twin" that can describe all the physical features of the surveyed item. Unfortunately, CH artifacts can be problematic for digital acquisition due to unstructured, mono- chrome, translucent, reflective, and/or self-resembling surfaces [20]. Taking advantage of previous studies as [21] and [22] an integrated winning methodology for the acquisition of high reflective items is here presented by using polarized filters.

3   Digital Acquisition of POIs for the AR Image Targets

Carried out in the CIVITAS project framework, the digitization of "Palazzo Ducale" in Urbino is meant as the first challenge for its safety, knowledge, and management, as described in [23]. Hence, the acquisition of the "Studiolo" was carried out as part of a complex workflow aiming at the museum challenge to document and preserve CH but also enhance visitors experience. For this specific element, also the conventional combination of different sensor and technology cannot lead to a proper result. In fact, the wood marqueteries covering the walls were restored many times, and the appli- cation of a transparent protective varnish made them high reflective. Several tests, considering the different aims of this survey, both documentation and fruition, lead us to the choice of the integration of TLS and DP using polarized light photography. TLS was used to accurately document the geometry, on the other hand DP implemented the high-resolution texture necessary to generate the image targets for the AR app. The use of polarized light has been adopted to eliminate the external reflectance generated by the light on the surface, enhancing the final image quality. By polarizing, light is transmitted in a determined angle convergent with that of the polarization axis' direction. If an object is illuminated with a polarized light, so that vibrates in only one plane, the superficial reflection remains in the same vibration plane. Putting a cross polarized filter in front of the camera lens, the superficial reflection could be completely blocked. For this acquisition, the camera was used in combination with a light equipped with a linear laminated film and the light beam oriented at 45° in relation to the surveyed object. The adequate setting of the polarizing filter put in front of the lens generated the cross polarization. All the pictures were taken in RAW file format then processed using Adobe Camera Raw to adjust the white balance, contrast and other image parameters. Converted the postprocessed files into jpg, these were aligned using the photogrammetry software Agisoft Photoscan. The point cloud obtained via TLS was in parallel processed to get a 3D mesh to be imported in the same software environment. Aligned pictures, reoriented in the mesh reference system using GCP manually added, were projected on it to generate a high-resolution texture. So, the orthoimages were extrapolated to be used as image targets for the AR app (Fig. 2).

The obtained results still show weak points. Some pictures present a light spot due to the incorrect positioning of the light. Moving two different tripods, one for the light and one for the camera, makes difficult to always keep the right illumination-acquisition distance and angle. So, the first improvement will be the use of a single support for both light and camera. The illumination lacks uniformity. For better results two lights, one on the left and one the right of the camera, will be used to ensure more homogeneity. Although the POI acquired images will be enhanced, those obtained at this stage could be also used to test how their quality affects targets recognition.

4 The AR-Based Mobile App for the "Studiolo"

Realized between 1473 and 1476 the "Studiolo" is considered a Renaissance master-piece. Meanwhile all the other walls were stripped of their decorations, it is the only internal space of "Palazzo Ducale" that keeps its original design, showing the taste at the time of the Urbino court and its magnificence.

Fig. 2. Orthoimages of the POIs to be used as image targets for the AR app.

The "Studiolo" is located on the noble floor of the building and was the private study of the Duke. The walls can be divided in two orders, the upper one with the portraits of great men present and past and the lower one characterized by wood marqueteries. Here the Duke is both celebrated as man of peace and man of war. Iconographic symbols as the sword and the baton, the armor, the flames and the blunderbuss refer to war, on the other hand books, the sandglass, the cardinal virtues and the representation of the Duke himself, with the tip of the spear kept down, are symbols of peace. All the depicted elements are a message to the visitor that could not get many of them due to the high complexity of the scene and the need of a deep knowledge of Renaissance History, Literature, Philosophy and Policy. At this stage of the work, POIs were detected only for this lower order. The sword, the sandglass, the representation of the Duke, a landscape, the lectern and some books were selected for their significance. Next step will be the creation of a storytelling connecting all these elements, also the portraits in the upper order, that will guide the visitor to the discovery of one of the Renaissance leading figure. A linear storytelling will be developed, from the entrance to the exit of the room, useful to avoid queues while ensuring the adequate visit time.

To communicate such contents, we propose an innovative fruition, exploiting the combination of image targets and SLAM into an AR-based mobile app. The challenges are (a) providing specific information once the POIs are recognized thanks to the image targets, and (b) suggesting other POIs inside the "Studiolo", understanding the user relative position thanks to the SLAM capabilities.

Figure 3 depicts the development process of the mobile app to enrich the visit of the "Studiolo". Once the images of the POIs are acquired and pre-processed as described in Sect. 3, they compose the database of image targets to be recognized at runtime, showing informative texts and small 3D models related to the recognized POIs. In fact, as highlighted in the "related works" section, the literature related to AR is plenty of works where image targets are references for computer graphics (2D and/or 3D) to be overlaid. Nevertheless, as showed in [24], to achieve a stable AR, image targets might not be enough. This is even more relevant if the goal is to exactly track the user's position, in order to suggest her/him other relevant POIs in the "Studiolo".

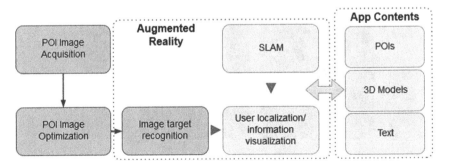

Fig. 3. Scheme for the AR app development: the SLAM allows performing the localization inside the "Studiolo" to suggests other POIs through the screen of the mobile device.

Given its effectiveness in the tracking of objects, we claim that SLAM can be used to locate the user's orientation inside the room and use 2D images such as arrows and icons superimposed on the mobile device display to direct the user's attention to other POIs. Hence, the proposed app consists of a main scene, a picture of the planimetry of the "Studiolo", from which the user can access to the AR mode, framing the "Studiolo" walls with the mobile device camera, or to a section with general information about the "Palazzo Ducale". Once the user selects the AR mode, the app activates the mobile device camera: when one of the POIs listed in Sect. 4 is detected thanks to the image target recognition, a descriptive text about the POI, and when its applicable, 3D animated models like the squirrel or the mandolin appear. When the image target is acquired, i.e. the POI is recognized, the app automatically switches to SLAM mode to perform instant tracking, assigning the visual elements related to the suggestion of other POIs to a virtual ground plane. This means that the user can move around, even framing different POIs, but the suggestions will maintain their original superimposed position even when the triggering POI is out of the field of view, preventing the need to recognize them again. Therefore, using an image target to recognize the POI and the SLAM to perform the object tracking leads to the achievement of a stable (and, if needed, real-scale) AR to provide a reliable immersive user-experience.

5   Conclusions

This paper proposed the development of an AR app integrating image targets and SLAM to support and guide a visitor in the fruition of the "Studiolo" inside the Palazzo Ducale in Urbino. Despite the presented research is in early stage and the app is a "proposal" to be implemented yet, the workflow explained in the paper has the potential to build a new kind of AR app mixing several sources and using different mapping technologies for the visualization of information superimposed to CH artifacts, even beyond the faced case study.

Of course, some limitations are also present in the described research. For example, ambient lighting and user confidence with other AR applications highly influence the results to achieve an effective interactive learning experience during the visit. Moreover, considering the implementation of the app, SLAM stresses the devices on a computation point of view. It recognizes and collects points of the real world with the camera so that if the user uses it for a long time the mobile device starts to overheat.

In addition to addressing the identified limitations, future works include:

1. The implementation of the proposed app and the tests inside the "Studiolo", aiming at understanding if the different lighting conditions available in the building might reduce the effectiveness of the image target recognition;
2. User tests, to assess both the validity of the application inside the "Studiolo" and the applicability of the proposed approach to CH fruition in general;
3. User tests, implementing the 3D model and the POIs in an Immersive Virtual Reality environment connected to an eye tracking device, to study visitors' behavior and attention to the interactive storytelling.

Acknowledgements. The research presented in this paper has been supported by the project "ChaIn for excellence of reflectiVe societies to exploit dIgital culTural heritAge and museumS" (CIVITAS), funded by Univpm.

References

1. Flavián, C., Ibáñez-Sánchez, S., Carlos, O.: The impact of virtual, augmented and mixed reality technologies on the customer experience. J. Bus. Res. **100**, 547–560 (2018)
2. Clini, P., Ruggeri, L., Angeloni, R., Sasso, M.: Interactive immersive virtual museum: digital documentation for virtual interaction. Int. Arch. Photogram. Remote Sens. Spatial Inf. Sci. **42**(2), 251–257 (2018)
3. Leue, M.C., tom Dieck, D., Jung, T.: Value of augmented reality at cultural heritage sites: a stakeholder approach. J. Destin. Mark. Manag. **6**(2), 110–117 (2017)
4. Clini, P., Frontoni, E., Quattrini, R., Pierdicca, R.: New augmented reality applications for learning by interacting. Archeomatica **8**(1), 28–33 (2017)
5. Quattrini, R., Pierdicca, R., Frontoni, E., Clini, P.: Mobile e Realtà Aumentata al Palazzo Ducale di Urbino: il Museo è digitale. Archeomatica **6**(1), 32–37 (2015)
6. Dal Poggetto, P.: La Galleria Nazionale delle Marche e le altre collezioni nel Palazzo Ducale di Urbino. Istituto poligrafico e Zecca dello Stato (2003)
7. Mortara, M., Bellotti, F., Fiucci, G., Houry-Panchetti, M., Petridis, P.: Learning cultural heritage by serious games. J. Cult. Heritage **15**(3), 318–325 (2014)
8. Hammady, R., Ma, M., Temple, N.: Augmented reality and gamification in heritage museums. In: Marsh, T., Ma, M., Oliveira, M.F., Baalsrud Hauge, J., Göbel, S. (eds.) JCSG 2016. LNCS, vol. 9894, pp. 181–187. Springer, Cham (2016). https://doi.org/10.1007/978-3-319-45841-0_17
9. Naspetti, S., Pierdicca, R., Mandolesi, S., Paolanti, M., Frontoni, E., Zanoli, R.: Automatic analysis of eye-tracking data for augmented reality applications: a prospective outlook. In: De Paolis, L.T., Mongelli, A. (eds.) AVR 2016. LNCS, vol. 9769, pp. 217–230. Springer, Cham (2016). https://doi.org/10.1007/978-3-319-40651-0_17
10. Empler, T.: Cultural heritage: displaying the forum of Nerva with new technologies. In: Digital Heritage International Congress, Digital Heritage 2015, pp. 581–586 (2015)
11. Canciani, M., Conigliaro, E., Grasso, M.D., Papalini, P., Saccone, M.: 3D survey and augmented reality for cultural heritage. The case study of aurelian wall at Castra Praetoria in Rome. Int. Arch. Photogram. Remote Sens. Spatial Inf. Sci. **41**, 931–937 (2016)
12. Botrugno, M.C., D'Errico, G., De Paolis, L.T.: Augmented reality and UAVs in archaeology: development of a location-based AR application. In: De Paolis, L.T., Bourdot, P., Mongelli, A. (eds.) AVR 2017. LNCS, vol. 10325, pp. 261–270. Springer, Cham (2017). https://doi.org/10.1007/978-3-319-60928-7_23
13. Clini, P., Quattrini, R., Frontoni, E., Pierdicca, R., Nespeca, R.: Real/not real: pseudo-holography and augmented reality applications for cultural heritage. In: Handbook of Research on Emerging Technologies for Digital Preservation and Information Modeling, pp. 201–227. IGI Global (2017)
14. Amin, D., Govilkar, S.: Comparative study of augmented reality SDKs. Int. J. Comput. Sci. Appl. **5**(1), 11–26 (2015)
15. Durrant-Whyte, H., Bailey, T.: Simultaneous localization and mapping (SLAM): part I. IEEE Robot. Autom. Mag. **13**(2), 99–110 (2006)
16. Bailey, T., Durrant-Whyte, H.: Simultaneous localization and mapping (SLAM): Part II. IEEE Robot. Autom. Mag. **13**(3), 108–117 (2006)

17. Gao, Q.H., Wan, T.R., Tang, W., Chen, L.: A stable and accurate marker-less augmented reality registration method. In: Proceedings of the International Conference on Cyberworlds, CW 2017, pp. 41–47 (2017)
18. Remondino, F.: Heritage recording and 3D modeling with photogrammetry and 3D scanning. Remote Sens. **3**(6), 1104–1138 (2011)
19. Huilin, L., Weizheng, L., Siqi, L., Lingxi, Z., Wenli, J., Zhang, Q.: The integration of terrestrial laser scanning and terrestrial and unmanned aerial vehicle digital photogrammetry for the documentation of Chinese classical gardens – a case study of Huanxiu Shanzhuang, Suzhou, China. J. Cult. Heritage **33**, 222–230 (2018)
20. Schaich, M.: Combined 3D scanning and photogrammetry surveys with 3D database support for archaeology & cultural heritage. A practice report on ArcTron's information system aSPECT3D. In: Proceedings of the Photogrammetric Week 2013, pp. 233–246 (2013)
21. Nicolae, C., Nocerino, E., Menna, F., Remondino, F.: Photogrammetry applied to problematic artefacts. Int. Arch. Photogram. Remote Sens. Spatial Inf. Sci. **40**(5), 451–456 (2014)
22. Cortón Noya, N., López García, Á., Carrera Ramírez, F.: Combining photogrammetry and photographic enhancment techniques for the recording of megalithic art in north-west Iberia. Digit. Appl. Archaeol. Cult. Heritage **2**(2), 89–101 (2015)
23. Nespeca, R.: Towards a 3D digital model for management and fruition of Ducal Palace at Urbino. Integr. Surv. Mob. Mapp. SCIRES-IT **9**(2) (2018)
24. Dragoni, A.F., Quattrini, R., Sernani, P., Ruggeri, L.: Real scale augmented reality. A novel paradigm for archaeological heritage fruition. In: Luigini, A. (ed.) EARTH 2018. AISC, vol. 919, pp. 659–670. Springer, Cham (2019). https://doi.org/10.1007/978-3-030-12240-9_68

Optimization of 3D Object Placement in Augmented Reality Settings in Museum Contexts

Alexander Ohlei[✉], Lennart Bundt[✉], David Bouck-Standen[✉], and Michael Herczeg[✉]

University of Lübeck, Lübeck 23562, Germany
{ohlei, bundt, bouck-standen,
herczeg}@imis.uni-luebeck.de

Abstract. Augmented Reality (AR) is a technology that can be used to provide personalized and contextualized information regarding physical objects in form of digital overlays. We use this technology in our research project Ambient Learning Spaces (ALS) to provide museum visitors with specific additional digital 3D information regarding the exhibits presented. With this technology, we enable museum curators to use a new form of transporting contextualized information without the need for additional physical space. However, the use of AR brings up new challenges for the creation and placement of digital contents into the museum space. In this context, we ran an anonymous survey on the use of AR in museums throughout Germany and studied responses of (N = 133) museum professionals. The results indicate that, although many museum professionals are interested in using AR technology, currently the integration is very costly and complex. This paper proposes a system we developed in a user-centered design process with a museum. This system provides an interface that helps museum professionals to cope with the complexity when placing and aligning digital 3D objects in their exhibition using mobile devices. Through this solution, visitors have the chance to experience the virtual objects spatially embedded in the exhibition by the curators themselves. In multiple user studies during the development phase we measured the usability of the interface. The findings show that the system provides a high degree of usability and can be applied effectively by museum professionals.

Keywords: Augmented reality · 3D object interaction · 3D object placement

1 Introduction

Augmented Reality (AR) is a technology that can be used in multiple settings to provide individualized and contextualized information regarding physical objects as digital information overlays [1]. These overlays can contain images, videos, and virtual 3D objects to support learning and understanding in the context of the physical object. To be able to see the digital content, a regular mobile device (e.g. a smartphone or tablet) can be used. It is also possible to use Head Mounted Displays (HMDs), like the Microsoft HoloLens or Magic Leap, to see the virtual content hands free. The release of

© Springer Nature Switzerland AG 2019
L. T. De Paolis and P. Bourdot (Eds.): AVR 2019, LNCS 11614, pp. 208–220, 2019.
https://doi.org/10.1007/978-3-030-25999-0_18

such HMDs, high-performance smartphones, as well as improved AR frameworks, has opened many possibilities for the development of new AR applications. However, AR content creation and content delivery are still time-consuming and complex tasks. In an industrial context, some companies have developed authoring tools like Vuforia Studio [2] which require no coding skills. In the museum context, only a few authoring tools exist.

In our research transfer project Ambient Learning Spaces (ALS) [3] we develop digitally enriched body- and space-oriented learning environments for schools and museums. In these environments, users can interact with multiple interconnected display devices in physical space to learn cooperatively. In the context of ALS, body- and space-oriented human-computer interaction in combination with Cross-Device Interaction (XDI) build a conceptual foundation. In the backend, the Network Environment for Multimedia Objects (NEMO) [4] is the platform for all ALS applications. Inside NEMO, all media created from physical objects are stored in a context-specific semantic model, which supports the use by ALS applications. All of these applications running on mobile and stationary devices access NEMO as a context repository.

Along with the backend, a suite of learning applications has been developed as frontend applications for ALS. One of these frontend applications is InfoGrid [5, 6]. Users can run InfoGrid on their smartphones and use it to see personalized and contextualized digital content superimposed over physical objects using AR technology. For museum curators, we provide the web-based ALS-Portal to add and edit content for InfoGrid and other ALS applications. We furthermore developed tools accessible through the ALS-Portal that enable museum curators to create and edit 3D objects [7].

In many cases, it is necessary that AR overlays fit to the surrounding space and need to be positioned relative to the physical environment. For this purpose, it is necessary that the translation, rotation, and scaling of the object can be adjusted to place it in spatial relation to its environment. This paper proposes an approach for the manipulation of virtual objects in AR environments on the smartphone. The manipulation takes place in an administration tool that is integrated into InfoGrid. We evaluated the functionality with students, research assistants and one of our museum project partners, the municipal Museum of Nature and Environment (MNU) in Luebeck, Germany.

2 Preliminary Museum Survey

To analyze the user group of museum professionals and to get a better understanding of the museum context, we ran an anonymous preliminary online survey. We prepared an online questionnaire using the LimeSurvey tool asking:

- Is the participant of the survey taking care of the design of the exhibition?
- How much experience does the participant have regarding AR?
- How is the participant's opinion about the implementation of AR at his/her museum?
- What kind of hardware- and software does the participant have access to?

- What does the participant think are the advantages and disadvantages of having AR at his or her museum?

We also added demographic questions. Answers consist of choosing option fields and free text fields, depending on the question. In July 2018, we sent e-mails to museum professionals introducing our research topic along with the link to take part in the survey. We collected feedback over a period of approximately one month and used the questionnaire as the only instrument for the survey.

2.1 Participants

We sent out e-mails to 800 museums in Germany and received feedback from 133 museum employees. 73 participants were female, 47 male, and 13 did not answer the question. The age of the participants ranged from 26 to 65 years. 39 participants were museum directors, 22 were curators, 55 had mixed job titles, and 17 did not answer the question. 26 participants always design exhibition, 46 often design exhibitions, 29 eventually exhibitions, 12 rarely design exhibitions, 5 never design exhibitions, and 14 did not answer the question.

2.2 Results

When asking for any experience regarding AR systems, the participants could answer with a Likert scale from one to five (see Fig. 1).

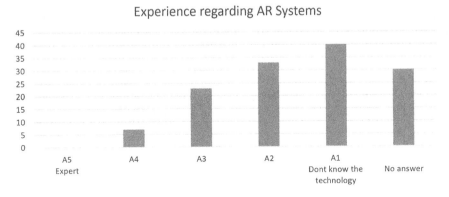

Fig. 1. Experience regarding AR systems of participants of the survey (N = 133). y-axis: number of participants. x-axis P

When asking whether the participants think the integration of AR into their exhibition makes sense we again offered a Likert scale from 1 to 5 (see Fig. 2).

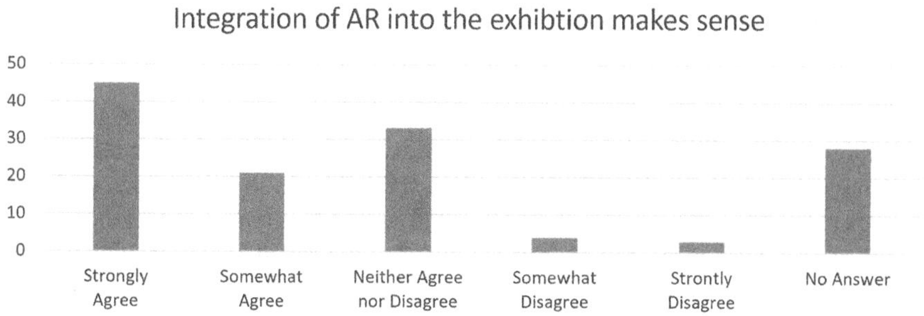

Integration of AR into the exhibtion makes sense

Fig. 2. Answers of the participants to the question if they think the integration of AR into their exhibition makes sense (N = 133). y-axis: number of participants.

When asking what kind of technology the participants is available in their museum, 18 responded that they are provided with a tablet, 42 are provided with a smartphone, 58 are provided with a laptop, 76 with a desktop PC. 38 participants did not answer the question.

When asking the participants for the advantages of integrating AR into their exhibition, they could answer with free text. Many responded that AR is a new and interesting form of giving information to their visitors. They think that especially young people could be attracted by offering AR contents and using animated 3D models is a nice way to present information vividly.

When asked for the disadvantages of implementing AR into their museum, it was again possible to answer with free text. Most participants responded that cost and technical complexity would be an issue. Some said that keeping the material up to date might be difficult. Some mentioned that a system failure would have a negative impact on the museum experience.

2.3 Discussion

Many participants see a positive impact of using AR in their exhibition and think that it makes sense to integrate AR technology into their museum. But the technical complexity is too high and costly for them to use this technology. Almost all museums have some computer equipment, e.g. in the form of a PC or Laptop.

3 Related Work

There is a variety of interaction methods, which are solutions for the rotation, translation, and scaling of virtual 3D Objects using smartphones. Most of them either use gestures on the touch screen of the smartphone or translate the smartphone's movement onto the virtual object.

Marzo et al. [8] describe multitouch gestures along with the movement of the smartphone to manipulate virtual objects. In their approach movements of the devices are used to translate objects and touch gestures are used to rotate the virtual object. Once the display registers a touch event the virtual object is fixed to the visual position on the screen. By moving the smartphone, the object can be moved until the touch of the display is not recognized anymore. Moving the finger across the touch screen initiates an arcball rotation as described by Heckbert [9].

Moving two fingers on the screen away from or towards each other, the virtual object moves closer or further away from the user. The results of user tests of this interaction methods show that this form of interaction is useful for rotations with large angles. Scaling of virtual objects is not possible with this interaction method.

Mossel et al. [10] developed an interaction technique called 3D touch, which allows for rotation and translation with six degrees of freedom (6DoF). Furthermore, the technique enables the scaling of the object on individual axes. The user can use buttons to choose the interaction method he wants to use. When one of the modes is selected the object is translated, rotated or scaled depending on the viewpoint of the mobile device. The user needs to change the perspective of the device towards the object if the object should be translated on the other axis besides the two visible ones. The interaction method can be used with one or two hands.

Another interaction method described by Mossel et al. [10] is called Hand Oriented Manipulation Extending Raycasting for Smartphones (Homer-S). This interaction technique only makes use of the device movement to rotate, translate and scale the virtual object. Similar to 3D Touch the individual manipulation is selected in a menu. Scaling with this interaction method is separated for each individual axis. This interaction method can also be used with one or two hands.

Cannavò et al. describe an interaction method called T4T [11]. It distinguishes between a cursor mode and a tuning mode. Selecting the tuning mode, the user can manipulate virtual objects using tangibles. Using the cursor mode, the user can select an object by pointing onto it for a couple of seconds. To cancel the selection the user can shake the smartphone. When an object is selected the user can use a menu to decide which interaction method he wants to use. It is possible to scale over all axes, scale over only one selected axis, translate over one plane, translate over the z-axis and rotate over individual axes. Scaling can be accomplished by using pinch gestures. The rotation works by moving two fingers around each other on the touch display of the smartphone.

Samini et al. [12] describe an interaction method called Relative-DPR that uses the movement of the device to manipulate the virtual object. In their approach, only one mode is available in which all movements of the device are transferred onto the virtual object's location. However, little research evaluates different interaction methods for the manipulation of virtual 3D objects in an AR smartphone app.

In our work, we assessed the usability of interaction techniques that can be used to translate, rotate and scale virtual 3D objects on mobile phones. Then we developed an AR application in which we implemented a combination of techniques with the best user rating. The selected methods are the rotation and translation via individual axes as described in 3D Touch [10]. We also implemented the scaling via pinch gestures as described in [10], but linked all axes together to select uniform scaling. We furthermore implemented the Arcball rotation as described by Heckbert [9] and the free hand translation as described by Marzo et al. [8]. The user can choose between different interaction methods by selecting the corresponding mode. We evaluated the final app in a quantitative user study using the SUS questionnaire.

4 System Description

The ALS project aims at developing digitally enriched body- and space-oriented learning environments for schools and museums. One of the applications that were developed is the frontend application InfoGrid. InfoGrid allows content creators in schools and museums to create AR tours using the web-based ALS-Portal. Inside the ALS-Portal our project partners can create individual tours consisting of multiple images that can be recognized by InfoGrid. These images are called image markers or simply markers. Each image marker can be assigned with a specific overlay type and a data file (see Fig. 3).

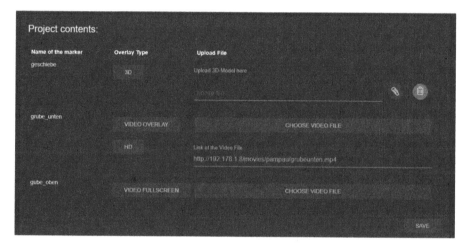

Fig. 3. ALS-Portal can be used to assign 3D objects to AR markers. The files are uploaded into our NEMO framework.

After saving the information through the ALS-Portal in our NEMO framework the InfoGrid smartphone app can be used to download the tour data in the institution where the markers are located. NEMO is a distributed system that allows media files to be stored in a local NEMO instance providing users with high-speed access to the respective files [4]. Once the tour data is downloaded InfoGrid can be used to scan the image markers and to experience the selected content. If the author decided to upload a 3D object file into the ALS-Portal the objects show up in a predefined scale and in an upright position towards the marker inside the app. In many cases, users want to present the virtual 3D object in an individual way. Therefore, we propose a system that can be used for translation and rotation on individual axes in three-dimensional space as well as uniform scaling using the smartphone app. Only the authors of the tours should be able to redefine the position and scaling of the 3D object, therefore the system should only be accessible after authentication. When the new position and scale is set correctly InfoGrid supports uploading all new parameters to the NEMO framework.

4.1 Prototype Testing

During the system development phase, we studied the manipulation of virtual objects by comparing three different manipulation methods.

The first manipulation method included the rotation and translation via touch gestures [8] and a uniform scaling via a slider. The user could either rotate the virtual object by using the arcball rotation with one finger or by using the two finger rotation [8]. The user could accomplish the translation by keeping the finger anywhere on the screen and while moving the device [8]. Additionally the user could translate the virtual objects forward or backward by pinch gestures [8]. Using this technique all manipulation tasks could be done simultaneously.

The second manipulation method was based on the first manipulation method. The difference was that the pinch gestures could be used for scaling [10] instead of moving the object to the front and to the back and the scaling slider was removed. The scaling was implemented to support uniform scaling.

The third manipulation method was based on the first interaction method as well. The difference was, that translation and rotation were separated [10]. The user had to choose between a rotation or translation mode selecting one of two buttons on the lower area of the smartphone display.

The purpose of the formative evaluation study was to find out the preferred method of users. Six participants, mainly students, took part in the study. Four participants were male and two female. The age range was from 22 to 54 years. We asked the participants to use the interaction method in a randomized order to place a randomly located virtual chair onto a fixed chair and to answer a questionnaire afterward.

The evaluation has shown that since the smartphone display is only 2-dimensional the missing depth information makes it hard to accomplish the task. To make sure that the object is placed correctly it was necessary to view the virtual objects from different angles by altering the perspective of the smartphone. Furthermore, it was noticed that the separation of translation, rotation, and scaling as described in [10] is preferred. We observed that this way the participants had less unintentional side effects such as unwanted object movement when the participant just wanted to rotate the object.

Therefore, we chose the third manipulation method with separated rotation and translation functions for the final application. The users preferred the scaling via pinch gestures to scaling via a slider. Thus, we adapted the third manipulation method to support scaling via multitouch gestures and removed the slider. We also noticed that the users preferred the option to choose between modes for fast interaction and for slower precise interaction. Thus, we added the modes for fast translation such as freehand translation and fast rotation such as the arcball rotation as well as for precise translation and rotation via individual axes.

4.2 Implementation of the Final Application

The final application combines approaches for fast and imprecise as well as slow and precise translation and rotation of the virtual 3D objects. The fast and imprecise interaction is based on the work of Marzo et al. [8]. If the user touches the screen of the smartphone the object can be translated in a constant relation regarding the smartphone (see Fig. 4).

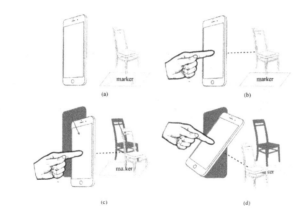

Fig. 4. (a) The virtual 3D object is shown on the marker. (b) The user touches the smartphone's display and initiates the translation interaction. (c) The user moves the smartphone and the virtual 3D object moves in a fixed relation regarding the smartphone. (d) The user rotates the smartphone. The virtual 3D object moves to the new area without reflecting the rotation angle.

The approach has been changed so that the user has to keep the free translation (see Fig. 5) button pressed to translate the object instead of touching any place on the display of the smartphone. Our formative evaluation has shown that without the button the users produced more errors. Our formative evaluation has also shown that users prefer that this interaction method can be used to only move the object but not to change the orientation at the same time.

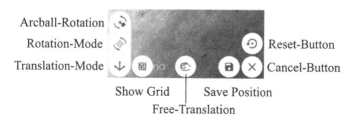

Fig. 5. Functions of the manipulation interface of the InfoGrid app.

To use the precise translation, the user has to activate the translation mode (see Figs. 5 and 6, left). Once activated translation axes are displayed on the smartphone's display. Touching the axes and moving the finger along the lines moves the object on that specific axis (see Fig. 6, left). Furthermore, the rotation mode (see Figs. 5 and 6, right) allows for the precise rotation. Using this mode, the smartphone displays 3 orthogonal circles around the virtual 3D object. Touching one of the circles, the user can rotate the object around the selected axis by moving his finger on the touchpad (see Fig. 6, right). In any mode, the user can resize the virtual 3D object by using pinch gestures. By moving the fingers away from each other the object gets bigger and by moving the finger closer towards each other, the object becomes smaller.

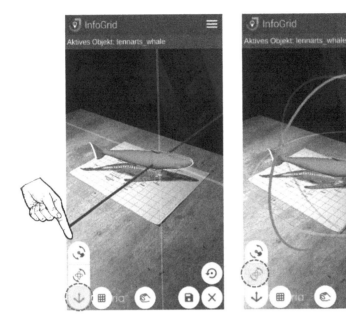

Fig. 6. (left) Smartphone display with activated translation mode (lowest button on the left). (right) Smartphone display with activated rotation mode (middle button on the left), additionally, the ground plane is overlaid with a grid to support the users understanding, if the object is located above the marker.

An arcball interaction mode was also implemented based on the ideas of Heckbert [9], to allow a quicker interaction (see Fig. 5). The rotation works by swiping with a single finger on the touchpad of the smartphone. To cope with the problem of having no depth information as noticed in the formative evaluation we decided to add a virtual grid on the ground plane defined by the marker (see Fig. 6). If the user moves the object below the ground plane, it is possible to see the intersection of the grid. This way, the user can decide if this is correct or not. If the grid interferes with the display of the object it can also be disabled.

5   User Studies

To evaluate the usability of the system we made a quantitative user study with student participants and a qualitative user study with employees of the Museum of Nature and Environment in Lübeck.

5.1   Quantitative User Study

Finally, we measured the usability of the new manipulation functionality using the SUS questionnaire [13] with 19 student participants. To run the study, we prepared 3 tasks in the usability lab of our institute that participants should solve using a Samsung Galaxy

S7 smartphone. The first task was to place a virtual 3D sculpture onto the center of a marker. Initially, the sculpture was placed in a predefined but seemingly meaningless position. The second task was to place a 3D model of a whale onto a 2D image of the whale. The third task was to place a virtual 3D model of London's Big Ben onto a physical model of the Big Ben so that they match. The participants could decide when they were finished with each task. The participants could also decide which interaction method they would like to use. After finishing all three tasks the participants were asked to fill out a SUS questionnaire and a free text field for further comments. Along with the questionnaire, the participants were observed solving the three tasks. 19 participants took part in the study. 15 participants were male and 5 female. The age range was between 20 and 32 years.

Results

The mean SUS-Score measured was 80.66 (SD = 13.76; N = 19; max. SUS-Score: 100) (see Table 1). Participants mentioned that the interface is clear and easy to use. Some participants needed time to understand how to use the manual translation button. They expected it to be a toggle button instead of a button that needs to be pressed all the time during interaction.

Table 1. Mean ± standard deviation of the SUS (statements originally provided in German). Values from 1–5 could be chosen. Statement (1, 3, 5, 7, 9): higher values are better; Statement (2, 4, 6, 8, 10): lower values are better.

ID	Statements	Total sample N; M(SD)
1	I think that I would like to use this system frequently	19; 3.68 (1.05)
2	I found the system unnecessarily complex	19; 1.57 (0.60)
3	I thought the system was easy to use	19; 3.94 (0.91)
4	I think that I would need the support of a technical person to be able to use this system	19; 1.63 (0.96)
5	I found the various functions in this system were well integrated	19; 4.32 (0.58)
6	I thought there was too much inconsistency in this system	19; 1.47 (0.77)
7	I would imagine that most people would learn to use this system very quickly	19; 4.00 (1.01)
8	I found the system very cumbersome to use	19; 1.42 (0.68)
9	I felt very confident using the system	19; 4.00 (1.05)
10	I needed to learn a lot of things before I could get going with this system	19; 1.58 (0.61)

Abbreviations: N, number of participants, M, mean; SD, standard deviation; SUS, System Usability Scale

5.2 Qualitative User Study

We conducted a qualitative user study using the think-aloud method along with an interview after the use of the app. The study was carried out inside the Museum of Nature and Environment (see Fig. 7) with one participant of the museum's staff.

Fig. 7. An employee of the Museum of Nature and Environment using the new manipulation interface to position a virtual whale onto a physical skeleton so visitors can see it properly aligned.

The task for the employee was to place a virtual whale model onto the physical bones in the exhibition hall. The manipulation functions were explained to her before using the app. Her thoughts (think aloud) during the use of the app, as well as answers to the interview questions, were written down. She mentioned that it was not clear at the beginning of the study that the 3D object had to be selected before the interaction with it became possible. She also mentioned that the rotation and scaling work very well and very precise.

During the interview, she mentioned that she thinks that users learn the functionality quickly and that it would make sense to her to use the app regularly. She also thinks that the manipulation works quickly as soon as one knows how to use the different functions. Finally, she mentioned that it is good to see what the museum's visitors will see when they use their app.

5.3   Discussion

We built the system so that it can be used quite easily, but some users needed some time and explanation to understand how they can use some functions. Therefore, we think it makes sense to provide users with a tutorial explaining all available functions.

The resulting SUS Score is interpreted as *good* based on the findings of Bangor et al. [14]. The SUS Score and the feedback of the museum employee indicate that the app provides good usability in the museum context and that employees will be able to work with it.

6   Conclusion

Manipulation of virtual 3D objects in AR scenes is a fundamental task when the objects should be presented in relation to physical objects. Each interaction technique on smartphones has different advantages and disadvantages regarding precision, speed and user satisfaction. Therefore, it seems helpful to offer multiple interaction forms for the user to reach the highest user satisfaction depending on the task and objects to manipulate.

In this paper, we described a system that has been designed and evaluated that can be used effectively to place virtual 3D objects in relation to physical markers. The study showed that users can successfully solve different tasks with the implemented functionality.

Acknowledgments. The authors thank the participants who spent their time for the study. The German Research Foundation (DFG) funds the ongoing project.

References

1. Bouck-Standen, D., Ohlei, A., Winkler, T., Herczeg, M.: Narrative semantic media for contexual individualization of ambient learning spaces. In: Iaria CENTRIC 2018 (2018, in press)
2. PTC: Vuforia Studio. https://www.ptc.com/en/products/augmented-reality/vuforia-studio. Accessed 01 Apr 2019
3. Winkler, T., Scharf, F., Hahn, C., Herczeg, M.: Ambient learning spaces. Education in a Technological World: Communicating Current and Emerging Research and Technological Efforts, pp. 56–67 (2011)
4. Bouck-Standen, D.: Erstellung einer API zur Anbindung des Network Environment for Multimedia Objects an die Ambient Learning Spaces (2016)
5. Ohlei, A., Bouck-Standen, D., Winkler, T., Herczeg, M.: InfoGrid: an approach for curators to digitally enrich their exhibitions. In: Mensch und Computer 2018-Workshopband (2018)
6. Ohlei, A., Bouck-Standen, D., Winkler, T., Herczeg, M.: InfoGrid: acceptance and usability of augmented reality for mobiles in real museum context. In: Mensch und Computer 2018-Workshopband (2018)
7. Bouck-Standen, D., Ohlei, A., Daibert, V., Winkler, T., Herczeg, M.: NEMO converter 3D: reconstruction of 3D objects from photo and video footage for ambient learning spaces. In: Mauri, J.L., Gersbeck-Schierholz, B. (eds.) AMBIENT 2017; The Seventh International Conference on Ambient Computing, Applications, Services and Technologies, Barcelona, Spain, pp. 6–12. IARIA (2017)
8. Marzo, A., Bossavit, B., Hachet, M.: Combining multi-touch input and device movement for 3D manipulations in mobile augmented reality environments. In: Proceedings of the 2nd ACM Symposium on Spatial User Interaction - SUI 2014, pp. 13–16 (2014). https://doi.org/10.1108/13595474200100023
9. Heckbert, P.S.: Graphics Gems IV. Academic Press, London (1994)
10. Mossel, A., Venditti, B., Kaufmann, H.: 3DTouch and HOMER-S: intuitive manipulation techniques for one-handed handheld augmented reality. In: Proceedings of the Virtual Reality International Conference: Laval Virtual, p. 12 (2013). https://doi.org/10.1145/2466816.2466829
11. Cannavò, A., et al.: T4T: tangible interface for tuning 3D object manipulation tools. In: 2017 IEEE Symposium on 3D User Interfaces, 3DUI 2017 – Proceedings, pp. 266–267 (2017). https://doi.org/10.1109/3dui.2017.7893374

12. Samini, A., Palmerius, K.L.: A study on improving close and distant device movement pose manipulation for hand-held augmented reality. In: Proceedings of the 22nd ACM Conference on Virtual Reality Software and Technology - VRST 2016, pp. 121–128 (2016). https://doi.org/10.1145/2993369.2993380
13. Brooke, J.: SUS-A quick and dirty usability scale. In: Usability Evaluation in Industry. CRC Press (1996)
14. Bangor, A., Kortum, P., Miller, J.: Determining what individual SUS scores mean: adding an adjective rating scale. J. Usability Stud. **4**, 114–123 (2009)

Transmedia Digital Storytelling for Cultural Heritage Visiting Enhanced Experience

Angelo Corallo, Marco Esposito, Manuela Marra,
and Claudio Pascarelli$^{(\boxtimes)}$

Department of Innovation Engineering, University of Salento, Lecce, Italy
{angelo.corallo, m.esposito, manuela.marra,
claudio.pascarelli}@unisalento.it

Abstract. The Italian cultural heritage is the richest in the world and its attractiveness is still far from a complete exploitation. The technological innovation is an opportunity to enhance the valorisation process and cultural institutions are willing to exploit this potential in order to attract even more visitors. New communication paradigms are needed to satisfy the emerging need for cultural knowledge and experience of citizens and tourists. This paper introduces an innovative technological and methodological framework aimed at facilitating the collaborative creation and sharing of cultural narrative experiences. The potential benefits and implication related to the framework development are described. The framework starts from the research and systematization of cultural heritage contents, which are then digitalized and virtually reconstructed in order to create interactive and immersive experiences.

Keywords: Cultural heritage · Transmediality · Digital Storytelling

1 Introduction

The Italian Cultural Heritage is well known thanks to his extraordinary abundance (4.976 museums, archaeological areas and monuments, 1 per 12.000 inhabitants) but it has a not yet fully exploited its potential. Trends, as regard profit and entry tickets, are positive, but no museums appear among the 10 most visited museums in the world, one out of three receives less than 1000 guests a year and the 70% of Italian people do not visit any museum [1]. A sample analysis, led by the Osservatorio Innovazione Digitale nei Beni e Attività Culturali [2] in 2016, shows that 52% of museums possess at least an account on common social networks. Nevertheless, digital services related to the enjoyment of museum collections, both online (e.g. online catalogues and virtual visiting) and onsite (as QR-code and proximity systems, mobile apps) are poor. Currently, therefore, cultural institutions have to face the challenge of sharing their own artistic and cultural heritage in a new way, closer to citizens and tourists demand for knowledge.

Enhancement of cultural assets field and tourism can be a strategical aspect in order to improve national systems competitiveness. In Italy, the rich cultural heritage is not managed effectively and efficiently, although it is of extraordinary importance with regard to local economic growth.

© Springer Nature Switzerland AG 2019
L. T. De Paolis and P. Bourdot (Eds.): AVR 2019, LNCS 11614, pp. 221–229, 2019.
https://doi.org/10.1007/978-3-030-25999-0_19

In Salento, a geographic region in the South of Italy rich of old artistic and cultural treasures, demand for tourism seems to be still referred only to sea places and to the capital city of Lecce. Although its growth from the nineties, tourism is limited to summer months. This aspect means a bad exploitation of the artistic, architectural, landscape, cultural, archaeological and historical heritage, which has to be overcome through innovative business models, made possible thanks to the implementation of new engagement technologies. Therefore, investments in the innovation, communication and organization fields are needed.

Nowadays, connections and integration among sites of archaeological interest serve as a showcase and they are often upgraded with unacceptable delays and the related communication is often missing.

Development and spread of new ICT technologies (sensors, virtual reality, wearable device, cloud technologies) applied to cultural heritage promotion increase the earning opportunities for these business operators, since some of the opportunities efficiently match with culture diffusion requirements. These ones become opportunities for the development of first-person experience technologies, which could guarantee a strong user engagement, beyond opening up new paths in museum didactics. This field, in avant-garde countries (as regards ICT technologies), is characterized by an escalation of projects and initiatives, such as the "Approachable" Statens Museum for Kunst in Denmark [3], the digitalization program at the Smithsonian in Washington [4], the digital reconstruction program applied to cultural heritage led by the Harvard University Semitic Museum [5], and so many others.

Another clear indicator highlighting the growing need for innovation consists of the drop in the number of young people visiting these museums for educational purposes in 2016, which dropped by over 6% from the previous year [6]. which means a progressive disaffection for culture among young people. This information is even more surprising, if compared to data related to different communication forms centred on user's involvement. For example, videogames as Assassin's Creed, with over 100 million copies sold in ten years of history [7] became the first knowledge vector to get in touch people with historical locations, characters and facts actually happened in the past. It seems clear that putting users in a highly emotional and interactive context makes processes like learning and involvement easier; this remarks the need for a renewal, oriented towards new generations, concerning communication of the cultural heritage of a territory.

Gamification has been implemented in information campaigns, becoming a real marketing strategy, useful in communicating a positive image and in increasing the visibility of contents, thanks to the viral diffusion of the game and the related desire for sharing satisfied by social media [8].

Then, the right solution can be to define a model of territory enjoyment, based on highly relevant contents, innovative technologies and gamification logics. Generational change in sharing the cultural heritage passes therefore by the new "media", since they use a language oriented towards empathising comprehension perceptive aspects. Contents digitalisation is the backbone of this new scenario and Virtual (VR) and Augmented Reality (AR) represents the key enabling technologies. VR and AR represent in fact the perfect tools for creating new contents and enriching the visitor's experience, telling stories on cultural heritage and letting them "speaking".

The Cultural Heritage Engineering Revolution (CHER) project, co-funded by the Apulia Region (Italy) and the European Union, started in 2018 and, involving cultural associations, small enterprises and a university, aims at defining an innovative methodology for content creation based on a modular Digital Storytelling model and adopting innovative technologies as touchable videos and Virtual Reality. The scope is to innovate the way of sharing cultural heritages, in order to make the cultural offer matching with tourists and citizens' real demand.

In this article, which extends and improves a previous work [9], an innovative framework is proposed and the project preliminary results are discussed.

In the following paragraph a short overview on ICT in the cultural heritage scenario is provided.

In the third section the preliminary results of the CHER project are presented and their expected future impact are discussed in section four.

Conclusions end the paper.

2   ICT and Cultural Heritage: An Overview

Digitally capturing cultural heritage resources have become nowadays a common practice. Information Technologies have made possible many important changes in the practices of research, narration and conservation of cultural heritage.

Rigorous scientific documentation induce to acquire and assimilate large amounts of data derived from the interaction with research objects. For this reason, complex databases are prepared to store all the metadata, relational schemas are defined to enable fast and easy communication with statistics and data analysis applications, and some more examples could be found.

Studying the past means to tell antiquity to scholars, scientific community and, as well, to the community of citizens; heritage professionals have to speak to a large and varied audience. Recording physical characteristics of an archaeological sites or of historic structures or, of a precious ancient book is a cornerstone of their conservation, whatever it means actively maintaining them or making a posterity record. The information produced by such activity potentially would support decision-making of property owners, site managers, public officials, and conservators around the world, as well as, to present the historic knowledge and values of these resources.

Digitisation, intended as a copy of a physical original, e.g. the scan of paper objects and documents or the digital image of a painting, is commonly used in different type of cultural institutions such as historic archives, libraries and picture gallery [10]. The digitisation of information (such as size, date, origin, title, description, context) resulting from earlier documentation or from personal knowledge generate a huge set of metadata which are useful to identify, describe, understand and value heritage material objects.

Gamification and interaction whit customers in a virtual environment represent the new frontier of cultural communication. A significant experience has been held in the National Archaeological Museum of Naples. The video-game Father and Son (downloadable for free) gained an excellent result in audience development (more than 1 million users from differ countries have downloaded the game and tens of thousands users in European, Asian and both Americas have learned the existence of the National Archaeological Museum in Naples [11]) and, thanks to this videogame, cultural contents spread out the physical boundaries of the museum and reached a very large and expanded public, that represent a new challenge for cultural accessibility (Fig. 1).

Fig. 1. Image from Father and Son (Developer: TuoMuseo).

VR and 3D modeling have been largely applied in the field of archaeology for more than three decades [12]. Much has been written on the potential of these tools for archaeological research [13] and most of projects involving these technologies focus on the communicational aspects of 3D modeling in various media, or on the technological improvements of 3D modeling tools and their use as a tool for scientific archaeological investigation [14].

The application of Virtual Reality to Cultural Heritage was therefore born for a dual purpose: to achieve functional goals and to spread applications of pure fascination. The visualization of places forbidden to the public or no longer existing are examples of applications that aim to satisfy the first objective; the splendidly executed virtual reconstructions, which are able to enchant the public from the emotional point of view, instead, respond to the second objective.

The current trend aims to combine these two aspects, with the objective of realizing virtual scientific reconstructions that are, at the same time, beautiful, increasing the amount of additional content made available to users.

VR becomes a visual portal to digitized resources, a tool for managing complexity for both researchers and general users.

With the continuation of this fruitful relationship, VR and Cultural Heritage begin to combine emotional aspects in the applications, trying to transfer knowledge by appealing not only to the rational but also to the emotional part of the users. The aim is to increase the accessibility to the cultural heritage for the general public, promoting a better sharing and communication of archaeological and historical information. In this sense, a recent experience of virtualizing cultural heritage of the Apulia Region is offered by the DiCET Project [15], where Augmented Reality solutions made available virtual tours of the main ancient monuments of the old town of Lecce during on-site visit.

In recent years, academic research started to analyse the evolution of narrative and storytelling paradigms through the exploitation of heterogeneous media. Henry Jenkins identified such phenomenon as "Transmedia Storytelling: a flow of contents across multiple media channels" [16].

Transmedia Storytelling concept is used when a story or a tale is told by different authors, which share and co-create contents to deliver through different media platforms. Transmedia designers and museum exhibition designers have common interests, starting from the common need to attract different types of people and communities by spreading stories across different media in the view of catching different audience needs [17]. At the same time, they both need to overcome the linearity of narration by offering different points of entry into the storyworld and stimulating the exploration of different fragments of content across multiple media.

Museums are not newcomers to transmedia storytelling: this new way of generating and disseminating content is something that museums have been doing for many decades. For example, when a museum sets up a temporary exhibition, it is usually accompanied by the edition of a catalogue. Exhibition and catalogue are two complementary ways of telling the same story. The difference stays in the way of enjoy it: In the exhibition, the visitor moves through the different section of the exhibition, in the catalogue they enjoy the story by turning and reading pages. Although applied in a very basic and limited way, transmedia storytelling has been a part of museums for years [17].

The combined exploitation of different media and technologies pave the way to innovative approach in cultural heritage dissemination and valorisation.

3   The CHER Framework

The CHER (Cultural Heritage Engineering Revolution) framework aims at modernizing the enjoyment and promotion of cultural heritages, spreading innovative technologies and competences in the creative and cultural industries. It proposes an innovative approach, able to engage visitors through an interactive formative path, which mixes gamification techniques, with video, mobile and immersive technologies.

The CHER framework final demonstrator will be an interactive story which rebuilds ancient worlds and elements, in an engaging and exciting way thanks to the gamification and fascinating technologies.

Elements, whose CHER framework consists of, are:

- A set of cultural heritage digitalized contents that can be combined to create ever new experiences;
- A modular Transmedia Digital Storytelling approach based on gamification logics;
- Mobile gaming, Interactive videos and Virtual Reality technologies.

All the elements composing CHER are combined as reported in the following figure (Fig. 2).

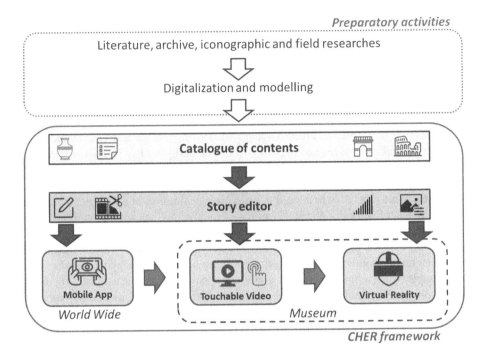

Fig. 2. CHER framework

Central element of the CHER framework is the "Catalogue of contents", collected through literature, archive, iconographic and field researches. These contents are systemised, digitalised and virtually rebuild in order to make them exploitable for a gaming experience. Possible CHER framework contents are buildings, landscapes, clothes, characters, sounds and artefacts.

Contents can be combined for creating different stories by mean of the "Editor". These tool allow intermediate users (e.g. museum curator) to customize some aspects of the gaming experience, which do not have a destructive impacts on the structure of the whole experience, which is instead pre-built. These can include, for example, a soundtrack, a model of an artwork, a minor dialogue, the background of a scenario, etc.

The framework supports the narrative creation, through a transmedia digital storytelling model application. With the purpose of maximising user's involvement in the cultural heritage diffusion, the CHER framework uses three technological directories: mobile gaming, Virtual Reality and Touchable (or interactive) video.

The Mobile gaming is an engagement method already exploited in the museum context, as stated in Sect. 2, since it is reachable by a worldwide audience. In the CHER framework it represent the ideal backbone for the transmedia storytelling approach.

Mobile gaming has a double purpose in CHER: to convey a large part of the story, that for reasons of time and opportunity cannot be completely narrated in a defined physical place, such as a museum, and to encourage the visit of the point of interest (e.g. museum), where the user can take advantage and complete the experience thanks to touchable videos and VR, being, at same time, physically immersed in the scenic context in which the story take place.

Touchable videos are interactive videos that enable the user to interact with the video itself and to access extensive contents. In order to make easier touchable contents individuation and overcome the main weakness of existing application, techniques belonging to the field of cinema, such as slow motion, slow zoom shot, Steadicam shots can be applied.

Virtual Reality allows developing and realizing the scenography able to make immersive final user's experience. The CHER framework involves 3D reconstruction of virtual scenery, which replay historical details, such as building materials, tools and artefacts features, characters' clothes and landscape morphology, to create a really realistic but, at the same time, fascinating experience for the final user.

4 Expected Impact

CHER Framework falls under the cultural contents digitalization, base of new best practices for culture spread. In particular, it aims to guarantee a certain kind of sensory and immersive experience that could change the cultural diffusion, defining different scenarios as path customization (Museum a la carte), contents multi-level follow-up (Matrioska Museum) and the playful experience-based aspect (Play Museum) based on interactivity and multimediality.

Using 3D-modeling, reconstruction and computer animation enhance visitor cognitive abilities. It is a way to efficiently create "righteous" learning processes, stimulating the need for a subjective discovery and consequently enabling new business opportunities. Supposing, for instance, a certain digital services improvement in the Italian museums context, based on relative investigations [18], it has been evaluated that visitors could raise from 180 thousands to 2 million units. This would take, obviously, to more economic incomes, ranging between 700 thousands and beyond 7 million euros. It is therefore expected that a quality enhancement of cultural heritage sharing processes should increase the territory attractiveness with indirect consequences in economics and occupations.

CHER framework requires the connection of a series of different competences, enabling strict synergies and continuous knowledge exchanges and fastening professionalism development, as the storyteller's one, main character in communication and marketing. It should also be considered that an upgrade of cultural heritage enjoyment mode would increase territories attractiveness with indirect consequences on tourism field.

The modular, and so customizable, structure, of CHER framework, locates it across different areas of application: cultural promotion but also for tourism services, fashion and local handicraft.

5   Conclusions

The paper proposes a technological and methodological framework that aims at innovating cultural heritages valorisation by combining mobile gaming, interactive videos and virtual reality, in order to increase users' engagement and, thus, the museum exhibitions attractiveness.

The CHER framework innovates the way cultural heritages are promoted and experienced, moving from current "passive" exhibitions to more attractive ones, where users can actively interact with historical contents, by mean of gamification techniques, immerging themselves into the narrative experiences.

The proposed interactive path starts with a mobile app, in order to catch an audience as wider as possible. By mean of game elements and an appealing story, the user starts to be involved in the historical scenario where the cultural heritages to be promoted belong. Once the user is sufficiently engaged, at the end of the mobile experience, which is still self-conclusive, he is invited to visit the cultural containers (e.g. museums) where touchable videos and virtual reality are used to further develop and extent the interactive information path.

The CHER framework also proposes a modular and flexible system of aggregating contents (3D models, artworks, characters, etc.) allowing intermediate users to customize the experiences, in order to produce ever new and updated stories.

The proposed framework, which will be piloted applied for cultural promotion, could also be applied for delivering services in other business areas such as tourism and fashion. For these exploitation potentials, however, further researches are required.

Acknowledgements. This research was funded by the European Union and the Regione Puglia Local Government through call Innonetwork, project Cultural Heritage Engineering Revolution (CHER).

Bibliography

1. ISTAT, I musei, le aree archeologiche e i monumenti in Italia-Anno 2015 (2016)
2. Osservatorio Innovazione Digitale nei Beni e Attività Culturali. https://www.osservatori.net/it_it/osservatori/osservatori/innovazione-digitale-nei-beni-e-attivita-culturali
3. Il museo "avvicinabile": la strategia social dello Statens Museum for Kunst – Danimarca. http://www.svegliamuseo.com/it/il-museoavvicinabile-la-strategia-social-dello-statensmuseum-for-kunst-danimarca/
4. Impact! The Importance of Digitization at Smithsonian. https://dpo.si.edu/blog/impact-importancedigitization-smithsonian
5. Semitic Museum Fundraises to Increase Digitization. http://www.thecrimson.com/article/2016/3/7/semitic-mueseum-fas-fundraising
6. British museums and art galleries hit by 1.4m fall in visitors. https://www.theguardian.com/culture/2017/feb/02/british-museums-art-galleries-hit-by-2m-fall-visitors
7. Assassin's Creed Franchise Reaches 100 Million Copies Sold. https://www.gamespot.com/articles/assassins-creed-franchise-reaches-100-million-copi/1100-6443544
8. A Practitioner's Guide to Gamification of Education. https://inside.rotman.utoronto.ca/behaviouraleconomicsinaction/files/2013/09/GuideGamificationEducationDec2013.pdf

9. Corallo, A., Esposito, M., Lazoi, M., Marra, M., Sammarco, M.: Innovating cultural heritage promotion through virtual and interactive technologies. In: Proceedings of 3th IMEKO TC4 International Conference on Metrology for Archaeology and Cultural Heritage (MetroArcheo 2017), Lecce, Italy, 23–25 October 2017 (2017)
10. Guccio, C., Martorana, M.F., Mazza, I., Rizzo, I.: Technology and public access to cultural heritage: the italian experience on ICT for public historical archives. In: Borowiecki, K.J., Forbes, N., Fresa, A. (eds.) Cultural Heritage in a Changing World, pp. 55–75. Springer, Cham (2016). https://doi.org/10.1007/978-3-319-29544-2_4
11. Solima, L.: Museums, accessibility and audience development (2017)
12. Remondino, F., Campana, S.: 3D Recording and Modelling in Archaeology and Cultural Heritage. BAR International Series, vol. 2598, pp. 111–127 (2014)
13. Niccolucci, F. (ed.): Virtual Archaeology: Proceedings of the VAST Euroconference, Arezzo, 24–25 November 2000, vol. 1. Archaeopress (2002)
14. Hermon, S., Nikodem, J.: 3D modelling as a scientific research tool in archaeology. In: CAA Conference Proceedings, Berlin, April 2007
15. Gabellone, F.: Integrated technologies for museum communication and interactive apps in the PON DiCet project. In: De Paolis, L.T., Mongelli, A. (eds.) AVR 2015. LNCS, vol. 9254, pp. 3–16. Springer, Cham (2015). https://doi.org/10.1007/978-3-319-22888-4_1
16. Jenkins, H.: Transmedia Storytelling: Moving Characters from Books to Films to Video Games Can Make Them Stronger and More Compelling. MIT Technology Review (2003)
17. Mateos-Rusillo, S.M., Gifreu-Castells, A.: Transmedia storytelling and its natural application in museums. The case of the Bosch project at the Museo Nacional del Prado. Curator: Museum J. **61**(2), 301–313 (2018)
18. Bem Research – Rapporto sull'eTourism 2016. https://www.bemresearch.it/wp-content/uploads/2016/07/Rapporto-e-tourism-2016-bem-research.pdf

Assessment of Virtual Guides' Credibility in Virtual Museum Environments

Stella Sylaiou[1](✉), Vlasios Kasapakis[2], Elena Dzardanova[1], and Damianos Gavalas[1]

[1] Department of Product and Systems Design Engineering,
University of the Aegean, Syros, Greece
sylaiou@gmail.com
[2] Department of Cultural Technology and Communication,
University of the Aegean, Mytilini, Greece

Abstract. Immersive virtual museum environments can play a crucial role for communicating cultural information in an engaging and educational way, especially when storytelling is involved. In this paper we explore and assess the impact that the status of three types of avatars (embodiments of a museum curator, a museum security guard and a museum visitor, respectively) may have on the credibility of their storytelling and the emotions they evoke to virtual visitors. Preliminary results derived from an experiment provide evidence that supports our initial hypothesis that the status of the avatar may indeed influence their credibility and the participants' emotions of sadness/worry, distress, shame, anger, relief and admiration.

Keywords: Virtual museum · Avatars · Virtual guide · Social presence · Credibility · Emotions

1 Introduction

Virtual Museums (VMs) are digital spaces that draw on the characteristics of a museum, in order to complement, enhance, or augment the museum through personalization, interactivity, user experience and richness of content [1]. They are mainly used as effective solutions for communicating cultural content and context via an entertaining and educational approach. With the help of ICT, they tell stories, they address the problem of accessibility, while also providing engaging experiences for different audiences [2]. Virtual museums adoption and experimentation range from interactive to immersive experiences involving cutting-edge technologies, such as Virtual, Mixed and Augmented Reality [3], and new concepts, such as virtual embodiment and storytelling by avatars in forms of Virtual Humans (VH) [4]. VH research explores social interactions between real and virtual humans [5]. The personality and the behavior of a VH can influence the quality of the virtual experience, make the stories presented more credible and perhaps influence the perceptions of the real human.

In this paper, we investigate the perceived credibility of three virtual avatars possessing different characteristics (with respect to their professional and social status) in an immersive virtual museum environment and the emotions they arouse. Within the

L. T. De Paolis and P. Bourdot (Eds.): AVR 2019, LNCS 11614, pp. 230–238, 2019.
https://doi.org/10.1007/978-3-030-25999-0_20

context of this experiment the term credibility is not aligned to notions of truthfulness or trust, but refers to the avatars' ability to engage the responders, mostly at an emotional level. A dramatic story about a museum exhibit, the sculpture of Arria and Pætus, is narrated by two virtual avatars that have different social distance and by one that has the same social distance from the participant. The remainder of the paper is structured as following: Sect. 2 explores the related work on social presence and how various factors may influence the credibility and the persuasiveness of avatars. Section 3 details the experimental procedure. Section 4 presents and discusses the research results. Section 5 reports on our conclusions and suggests directions for future research.

2 Related Work

According to the media equation theory, media interaction is essentially no different to the interaction with an actual person in psychological terms. As a general communication theory delineated by Reeves and Nass [6] it provides insights into the fact that people tend to interact under the same mindset irrespective of whether they are dealing with, e.g., an avatar, a VH, or a real human. As Liew and Tan [7] put it, 'people will unconsciously treat media technologies as social beings, such that people's behavioral and emotional responses to computers would adhere to human-to-human social norms' [6, 8]. This understanding forms the basis of a discussion on credibility issues and power-relation balances which affect the interaction between the participants in the experiment and the three avatars, in the sense that the psychological as well as social mechanisms that are in place regarding these factors are no different in a virtual environment as they are in 'real life' situations. Hence, the responses of immersed users interacting with VHs are congruent if not identical to those that would ensue from interaction with real persons having the equivalent roles. This brings forth the issue of whether the avatars are indeed persuasive as entities representative of specific roles and/or social status, namely those of a curator, of a museum security guard and a visitor.

Appearance and congruence between the attire of an avatar and the assumed social role is a key factor as Parmar et al. [9] posit in their examination of responses to avatars of physicians with clothing ranging from casual to formal and strictly professional. This illustrates the fact that the more the image of the avatars adheres to the stereotype associated with their assumed professional role, the better they have been accepted with respect to credibility. Interestingly enough, the responses in terms of avatar persuasiveness from the persons involved in the experiment have been more positive towards the avatars stereotypically dressed (with attire most appropriate and expected from a physician) irrespective of how 'uncool' or 'different to them' the avatars have been in style (ibid). This shows that adherence to appearances signifying specialization are preferable even if they generate social distance between the style/appearances of the people involved in the experiment and, by extension, the users of a Virtual Reality (VR) resource where avatars communicate verbally an account or information from a position of those who possess specialists' knowledge. Embodied virtual agents are computer-generated visual characters that simulate assistants [10] and due to their exponentially increasing use (mostly in commercial context), a substantial body of research on their potential, optimal characteristics and overall function is undertaken. As Liew and Tan [7] put it 'they are capable of conveying verbal and nonverbal cues through animated facial expression, body gestures, and text-to-speech dialogues' thus humanizing and rendering emotionally engaging and agreeable experience of the human-avatar interface [11–14]. Although this research has focused on commercially-oriented context, the findings and insights are applicable to human-avatar interfaces where persuasiveness and credibility is of paramount importance as is the case with the experiment findings on which our paper hinges on.

Embodied virtual agents conveying information as acting and knowledgeable subjects (as it is the case with the herein presented experiment) are not only perceived as able to persuade, in accordance to their specialization and relevance to the topic they present/convey, but all the more, their assumed specialization affects the very way they are perceived: peoples' perception tends to ignore elements that are incongruent or diverging from their perceived profile. Respectively, elements, information or aspects of their accounts which enhance or correspond to their perceived specialization are subconsciously foregrounded and given more attention by the experiment participants. This phenomenon which filters out - to an extend - what contradicts the expectations that human actors have form avatars or embodied virtual agents, develops in a heuristic way; thus, it changes reality that is contradicting expectations to the extent of 'ignoring information that is/are seemingly dissonant with the stereotype' [7]. In other words, the minor differences in register, tone and vocabulary used by the three avatars in our experiment which are expected to make their roles more persuasive as such, will be further altered upon perception to match the stereotypes of their assumed roles/identities. The fact that the museum visitor has no professional profile whatso-ever, is hypothesized that will decrease that avatar's ability to take advantage of the added credibility automatically bestowed and subconsciously enhanced when author-itative, erudite and specialist embodied virtual agents are addressing human subjects. The added symbolic and social capital associated with the role of the curator is expected to attract more persuasiveness and attention.

However, given the nature of the emotionally charged account, there could be a question as to whether specialist knowledge will prevail over a more humanized and more socially similar and thus approachable generic avatar in the form of a museum visitor, which might elicit the element of perceived identification with the visitors of the virtual museum/exhibit who partake in the experiment. The hypothesis is that, given the prevalence of specialization over social proximity as illustrated by Parmar et al. [9], this is unlikely to happen even though the topic is not directly related to the trans-ference of technical information or guidance but is more relevant to subjective emo-tional states generated by the touching story immortalized in the form of the statue presented in the virtual museum environment. Carrozzino et al. [15] explored a virtual museum with three alternatives of storytelling, including one featuring VH, and compared the research results in terms of engagement and understanding of the pro-posed content. According to them 'findings confirm the hypothesis that an embodied virtual agent is able to stimulate attention and involvement, and contributes to a better content delivery and learning [ibid, p. 301]. Despite the fact that this is not surprising as it confirms the initial hypothesis of their research, it nevertheless consolidates the importance, value and potential of VHs as guides in virtual museum settings, which in turn, makes the research on how to optimize VHs in such contexts imperative.

The Expectancy Violations Theory [16] posits that when the expectations form an agent with specific social characteristics such as assumed expertise are not met (e.g. when an expert provides unconvincing information or even appears visually incongruent with her status or role) then strong negative emotions are elicited and distrust ensues as result. The opposite applies though: according to research [ibid], when expectations are exceeded as it is the case when a non-specialist provides accounts showing surprisingly high expertise that is beyond their, e.g., professional background, then an exceedingly positive response in terms of liking, credibility and trust might be generated at the human actors' level. Therefore, this factor can influence the response towards the only non-specialist avatar who nevertheless provides an account essentially of the same information-value as with those of the more expert/relevant roles. In other words, when an avatar which is supposedly, e.g. an art historian does not look or behave like one will engender disappointment, whilst an avatar which is not an art historian provides guidance of surprisingly high quality for a non-specialist might generate empathy and trust -a factor that complicates, yet, adds to the interest and value of this experiment.

3   Materials and Methods

3.1   Apparatus and Visual Content

The experiment has been conducted in the Laboratory of Virtual Reality of the Department of Cultural Technology and Communication, University of Aegean, Greece. The hardware used in this study included: (a) headphones, (b) a VR - Ready PC with Nvidia Geforce GTX 970 card, (c) an Oculus Rift™ [17], a Virtual Reality headset, with a field of view (FoV) of 100° that supported participants' head movement tracking, in order to facilitate them to look around the virtual museum based on their head movement (the frame rate has been 80 frames per second [fps] and the resolution 960 × 1080 per eye), (d) an Xbox 360™ controller, and (e) headphones for hearing the voice of the virtual guides. The experimental setup is presented in detail in [18]. For the case study, we have created a virtual museum room with one 3D exhibit, an artwork of the 17th century that depicts Arria et Pætus from Louvre Museum[1]. We prepared three variants of the same scenario during the visit to the virtual museum. In particular, three different high-detail human avatars narrate the story of Arria et Pætus: a female **curator** of the exhibition; a male **museum security guard** of the exhibition; a female **visitor** of the exhibition. Avatars behave accordingly to their status in order to engage the virtual visitors. To increase the level of realism we have added movement of the body, hands, and lips sync according to the spoken text. Additionally, the avatars make direct eye contact with the participants.

---

[1] The 3D exhibit has been downloaded from: https://sketchfab.com/models/
e5dc1871b7654429b883b9e04c8418c4.

3.2   Participants

Forty-five (45) volunteers (22 males, 23 females, age range: 22–41) mainly post-graduate students from the University of Aegean, Department of Cultural Technology and Communication, participated in the experiment. Three experiments have been conducted and homogeneous groups of 15 users that belong to the same age group and possess similar cultural background, knowledge and VR experiences (medium knowledge in History of Art and 1–3 immersive experiences) participated in each of them. In this way (i.e., through assigning different experimental scenarios to disjoint participant groups) we have avoided: (a) biased attitudes and opinions - towards the different virtual guides - induced due to the repetition of the same experimental scenarios (i.e., earlier VR experiences would affect the subsequent ones); (b) fatigue (due to participating in many experimental sessions and having to fill-in many questionnaires) which would compromise the validity of participants' feedback. It is noted that participants have been ignorant as to the purpose of the experiment.

3.3   Experimental Procedure

Three steps have been undertaken: briefing about the experiment (users start with a plan of the tasks to be accomplished), exploration (users explore freely the virtual space and test their capabilities and affordances of the virtual environment) and assessment (users interpret the avatars' stories and assess their credibility). The experiment has been conducted in a comfortable museum room with one door and only one exhibit, the sculpture that depicts the story of Arria and Pætus, artificially lighted. We decided to keep the room as simple as possible, without windows and other exhibits, to avoid complexity that could distract participants' attention. Firstly, each participant is invited to complete a demographics questionnaire with information about their age, gender, education, knowledge about art, number of immersive experiences, whether they visit museums and how often. Then, they are briefly informed about the context of the experiment (less than 1').

Then the participant may freely explore the virtual museum room. The narration opening is not sudden, since it starts after the eye contact of the immersed user with the avatar. When the user approaches the exhibit, an avatar is introducing itself and starts narrating a story about the exhibit (duration of 1–2 min). Specifically, the avatar narrates the dramatic story of Arria et Pætus in ancient Rome (AD 42). Pætus, a Roman senator, was condemned to death for his role in a revolt against emperor Claudius. A suicide would be a noble death; however, he was afraid to commit it. His wife, Arria visited her imprisoned husband, stabbed herself first to encourage him and then passed the dagger to her husband saying *Pætus, non dolet!* (Pætus, it doesn't hurt!) [19].

On each experimental session, one of the three avatars assumes the role of a guide and starts his/her narration. The avatar of the curator conveys the story of the couple in a flat way and she emphasizes to the historical information, formal analysis of the artwork, and the description of the sculpture's features using scientific terms. The avatar of the museum security guard tells the story of the couple using simple words underlining the courage and the internal power of Arria. The avatar of the visitor provides some personal information about the couple (e.g. the fact that their son had died, while Pætus was in prison, but Arria never told him the truth, in order not to worsen his sentimental condition) highlighting the emotions, such as love, sadness, fear etc. Having completed the immersive experience, each participant filled in a second questionnaire to specify his/her impressions about the story s/he has heard and more specifically, in what degree s/he experienced the emotions of sadness/worry, distress, shame, anger, relief and admiration, how convincing the story was, and if s/he considers that the narrator had broad and/or deep knowledge of the artwork.

3.4   Methods

The presented study investigates the impact that the status of three types of avatars (embodiments of a museum curator, a museum security guard and a museum visitor, respectively) has on the credibility of their storytelling and the emotions (sadness/worry, distress, shame, anger, relief and admiration) they evoke to virtual visitors. More specifically, our research explores whether there is: (a) a correlation between the three avatars, (b) a correlation between emotions, (c) a correlation between the avatars and emotions. The statistical package SPSS v.21 has been used for the data analysis. The primary aim of the analysis was to study the influence/effect of each of the three avatars to the six emotions under study (1): Sadness/worry, (2): Distress, (3): Shame, (4): Anger, (5): Relief, (6) Admiration. For this purpose, the repeated measures design ANOVA[2] [20] has been used. If sphericity is violated, then automatically correction methods of F-ratios are applied. We have considered the Greenhouse–Geisser correction. All the tests have used a significant level of .05.

4   Results and Discussion

According to the descriptive statistics of the sample, the emotions of shame and anger appear lower mean scores, whereas admiration is the predominant emotion regardless the type of the avatar. The results of repeated measures ANOVA, with corrected F values (Greenhouse-Geisser), have shown that there is no important effect of the type of the avatar ($p = 0.269$). However, there is a significant effect of the type of emotion ($p = 0.000$) and of the interaction between the avatar and the type of emotion ($p = 0.031$).

(a) *Correlation among the avatars*
    The pairwise comparisons with the Bonferroni correction for the three avatars show that there is no significant statistical difference between them ($p > 0.05$), even though in Fig. 1 the curator seems to have the lowest mean of emotions from the other two avatars (museum security guard and visitor).

(b) *Correlation between emotions*
    In the sample we observe that the greatest mean is in admiration (4.178) and the lowest in shame (2.000). Then, we examine correlations between emotions,

---

[2] Repeated measures design is a research design that involves multiple measures of the same variable taken on the same or matched subjects either under different conditions or over two or more time periods. ANOVA is a commonly used statistical approach to repeated measure designs.

independently from the type of avatar factor. Most of the emotions are positively correlated between them, besides the admiration that is negatively correlated with all the emotions, except the emotion of sadness/worry.

(c) *Correlation between avatars and emotions*

As it is shown in Fig. 1, there is an interesting observation that the effect of the three avatars' narrations to participants' emotions, as well as the changes in each emotion depends on the type of the avatar.

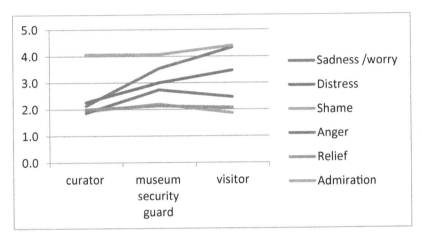

Fig. 1. Interaction graph between avatars (vertical axis) and types of emotions (horizontal axis)

Thus, it is clear that the curator yields the lowest scores on all types of emotions. Our initial hypothesis has been confirmed and we conclude that the museum curator is perceived as the most indifferent narrator. Her narration may attract more persuasiveness and attention. However, the fact that she chooses to highlight the historical information, formal analysis of the artwork, and the description of the sculpture's features using scientific terms, does not have any effect on participants' emotions.

It seems that the museum security guard's and the visitor's stories arouse participants' emotions. The museum security guard's description provokes more intense feelings, compared to the curator, apart from admiration, in which they are on the same level. Finally, the visitor's narrative, that discloses personal information about the couple, has an even greater impact on increasing sadness, despair, and admiration. But in feelings like shame, anger and relief, the curator and the visitor score lower than those of the museum security guard. In order to better interpret this difference between the three narrators, the question "What impressed you most in the story you heard?" has provided insightful results. It was observed that in the description of the museum security guard, words such as "drama", "sacrifice", "suicide" have been spoken, while during the narrative of the visitor, emphasis was mostly placed on "drama", while some participants stated that they were attracted by the graphics and the design. This means that the character of the curator was fond rather indifferent, the museum security guard very dramatic and the visitor attractive and more natural.

5 Conclusions

This paper presents the initial results of the experiment concerning the credibility of various types of virtual guides in a virtual museum environment. These high-detail human avatars (a female museum curator; a male museum security guard; and a female visitor of the exhibition) narrate in different ways (using scientific terms, underlining the courage, and highlighting the emotions respectively) the dramatic story of the sculpture that depicts Arria et Pætus. We have tested the emotions of sadness/worry, distress, shame, anger, relief and admiration they have felt in each case, which provided insights in respect to how convincing every narrator-avatar was and more importantly what specific effect every VH seemed to have in terms of persuasion which accrues from avatars' potential for emotional engagement. Some interesting information has been revealed concerning the fact that the narration of the museum curator has not evoked strong emotions to the participants. Although admiration (for the brave action of Arria) is high irrespective of virtual narrator, emotions such as anger, and especially distress (sense of despair and helplessness), sadness/worry (feelings of sadness) were far more common and prevalent when the narrator was the virtual museum visitor. This means that when the narrator is perceived as a specialist, feelings of admiration or appreciation are almost equally present, but the narrator who is closer to the status of the responders (something reflected to her more communicative style that does not convey higher status or level of expertise), elicit stronger emotions of empathy and personal emotional involvement. These differentiations can inform practices of virtual museums and help them to optimise the characteristics of the virtual guides they might wish to use, according to what they want to achieve in terms of audience engagement. As future research directions they are planned more experiments that will use more experimental variables, like additional visitors, duration of narration, while they will be recorded and analysed more quantitative metrics, like proximity and duration of eye contact among the user and the virtual guide.

Acknowledgment. This research is co-financed by Greece and the European Union (European Social Fund - ESF) through the Operational Programme "Human Resources Development, Education and Lifelong Learning 2014–2020" in the context of the project "Social Interaction in Virtual Reality Environments" (MIS 5004223).

References

1. The ViMM Definition of a Virtual Museum, ViMM. https://www.vi-mm.eu/2018/01/10/the-vimm-definition-of-a-virtual-museum/
2. Sylaiou, S., Liarokapis, F., Kotsakis, K., Patias, P.: Virtual museums, a survey and some issues for consideration. J. Cult. Heritage **10**(4), 520–528 (2009)
3. Sylaiou, S., Kasapakis, V., Gavalas, D., Dzardanova, E.: Leveraging mixed reality technologies to enhance museum visitor experiences. In: Proceedings of the 9th IEEE International Conference on Intelligent Systems (2018)
4. Papagiannakis, G., et al.: Mixed reality, gamified presence, and storytelling for virtual museums. In: Lee, N. (ed.) Encyclopedia of Computer Graphics and Games, pp. 1–13. Springer, Cham (2018). https://doi.org/10.1007/978-3-319-08234-9

5. Kim, K., Maloney, D., Bruder, G., Bailenson, J.N., Welch, G.F.: The effects of virtual human's spatial and behavioral coherence with physical objects on social presence in AR. Comput. Animat. Virtual Worlds **28**(3–4), e1771 (2017)

6. Reeves, B., Nass, C.: The Media Equation: How People Treat Computers, Television, and New Media like Real People and Places. Cambridge University Press, Cambridge (1996)

7. Liew, T.W., Tan, S.M.: Exploring the effects of specialist versus generalist embodied virtual agents in a multi-product category online store. Telematics Inform. **35**(1), 122–135 (2018)

8. Isbister, K., Nass, C.: Consistency of personality in interactive characters: verbal cues, non-verbal cues, and user characteristics. Int. J. Hum.-Comput Stud. **53**(2), 251–267 (2000)

9. Parmar, D., Olafsson, S., Utami, D., Bickmore, T.: Looking the part: the effect of attire and setting on perceptions of a virtual health counselor. In: Proceedings of the 18th International Conference on Intelligent Virtual Agents, pp. 301–306. ACM (2018)

10. Holzwarth, M., Janiszewski, C., Neumann, M.M.: The influence of avatars on online consumer shopping behavior. J. Mark. **70**(4), 19–36 (2006)

11. Beldad, A., Hegner, S., Hoppen, J.: The effect of virtual sales agent (VSA) gender–product gender congruence on product advice credibility, trust in VSA and online vendor, and purchase intention. Comput. Hum. Behav. **60**, 62–72 (2016)

12. McGoldrick, P.J., Keeling, K.A., Beatty, S.F.: A typology of roles for avatars in online retailing. J. Mark. Manag. **24**(3–4), 433–461 (2008)

13. Qiu, L., Benbasat, I.: Evaluating anthropomorphic product recommendation agents: a social relationship perspective to designing information systems. J. Manag. Inf. Syst. **25**(4), 145–182 (2009)

14. Jin, S.A.A., Bolebruch, J.: Avatar-based advertising in second life: the role of presence and attractiveness of virtual spokespersons. J. Interact. Advert. **10**(1), 51–60 (2009)

15. Carrozzino, M., Colombo, M., Tecchia, F., Evangelista, C., Bergamasco, M.: Comparing different storytelling approaches for virtual guides in digital immersive museums. In: De Paolis, L.T., Bourdot, P. (eds.) AVR 2018. LNCS, vol. 10851, pp. 292–302. Springer, Cham (2018). https://doi.org/10.1007/978-3-319-95282-6_22

16. Afifi, W.A., Burgoon, J.K.: The impact of violations on uncertainty and the consequences for attractiveness. Hum. Commun. Res. **26**(2), 203–233 (2000)

17. Oculus Rift™. https://www.oculus.com/

18. Kasapakis, V., Gavalas, D., Dzardanova, E.: Creating room-scale interactive mixed-reality worlds using off-the-shelf technologies. In: Cheok, A.D., Inami, M., Romão, T. (eds.) ACE 2017. LNCS, vol. 10714, pp. 1–13. Springer, Cham (2018). https://doi.org/10.1007/978-3-319-76270-8_1

19. Arria and Pætus, Louvre Museum. https://www.louvre.fr/en/oeuvre-notices/arria-and-paetus

20. Keselman, H.J., Algina, J., Kowalchuk, R.K.: The analysis of repeated measures designs: a review. Br. J. Math. Stat. Psychol. **54**(1), 1–20 (2001)

Immersive Virtual System for the Operation of Tourist Circuits

Aldrin G. Acosta[1(✉)], Víctor H. Andaluz[1(✉)],
Hugo Oswaldo Moreno[2(✉)], Mauricio Tamayo[3(✉)],
Giovanny Cuzco[4(✉)], Mayra L. Villarroel[1,2,3,4],
and Jaime A. Santana[1,2,3,4]

[1] Universidad de las Fuerzas Armadas ESPE, Sangolquí, Ecuador
{agacosta,vhandaluzl}@espe.edu.ec
[2] Escuela Superior Politécnica de Chimborazo, Riobamba, Ecuador
h_moreno@espoch.edu.ec
[3] Universidad Técnica de Ambato, Ambato, Ecuador
fm.tamayo@uta.edu.ec
[4] Universidad Nacional de Chimborazo, Riobamba, Ecuador
gcuzco@unach.edu.ec

Abstract. This article presents the results of the development of an immersive VR application that serves as a virtual guide to the tourist within the circuits programmed in a visit design. It presents the structure of programming and 3D digitization process through which virtual scenarios were built through the APP allow to enjoy the immersive experiences within the tours programmed by the user according to the attractions, services and facilities of the destination; allowing to select the type of tourism, choose the site you want to visit and the time you have to make the visit, showing the location within the destination and projecting the circuit that may travel with the points where the tourist avatar will make, generating an interactive experience to tourists in scenarios and virtual tours marketed as new products making the destination more competitive and categorizing them as spaces with facilities for the development of e-tourism.

Keywords: Immersive experience · 3D unity · Virtual guide · Tourist circuit

1 Introduction

The use and incorporation of new technologies in the tourism sector have contributed significantly to the growth of tourism, which is reflected in the 10.2% of the global GDP, representing $ 7.6 trillion, figures generated by the movement of 1200 million tourists around the world in 2016; by 2017 tourist flows reached 1323 million representing a 7% growth over the previous year. Europe with 50.71% was the continent with more visits, while in America the number of tourists was 207 million which represents 15.65%. Figures showing the positioning of the tourism industry and adaptation to changes and technological trends [1, 2].

This great demand of tourists looks for destinations that offer a wide and diversified offer of tourist circuits that generate recreational experiences among which there are

L. T. De Paolis and P. Bourdot (Eds.): AVR 2019, LNCS 11614, pp. 239–255, 2019.
https://doi.org/10.1007/978-3-030-25999-0_21

activities and services that supported with elements and technological applications make the visit and stay more dynamic. This has been achieved since tourism companies have incorporated in their processes of management, operation, promotion and marketing the use of information and communication technologies (ICTs) encouraging new ways of tourism, emerging and solidified digital services driven by networks, systems and applications that make more dynamic a destination and its circuits, generating radical changes [3, 4].

These substantial changes are a consequence of the use of immersive technologies that have propitiated new tourist products through virtual reality (VR) which generates an enveloping experience resulting from the capacity of physical and psychological simulation that is achieved by recreating a credible scenario and provoking a tourist avatar in which the tourist manages to simulate the use of the five senses in a sensory way at the moment he explores and interacts with the environment recreated in 3D. For this reason tourists have shown great interest in virtual guides and immersive experiences that make more attractive and interactive tours and sites of visit thus making the destination more attractive and competitive [5–7].

In this context, the Spanish government, through the Sociedad Estatal para la Gestión de la Innovación y las Tecnologías Turísticas SEGITUR, has developed the Tourist Intelligence System that converts the data generated by tourists into a digital information base that is put at the service of companies, operators and tourism managers so that they can make decisions in real time and efficiently manage the destination by incorporating digital services and immersive technologies in tourist visits. The business of immersive technologies in Spain has grown 86.6% in the last two years, underpinning it in the market for the development of intelligent technological systems [8, 9, 10].

The immersive technologies used in a tourist visit allow the tourist to interact in a reconstructed environment with 3D models implementing objects such as facades, roads, species of the place, etc.; and assigning characteristics specific to the site of visit (climate, gravity, seasons, etc.), from which immersive virtual environments and systems are created that generate the recreational experience and configure tourist circuits that are constituted in routes that induce the displacement of an initial point of attraction towards another point of visit, showing the attractiveness of the route as well as the hotel facilities and equipment that exists along it, and that is established in a new offer of tourist products and services with technological support [11].

Technology and tourist activities are factors that make a destination competitive and that according to the study of countries with the highest economic competitiveness in the world, to select a site where to arrive and make tourism, it must be considered that so productive in terms of services and attractions, is the area or region chosen, where technology and information are essential factors to achieve the productivity of a territory and boost the profitability of services of accommodation, food, guidance and transport, which promoted within a circuit and tourist package are constituted in the offer of a destination [12–14].

This article is divided into 5 sections: 1. Introduction, which contextualizes the actions of tourism and the VR approach as part of technological trends, presenting the problems of research; 2. Description of the System, which refers to the structure of the system related to tourism with the recreational experience generated by VR.

Development of the System, which explains the procedure for promoting 2D and 3D virtual environments; 4. Results, where the experimental performance of the application of VR oriented to the virtualization of tourist environments is presented; and 5. Conclusions of the investigation.

2   System Description

When talking about tourism, it is inevitable to look at its functioning as a system where tourists move around using routes and circuits that connect them not only with the most attractive resource in the locality, but also with the one that generates a recreational experience depending on the destination through which they traveled and are visiting. This determines that the tourist system is configured by three main elements that are: tourist - circuit/routes - destination, and that for the present investigation are constituted in the components on which the system of blocks is structured and of 3D programming that mark the relational line to create the recreational experience in VR [15, 16]. See Fig. 1.

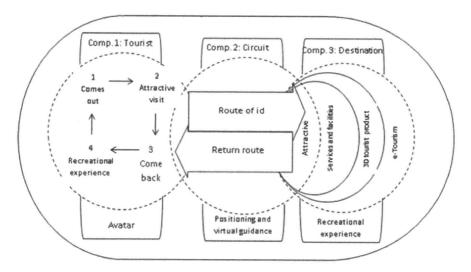

Fig. 1. Relational system to structure the recreational experience VR.

These three components are those that will intervene with immersive technologies, to highlight the characteristics of the destination, as well as the path that leads to and within it, and on which the 3D interaction is generated for the tourist to make the tourist avatar which will be shown as the e-tourism product of the visit destination.

Component 1: Tourist. The user of the application is the tourist who travels and recovers the destination through the circuit, and is the consumer of the VR service and who identifies himself as the Avatar within the recreated scenarios virtually on whom the recreational experience is generated.

Component 2: Circuit. The environment in which the tourist is virtually guided, acceding to a database that facilitates his travel and points out his positioning leading him to the destination attraction where he can make the tourist avatar.

Component 3: Destination. The area where the e-tourism circuits that are constituted in the 3D products generated by the virtualization of attractions, services and facilities creates the recreational experience of the tourist in the destination of visit.

With this structure it is possible to enhance the tourist resources existing in the destination with informative contents and graphics that show in the reconstructed scenarios highlight the characteristics of the resource, making it more interactive and turning it into a virtual guide that will allow you to know the environment and choose an adjusted service to your preferences and personality; increasing interest in those destinations that incorporate technological facilities in their tourism products [17] (Fig. 2).

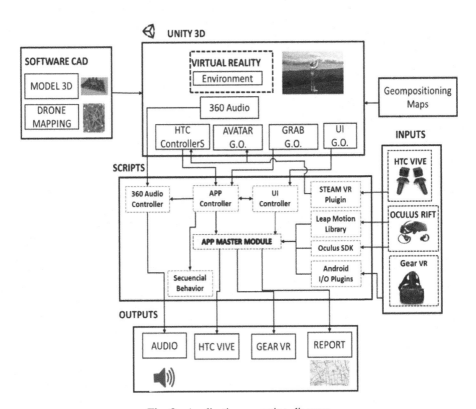

Fig. 2. Application operating diagram.

CAD software allows you to digitize virtualize the environment through the use of 3D models with their corresponding objects such as: facades, trees, shrubs, roads, windows, and native species of each place, which are implemented in the simulation scenario for the suitable immersive environment and will have a realistic user experience.

The Scene VR contains the resources obtained from CAD software represented by generic blocks called Games Objects GO, these elements are assigned characteristics of the environment to be simulated as: climate, gravity, season of the year, among other physical characteristics, in the main scene it is deployed the execution and use components of the application that allow the user to enter the touristic category and the estimated time of travel, with these data the scene generates a tourist route by which the user will be guided in his journey.

The set of Scripts contains the code to manage the data generated by the user. To interact with the input devices, the Steam VR Plugin and Oculus Library modules are used to interact with HTC VIVE, Oculus Rift and GearVR devices. In order to have a geo positioning in the virtual environment, a web service is consumed, which returns the latitude and longitude values of the tourist route generated to reach the user's destination

The Inputs contain the devices to access the simulated virtual reality environment, the devices are selected according to the generated route with the data entered by the user, so: HTC VIVE and OCULUS RIFT is used when the work space is large and fine traking is required for mobilization in the virtual environment and Gear VR is ideal when mobility and easy access are required.

In the Outputs you get the spatial audio output to have a better experience between the user and the simulated environment, a general module that indicates the route generated according to the category and estimated time.

3   Virtual System Development

The development of the proposed system is presented in Fig. 3, in which three main stages are considered in which a specific task is defined, in addition, of one or several computational packages that allow executing the tasks of the workflow: (i) Layer 1 in this layer is the construction of 3D models and the digitization of tourist surroundings; (ii) Layer 2 is in charge of modifying the characteristics of axes, orientation of the 3D object and modification of meshes of the point cloud of the digitized model of the environment; and finally (iii) Layer 3 has as objective the graphic edition that recreates the environment and the codification that establishes the behavior of each environment, giving as final product a build of the application.

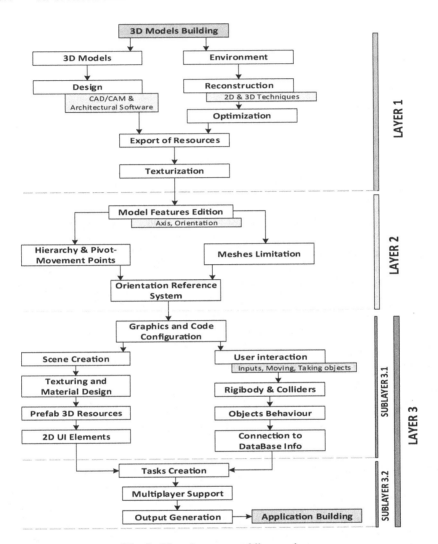

Fig. 3. Virtual system multilayer scheme.

(i) **Build 3D Models**, in this block we mainly consider two types of 3D models that are built specifically for the application: 3D objects and tourist environments. 3D objects are built using architectural software packages (sculptures and urban resources) and CAD/CAM programs (replicas of 3D mechanical objects). Tourism environments are obtained from real locations through 2D and 3D digital reconstruction techniques. Figure 4 shows the scheme implemented for the virtualization of environments and tourist attractions, which can be subdivided into four main stages:

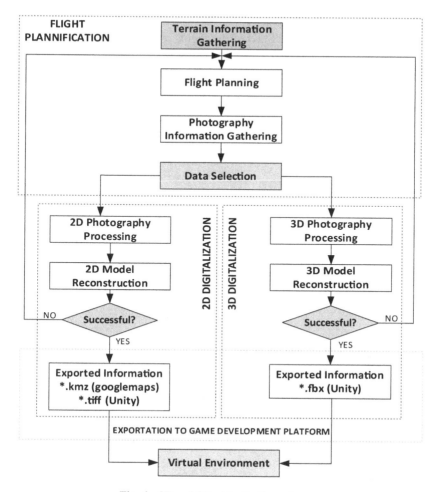

Fig. 4. 2D and 3D virtualization scheme.

(a) Flight planning will be carried out according to the selected sites and resources of the tourist destination to be intervened; the information is collected through an unmanned aerial vehicle, UAV, which consists of a viewing camera positioned at an angle of 90° for the 2D case and with different inclinations in a range of 45° to 80° for the 3D case. The planning of the autonomous flight of the UAV must consider the characteristics of the terrain, (maximum height of the existing objects are these monuments or buildings), flight space (perimeter of the area) and natural conditions (clear day with high natural luminosity and little concurrency of people); The result of the survey has a big data of flat pictorial representations, the same ones selected, will serve to perform the superposition of images to generate the 2D and 3D virtual model, see Figs. 5 and 6.

(b) 2D Virtualization allows to project flat images in two dimensions (X - Y) with updated georeferenced information of the site and tourist resources. The

(a) Planning for 2D digitization

(b) Planning for 3D digitization

Fig. 5. Autonomous flight planning for 2D and 3D the reconstruction.

information can be stored in a computing cloud in order to generate interactions with the Google Maps Earth search engine and in this way the user can in real time define routes and places of visit with geographic positioning data as well as general characteristics of the site (climate, height, among others). With the photographic information captured with a constant firing angle of 90°, the reconstruction software finds coincidences between edges of each one of the figures in order to generate a general model of all the captures. Figure 7 presents the two-dimensional reconstruction of the University of the Armed Forces ESPE, the result of processing in the Agisoft photo Scan software.

(c) 3D virtualization combines flat images obtained through different routes (double grid - circular and free flight) in order to generate graphic

Fig. 6. Uprising of information with different angles of inclination of the UAV vision camera.

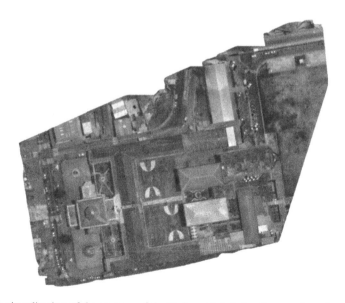

Fig. 7. 2D virtualization of the campus of the Universidad de las Fuerzas Armadas, Latacunga, Cultural Heritage of the Cotopaxi province -Ecuador.

(a) Dome of the administration building

(b) Facilities for classrooms and University Stadium

Fig. 8. 3D Scanning of Campus of the Universidad de las Fuerzas Armadas ESPE.

representations in three dimensions (X - Y - Z) that will serve to recreate existing tourist attractions or modify them virtually by means of the incorporation of objects that enrich the tourist resource and promote the characteristics of attraction for the visitor. For a three-dimensional reconstruction, photographs taken from different angles, 45° and 80°, are used in the same location, so that heights and depths can be estimated through the variation of camera rotation. Through the information acquired from the metadata of each image, the Autodesk ReCap Photo software reconstructs a three-dimensional scenario similar to the real one. Figure 8 shows 3D scanning.

(d) Optimization model consists in the unification of reconstructed 3D models with the 2D models of the digitized place to soften the mesh, eliminate noise created by the reconstruction, group the model layers, smooth, add colors and textures to have a real visual experience; using the SKETCHUP program in which you enter a .obj file of the reconstruction shown in Fig. 8 and obtain the optimized models indicated in Fig. 9

(a) 3D structure generated in the reconstruction

(b) Final result of optimization

Fig. 9. Optimization of the model generated in 3D scanning.

(e) Export to 3D graphics engine, consider the use of images from the 2D digitization block with kmz extension to use them georeferenced in Google Earth, which updates the view of the scanned area; while for use in Unity it is exported with a .tiff extension as a reference to the exact location of an architectural structure developed parallel to the image. Finally, the models of the 3D scanning stage are used in Unity with the extension fbx, resulting in a virtual environment on which the functionalities of the application are implemented, see Fig. 10. The captured photographs are not imported directly into the 3D graphic engine, but rather their two-dimensional and three-dimensional reconstruction. The re-construction software generates different files that can be imported into Unity 3D, being in * .tiff format and * fbx those delivered by agisoft and reCAP photo, respectively. The first * .tiff format is a superior image of the whole campus, in which there are no elevations, only flat photographs. The second *.fbx format includes elevations and can be used to navigate in the reconstructed environment. Figure 10 presents the export to Unity 3D.

Fig. 10. Export of the campus of the university to the unity 3D graphics engine.

(ii) **Edit Model Properties**, in this layer the hierarchy of elements is implemented by applying the parent-child model, the movement of the pivot point towards the concentric point of the object (facilitates direct manipulation of the object) for the pieces or set of pieces that make up a 3D model. In the case of the digitized model of the scenario, the limitation of meshes is done (deleting excess faces, edges and lines) for reasons of stylizing the chosen area. In addition, the orientation correction of the central and pivot axes of each model is made towards the global reference system.

(iii) **Graphic and coding,** this layer is developed in the Unity3D graphic engine, which is subdivided into two main sublayers; the sub-layer 3.1 considered sequential processes where graphic and code editors are used.

(a) *Graphic editor*, the layout of the scene is done by placing objects in the desired position, the texturing of the 3D objects within the diagram and the creation of materials for interaction with the user and with each other. As the objects take shape within the environment it is necessary to create prefabs to automate the changes created in a model to the other Game Objects descendants of the Pre-fab; the user interface is implemented with internal and/or external resources (to the resources granted by default in Unity), the text balloons show information on the status of the application and help when there is no response from the user for a long time.

(b) *Code editor* starts with the configuration of inputs, movement functions, locomotion and grabbing objects. Next, the characteristics of Rigibody and Colliders that give the model the ability to be affected by gravity and detect collisions respectively are added. The previous block allows you to program the behavior of each object including animations resulting from the user's interaction with said model. Finally, the connection to the database is made to consume dynamic data such as descriptions, messages, warnings, among others

4 Experimental Results

The application presents an initial welcome screen to the Tourist Destination. Here the user enters his name and the reason for his visit within which he selects the type of tourism that he wishes to carry out according to the following categories: Ecotourism and nature, sports and adventure, cultural, religious and gastronomic heritage, see Fig. 11.

Fig. 11. Welcome screen within the VR environment.

Once the category has been selected, the list of sites of touristic interest corresponding to the type of tourism is displayed, where the tourist chooses the site he wishes to visit, guided by the attributes, characteristics and peculiarities, as well as by recommendations they allow you to have a more focused perception of your profile. Here also choose the range of time available to make the visit (Fig. 12).

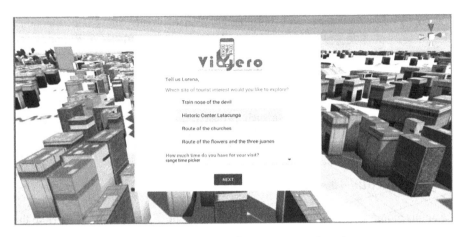

Fig. 12. Selection screen of category-type of tourism.

The application takes the preferences indicated by the user and complements them with geographic information obtained from the device showing its location within the destination and projecting the perimeter that could travel according to the selected time range.

So that the user obtains his route to follow according to the type of route and time of duration, he carries out the following steps: (a) Indicate the starting point of the trajectory in the structure of the 3D model; (b) Select the tourist destination that you want to know on the map that indicates the reconstructed tourist attractions; (c) Indicate the type of trajectory that is going to be carried out and the duration of the journey. See Fig. 13.

The With the data entered and from the nearest point the generated circuit is displayed so that the user can travel the guided route, as indicated in Fig. 14(a); while the tour is being done, the user visualizes the virtualized and digitized tourist structures obtaining an experience in Virtual Reality completely immersive to the reality that the tourist site has, look at Fig. 14(b).

(a) Selection of the initial point of the trajectory to generate

(b) Selection of tourist destination

(c) Indicate de type of tour and the time estimation

Fig. 13. Steps in edition of the digitized model.

(a) Route generated in the virtual map according to the entered data

(b) Immersive experience that the user has along the route

Fig. 14. Generated route and route made by the user.

5 Conclusions

In this article we present the development of an application and immersive virtual system for programming tourist circuits with VR. The results presented demonstrate the correct performance of the geographic information system and the attributes provided by the APP to tourists. Its use demonstrated that the application serves as a virtual guide for the tourist, showing the sites of tourist interest according to the category of visit selected and the route to follow according to the initial point of location marked. It was also evident the access to information that the tourist had in real time and the ease in downloading and programming the tourist circuit, which shows that the application provides an immersive experience, dynamic and interactive.

Acknowledgment. The authors would like to thank the Corporación Ecuatoriana para el Desarrollo de la Investigación y Academia – CEDIA for the financing given to research, development, and innovation, through the Grupos de Trabajo, GT, especially to the GT -

eTURISMO; also to Universidad de las Fuerzas Armadas ESPE, Universidad Técnica de Ambato, Escuela Superior Politécnica de Chimborazo, Universidad Nacional de Chimborazo, and Grupo de Investigación ARSI, for the support to develop this work.

References

1. The World Economic Forum within the framework of the Economic Growth and Social Inclusion System Initiative and the Future of Mobility System Initiative. The Travel & Tourism Competitiveness Report 2017. World Economic Forum, pp. 1–387 (2017)
2. Garcia, R.: Estadísticas de la Organización Mundial de Turismo. Aprende de turismo.org, p. 1 (2018)
3. López, T., González, R.: Mejora del uso de internet en la relación empresa turística-cliente. TURyDES, vol. 6, n° 15 (2013)
4. Mejía, M.: Nuevas tecnologías para el desarrollo de la industria turística en Guanajuato, México. ROTUR, Revista de Ocio y Turismo, pp. 35–43 (2015)
5. Guttentag, D.: Virtual reality: applications and implications for tourism. Tour. Manag. **31**, 637–651 (2010)
6. Fritz, F., Susperregui, A., Linaza, M.: Enhancing cultural tourism experiences with augmented reality technologies. Asociación VICOMTech, pp. 1–6 (2005)
7. Escartín, E.: La realidad virtual, una tecnologia educativa a nuestro alcance. Pixel-Bit Revista de Medios y Educación, pp. 5–21 (2000)
8. Spinar: Desarrollo de aplucaciones móviles para plataformas IOS Y ANDROID. Spinar, p. 1 (2018)
9. ABCDIARIO, J.M.S.: España quiere ser potencia mundial de la realidad virtual y aumentada, 25 October 2018. https://www.abc.es/tecnologia/informatica/software/abci-espana-quiere-potencia-mundial-realidad-virtual-y-aumentada-201810250337_noticia.html. Accessed 08 Mar 2019
10. SEGITTUR: SMARTTRAVEL (2018). https://www.smarttravel.news/2018/01/18/segittur-promueve-fitur-2018-uso-la-realidad-virtual-sector-turistico/. Accessed 08 Mar 2019
11. Dragos, S.: Travel & Tourism competitiveness: a study of world's top economic competitive countries. Procedia Econ. Financ. **15**, 1273–1280 (2014)
12. Alberca, P., Parte, L., Muñoz, A.: La incidencia del destino turístico en la eficiencia y la productividad de las empresas hoteleras. Estudios y Perspectivas en Turismo, pp. 159–179 (2012)
13. Hu, W., Wall, G.: Environmental management, environmental image and the competitive tourist attraction. J. Sustain. Tour. **13**(6), 617–635 (2005)
14. García, B.: Características diferenciales del producto turístico rural. Cuadernos de Turismo, pp. 113–133 (2005)
15. Osorio, F., Arnold, M.: Introducción a los Conceptos Básicos de la Teoría General de Sistemas. Cinta moebio **3**, 40–49 (1998)
16. Dionísio, F.: Una Teoría Dinámica del Sistema Turístico, pp. 15–64. Universidad del Algarve (2007)
17. Han, D., Jung, T., Gibson, A.: Implementing augmented reality (AR) in. Information and Communication Technologies in Tourism, pp. 1–13 (2014)
18. Spinar: Desarrollo de aplicaciones móviles para plataformas IOS Y ANDROID. Spinar, p. 1 (2018)

Virtual Museums as a New Type of Cyber-Physical-Social System

Louis Nisiotis$^{(\boxtimes)}$ ⬥, Lyuba Alboul ⬥, and Martin Beer ⬥

Sheffield Hallam University, Sheffield, UK
{L.nisiotis, l.alboul}@shu.ac.uk, mdb.shu@gmail.com

Abstract. Museums are institutions that primarily care for cultural heritage exhibition, preservation and conservation of historical artifacts. However, simply displaying artifacts and provide complex information to describe them is simply not sufficient to effectively engage museum visitors. To improve visitors engagement and their overall museum experience, the use of technology utilized by museums, introducing the concept of Virtual Museums. This paper discusses the use of Virtual Reality through the use of smart phone devices as a mean of a Cyber-Physical-Social system to support, improve and enhance the visitors' experience. The RoboSHU prototype, its current development stage and future work are presented, together with the future research directions of the research team.

Keywords: Virtual Museum · Virtual Reality · Smart Phone VR · Google Cardboard · Cyber-Physical-Social systems

1 Introduction

Museums are institutions responsible for cultural heritage preservation, artefacts exhibition, restoration and conservation, allowing everyone access and be educated about culture and history. The main aim of a museum is to allow visitors understand historical events that took place over time by providing accurate information supplemented with visual elements to engage and educate them. In modern days however, museum visitors demand more interactive, immersive and stimulating experience from what the traditional museology has to offer, which is mainly limited to displaying artefacts in glass cases accompanied by complex descriptions [1]. In the era of the 'museum experience' in which the visitor is seems as a consumer, visitors' satisfaction is crucial to support museums continuity [2]. With the recent advancements in technology, museums have started using smart phones, tablets and VR to support and enhance their visitor's experience, introducing the concept of Virtual Museums [3]. Virtual museums refer to the "*digital spatial environment, located in the WWW or in the exhibition, which reconstructs a real place and/or acts as a knowledge of a metaphor, and in which visitors can communicate, explore and modify spaces and digital or digitalized objects*" [4], and have drawn a lot of interest over the past few years [3, 5–9]. One of the recently introduced technologies is the use of Virtual Reality in general, and smart phones enabled VR in particular. This paper presents a Smart Phone VR prototype that has been developed to investigate the potentials of such

© Springer Nature Switzerland AG 2019
L. T. De Paolis and P. Bourdot (Eds.): AVR 2019, LNCS 11614, pp. 256–263, 2019.
https://doi.org/10.1007/978-3-030-25999-0_22

technology to support and enhance museum experience, aiming at introducing the concept of a Cyber-Physical-Social system that support social, interactive and immersive experience to visitors.

2 Theoretical Background

Virtual Museums support a mixture of traditional museum practices, utilizing a wide range of communication modes and current technological trends [1], which can customize visitors experience in a museum to improve their overall satisfaction [2, 10, 11]. Among the different technologies, the use of VR has been drawing a lot of attention and used in the field of cultural heritage, conservation, restoration, digital storytelling and education [12]. VR is a technology that involves a user interacting with a computer generated artificial 3D environment, in which the users movement is tracked in real time through sensors, and updating the visual input, sound and the environment [13]. VR technology is used in museums to display, reconstruct, as well as perform virtual restoration of artefacts, cultural heritage locations and archeological sites that may have been damaged or perhaps not even exist anymore [12, 14–16]. VR is identified capable of helping visitors adapt to information about artefacts and exhibits [17], and can provide highly immersive and realistic experiences when compared to tools and techniques used in traditional museology [18], due to the technology affordances of immersion and presence. The feeling of immersion refers to "*a form of spatiotemporal belonging in the world that is characterized by deep involvement in the present moment*" [19] and relates to the experience of a technology that is exchanging sensory input from reality with digitally generated input [20], expressing the full absorption of the user into a digital dimension, which stimulates interest, pleasure, cognitive and emotional engagement [21]. Presence is a similar notion, but distinct from immersion [22], and concerns "*the subjective experience of being in one place or environment, even when one is physically situated in another*" [23]. Presence is the subsequent reaction to immersion, which leads the users reaction to the virtual environment to be the same way as the real world [24]. These unique attributes allow the development of virtual experiences that may be difficult or even impossible to reconstruct in the real world, supporting motivation for technology adoption [25].

Since its initial introduction in the 1950's, VR was being a very expensive technology that was challenged by many technical issues and requirements. However, with the rapid development of technology it is now finally an affordable and mature customer-ready technology [26]. Despite some initial resistance to adopt [2], VR has been now increasingly used in museums to improve their visitors' experience and interactions with cultural heritage [12].

2.1 Technical Characteristics of VR

The technical characteristics of VR require the use of hardware equipment to generate the virtual environment and to display information, and there are a number of tools to support this approach. The CAVE (Cave Automatic Virtual Environment) for instance, is a supreme quality immersive technology featuring a room in which the walls, floor and ceiling are projection screens and the user can freely navigate and interact. Similar technologies are the Power Wall and Reality Deck [27] providing the highest level of immersion [30, 31]. Such systems are already in place to support cultural heritage education to allow students visit 'live' archeological sites (see [28]) However, these systems are very expensive and are not mobile as they require the utilization of dedicated spaces [20, 29]. The use of Head Mounted Displays (HMD) is a more affordable and mobile technology which provides highly immersive experience to users [29]. HMD such as Oculus Rift and HTC Vive offer head tracking and interactivity with the virtual environment, are customer level, and provide very high graphics resolution, simulation and quality of experience to users. For example, Oculus Rift [30] is a HMD which uses a positional tracker camera sensor to track the user's position as well as a magnetometer, gyroscope and accelerometer to accurately track the head movement. The limitation of this technology however relates to the need of a high spec computer to be connected with the HMD to generate the virtual experience. With the significant technological advancement over the past few years, the opportunity for using smart phone devices to generate VR experiences is also now possible and is a more compact and cost effective option. This technology is utilising the processing power and high quality screen of modern smart phone devices to generate VR experiences, and with the use of low cost display units such as the Google Cardboard, Daydream, Samsung Gear VR and others, can display good quality VR immersive experiences [31]. For instance, Google Cardboard [32] is a handheld device where the user puts his/her smart phone into a cardboard box with lenses, and the visual information is updated using the gyroscope and accelerometer information of the smart phone. However, unlike Oculus Rift, the Google Cardboard can track head rotation but not position.

The ability to transform smart phone devices into VR headsets offer affordable and portable ways to experience VR and open a wide range of possibilities of utilising this technology to support virtual museums and experience the past [33]. It has been identified that to date, there is only a small number of museums who have managed to explore the potentials of VR, mainly due to affordability of developing and executing a virtual environment [18]. Therefore, the use of smart phone enabled VR technology is available to be used as a more accessible and cost effective solution.

2.2   The Virtual Museum Prototype

In order to gain a better understanding of the affordances of VR in the topic of virtual museums, we have developed a prototype and we are experimenting with different technologies. The prototype is named RoboSHU[1] (Fig. 1) and aims at promoting the history of robotics in: (i) desktop 3D virtual world, and through (ii) VR technologies. RoboSHU features informational boards and exhibits designed by students, aiming at informing visitors about the history of robotics, provide information about the research conducted by the Sheffield Robotics group, and the current state of the literature. The virtual museum prototype was first presented in [34].

---

[1] http://virtualshu.com/roboshu.

Fig. 1. The RoboSHU.

RoboSHU was first implemented as part of an existing Multi User Virtual Environment (MUVE) named VirtualSHU. This is a multi-user 3D virtual world in which users can interact with the environments, its objects, and each other through the use of a graphical representation of their selves called the Avatar. VirtualSHU (Fig. 2) is used to support the delivery of a computing module at our university, for students' dissertation projects, and for research purposes (see [35–37]). The environment is developed using OpenSimulator, an open source MUVE platform, and can be experienced in Desktop 3D and VR mode. In the 3D desktop mode, users are using a computer and a monitor to experience the visual aspect of the virtual world, and use a keyboard and a mouse to interact with the environment and each other. RoboSHU is located on a dedicated area within the VirtualSHU environment, and visitors can visit, navigate, coexist, communicate and interact with other visitors and the virtual museum. Communication is established through the use of Instant Messages, nearby public messages and through Voice over IP.

In the VR mode, the environment can be experienced using the Oculus Rift HMD. Navigation and interaction can be done through the use of an Xbox 360 controller or with the keyboard and a mouse.

In addition to the 3D desktop and VR mode, we have ported the RoboSHU Virtual Museum into a smart phone enabled VR experience using Unity3D Game Engine [30]. We have developed an Android application, which is not yet publicly available; as it is still in a development stage (see Fig. 3).

Fig. 2. The VirtualSHU.

The application is targeting modern Android powered smart phone devices and the environment can be experienced with the use of the low cost Google Cardboard or similar low cost HMD device. The latest version of the Google Cardboard features single button functionality through an internal mechanism that generates a touchscreen input, which has been used for user navigation in the virtual environment. Interaction with the environment and artefacts is taking place using the graphical user interface of the virtual museum, in which the visitor has to focus their view for a few seconds in hotspot areas in order to interact with them.

Fig. 3. The RoboSHU through the Google Cardboard HMD.

3 Conclusions and Future Work

The concept of a Virtual Museum can be seen as a union of intertwined and interrelated spaces. 'Navigating' through those spaces, observing their exhibits, interacting and communicating with other visitors in those spaces will lead to forms of new immersive, interactive and personalized experiences to enhance our understanding of the world around us and our cultural roots. The Virtual Museum will be able to connect the audience to events and/or objects, or phenomena, separated either in time or in space or both, as well as provide 'rendezvous' among members of the audience, via various media and mediums.

In this paper we focus mostly on technical aspects of developing a VR museum prototype, however the concept of the Virtual Museum can be perceived in a much broader sense, as a type of a Cyber-Physical-Social society, that can be applied to a plethora of domains [38, 39].

For future work, we are concentrating on developing RoboSHU in several ways. First of all, we aim to include more exhibits and additional information relevant to the history of robotics to improve the educational efficacy of the virtual museum. Moreover, we are concentrating on providing additional functionality to the environment to provide greater user interactivity between users and the environment. To date, the handheld VR prototype is a single user experience, and we are experimenting in converting it to a multi user virtual world to support the concept of a Cyber-Physical-Social system. Furthermore we aim to conduct a series of evaluation studies to investigate the usability and technical aspects of the environment as well as the users' perception of presence and immersion during the VR experience. Another research direction aims at connecting 'virtual robots' that 'live' in the museum to some of real robots that we have in our Robotic lab.

Acknowledgements. We would like to acknowledge Enohor Igbeyi, a PhD student at the Centre for Automation and Robotics Research, Sheffield Hallam University, for her contribution to the design of RoboSHU.

References

1. Lorente, G.A., Kanellos, I.: What do we know about on-line museums? A study about current situation of virtual art museums. In: International Conference in Transforming Culture in the Digital Age, pp. 208–219 (2010)
2. Galdieri, R., Carrozzino, M.: Natural interaction in virtual reality for cultural heritage. In: Duguleană, M., Carrozzino, M., Gams, M., Tanea, I. (eds.) VRTCH 2018. CCIS, vol. 904, pp. 122–131. Springer, Cham (2019). https://doi.org/10.1007/978-3-030-05819-7_10
3. Jensen, L., Konradsen, F.: A review of the use of virtual reality head-mounted displays in education and training. Educ. Inf. Technol. **23**, 1515–1529 (2018)
4. Pujol, L., Lorente, A.: The virtual museum: a quest for the standard definition. Archaeol. Digital Era **40**, 46 (2012)
5. Shaw, J.: The virtual museum. Installation at Ars Electrónica. ZKM, Karlsruhe (1991)
6. Schweibenz, W.: The "virtual museum": new perspectives for museums to present objects and information using the internet as a knowledge base and communication system. ISI **34**, 185–200 (1998)
7. Carlucci, R.: Archeoguide: augmented reality-based cultural heritage on-site guide (2002)
8. Petridis, P., et al.: Exploring and interacting with virtual museums. In: Proceedings of Computer Applications and Quantitative Methods in Archaeology (CAA) (2005)
9. Styliani, S., Fotis, L., Kostas, K., Petros, P.: Virtual museums, a survey and some issues for consideration. J. Cult. Heritage **10**, 520–528 (2009)

10. Pagano, A., Armone, G., De Sanctis, E.: Virtual museums and audience studies: the case of "Keys to Rome" exhibition. In: 2015 Digital Heritage, pp. 373–376. IEEE (2015)
11. Choi, H.-S., Kim, S.-H.: A content service deployment plan for metaverse museum exhibitions—centering on the combination of beacons and HMDs. Int. J. Inf. Manag. 37, 1519–1527 (2017)
12. Carrozzino, M., Bergamasco, M.: Beyond virtual museums: experiencing immersive virtual reality in real museums. J. Cult. Heritage 11, 452–458 (2010)
13. Coelho, C., Tichon, J., Hine, T.J., Wallis, G., Riva, G.: Media presence and inner presence: the sense of presence in virtual reality technologies. In: From Communication to Presence: Cognition, Emotions and Culture Towards the Ultimate Communicative Experience, pp. 25–45. IOS Press, Amsterdam (2006)
14. Levoy, M.: The digital michelangelo project. In: 1999 Proceedings of Second International Conference on 3-D Digital Imaging and Modeling, pp. 2–11 (1999)
15. Sideris, A., Roussou, M.: Making a new world out of an old one: in search of a common language for archaeological immersive VR representation. In: Proceedings of 8th International Conference on Virtual Systems and Multimedia (VSMM), pp. 31–42 (2002)
16. Grün, A., Remondino, F., Zhang, L.: Reconstruction of the great Buddha of Bamiyan, Afghanistan. Int. Arch. Photogramm. Remote Sens. Spat. Inf. Sci. 34, 363–368 (2002)
17. Reffat, R.M., Nofal, E.M.: Effective communication with cultural heritage using virtual technologies. Int. Arch. Photogramm. Remote Sens. Spat. Inf. Sci. 5, W2 (2013)
18. Lepouras, G., Vassilakis, C.: Virtual museums for all: employing game technology for edutainment. Virtual Reality 8, 96–106 (2004)
19. Hansen, A.H., Mossberg, L.: Consumer immersion: a key to extraordinary experiences. In: Handbook on the Experience Economy, p. 209 (2013)
20. Freina, L., Ott, M.: A literature review on immersive virtual reality in education: state of the art and perspectives. In: eLearning and Software for Education (2015)
21. Sorensen, C.G.: Interface of immersion - exploring culture through immersive media strategy and multimodal interface. In: Proceedings of the DREAM Conference - The Transformative Museum, pp. 409–421 (2012)
22. Lombard, M., Ditton, T.: At the heart of it all: the concept of presence. J. Comput.-Mediated Commun. 3, JCMC321 (1997)
23. Witmer, B.G., Singer, M.J.: Measuring presence in virtual environments: a presence questionnaire. Presence: Teleoperators Virtual Environ. 7, 225–240 (1998)
24. Slater, M.: A note on presence terminology. Presence Connect 3, 1–5 (2003)
25. Mikropoulos, T.A., Natsis, A.: Educational virtual environments: a ten-year review of empirical research (1999–2009). Comput. Educ. 56, 769–780 (2011)
26. Fradika, H., Surjono, H.: ME science as mobile learning based on virtual reality. J. Phys: Conf. Ser. 1006(1), 012027 (2018). IOP Publishing
27. Lee, H., Tateyama, Y., Ogi, T.: Realistic visual environment for immersive projection display system. In: 2010 16th International Conference on Virtual Systems and Multimedia (VSMM), pp. 128–132. IEEE (2010)
28. Ott, M., Pozzi, F.: ICT and cultural heritage education: which added value? In: Lytras, M.D., Carroll, J.M., Damiani, E., Tennyson, R.D. (eds.) WSKS 2008. LNCS (LNAI), vol. 5288, pp. 131–138. Springer, Heidelberg (2008). https://doi.org/10.1007/978-3-540-87781-3_15
29. Gonizzi Barsanti, S., Caruso, G., Micoli, L.L., Covarrubias Rodriguez, M., Guidi, G.: 3D visualization of cultural heritage artefacts with virtual reality devices. In: 25th International CIPA Symposium 2015, pp. 165–172. Copernicus Gesellschaft mbH (2015)
30. Oculus Rift. https://www.oculus.com/
31. Cochrane, T.: Mobile VR in education: from the fringe to the mainstream. Int. J. Mob. Blended Learn. (IJMBL) 8, 44–60 (2016)

32. Google Cardboard. https://vr.google.com/cardboard/
33. Fabola, A., Miller, A., Fawcett, R.: Exploring the past with Google cardboard. In: 2015 Digital Heritage, pp. 277–284. IEEE (2015)
34. Alboul, L., Beer, M., Nisiotis, L.: Merging realities in space and time. In: Virtual Reality Summer School 2017, Lecce, Italy (2017)
35. Nisiotis, L., Kleanthous, S.: The development and evolution of transactive memory system over time in MUVEs. In: 10th Computer Science and Electronic Engineering Conference. IEEE (2018)
36. Nisiotis, L., Kleanthous Loizou, S., Beer, M., Uruchurtu, E.: The development of transactive memory systems in collaborative educational virtual worlds. In: Beck, D., et al. (eds.) iLRN 2017. CCIS, vol. 725, pp. 35–46. Springer, Cham (2017). https://doi.org/10.1007/978-3-319-60633-0_4
37. Nisiotis, L., Kleanthous Loizou, S., Beer, M., Uruchurtu, E.: The use of a cyber campus to support teaching and collaboration: an observation approach. In: The Immersive Learning Research Network (iLRN) Conference, 26–29 June 2017 (2017)
38. Alboul, L., Beer, M., Nisiotis, L.: Robotics and virtual reality gaming for cultural heritage preservation. In: Resilience and Sustainability of Cities in Hazardous Environments, pp. 335–345 (2019)
39. Alboul, L., Beer, M., Nisiotis, L.: Merging realities in space and time: towards a new cyber-physical eco-society. In: Dimitrova, M. (ed.) Cyber-Physical Systems for Social Applications. IGI Global, Hershey (2019)

Virtual Portals for a Smart Fruition of Historical and Archaeological Contexts

Doriana Cisternino, Carola Gatto, Giovanni D'Errico,
Valerio De Luca, Maria Cristina Barba, Giovanna Ilenia Paladini,
and Lucio Tommaso De Paolis[✉]

Department of Engineering for Innovation,
University of Salento, Lecce, Italy
{doriana.cisternino,carola.gatto,giovanni.derrico,valerio.deluca,
cristina.barba,ilenia.paladini,lucio.depaolis}@unisalento.it

Abstract. The experimentation of new effective visualization techniques provides viable technological solutions for the archaeological heritage fruition and the visitors involvement within cultural places (museums, gallery, archaeological sites).

This article discusses the feasibility of a project aimed at extending outdoor the fruition of archaeological sites, by exploiting mixed reality technology. In a previous work we focused on the development of a mobile application that exploits the Augmented Reality technology, limited to an indoor fruition. Therefore we decided to extend the app by introducing a strategy for an on-site outdoor exploration based on "virtual portals". Virtual portals are the medium that allows the transition from reality (present) to virtuality (past) and vice versa. They are located outdoor at some point of interest, so that user has the perception of passing through different space-temporal dimensions: this is actually what we want to provide as user personal experience. We considered two archaeological areas as case studies, both excavations sites of the University of Salento, as an instance of a wider network we aim to create. In this paper we give some hints about a feasible design for this project.

Keywords: Augmented reality · Mixed Reality · Cultural Heritage · Archaeology · Virtual portals

1 Introduction

In cultural heritage scenarios, visitors should never be considered as passive spectators: in order to establish a learning process with the historical contents, visitors should play an active role, in a sort of interactive learning by doing experience.

The experts talk about edutainment, a word coined by Bob Heyman in 1973 that comes from the fusion of the two words educational and entertainment [1]. Edutainment techniques often pass through the use of Information and Communications Technology (ICT). This allows the humanistic disciplines to achieve

L. T. De Paolis and P. Bourdot (Eds.): AVR 2019, LNCS 11614, pp. 264–273, 2019.
https://doi.org/10.1007/978-3-030-25999-0_23

higher results in terms of audience engagement. In this project we would like to stimulate the user interest through an interactive learning process.

It should be taken into account that, in an archaeological context, the lack of tangible elements brings the need for "augmenting" the reality. The aim is to extend the visible by presenting a world where real and virtual objects exist simultaneously. A helpful answer to this need comes from Virtual, Augmented and Mixed Reality technologies [2]:

- Virtual Reality (VR) isolates users from the real world and immerses them in computer-generated synthetic environments;
- Augmented Reality (AR) is a technology able to extend the visible, without replacing the reality with a synthetic one, but by presenting a world where physical and digital objects co-exist;
- Mixed Reality (MR) combines elements of both AR and VR to produce new environments and visualizations enabling virtual objects to be not just overlaid on the real world but able to interact with it.

Virtual reconstructions make historical concepts more accessible to the general public. In the edutainment paradigm [3], they allow the design of game oriented environments [4,5], which provide the users with engaging multi-channel and multi-sensory experiences [6]. In addition, MR environments allow the user having his/her own personalized interactive experience.

In a previous work [7], the aim was to experiment with the use of Augmented Reality technology for the archaeological sites of the Museo Diffuso Castello di Alceste, in San Vito dei Normanni (BR) and Fondo Giuliano, in Vaste (LE), both excavations of the University of Salento. The output was a mobile application (both for Android and iOS devices) able to transmit historical-archaeological content to a wide, non-specialized audience, through virtual reconstructions of the environment displayed in AR. Indeed the user is able to visualize the 3D reconstructions by simply pointing the camera of a smartphone at some ortho photographs of the site, used as image targets.

The limit of such application is that it has been developed and optimized for an indoor fruition, for instance inside a museum that collects archaeological remains of those sites. In order to have a significant impact in terms of user experience, the extension we are going to describe concerns a possible outdoor and on-site fruition by means of Mixed Reality. The idea is to provide the user with a mobile application aimed at bringing him/her inside a virtual reconstruction of the ancient archaeological site, by means of a "virtual portal" that permits a sort of a virtual journey into the past. The visualization of a virtual door in some points of the site allows the user to move from the real environment (the present) to the virtual one (the past).

1.1 Previous Indoor Application

According to the project and development of the previous app, the idea was to enhance the understanding of Apulian ancient culture. In particular we focused on two archaeological contexts:

– Museo Diffuso Castello di Alceste in San Vito dei Normanni (BR),
– Fondo Giuliano site in Vaste (LE).

We analyzed the historical context according to the data emerged during the digs campaigns [8,9]. In this way we defined, adapted and digitally optimized the contents that are actually object of the application and that the user can visualize by using augmented reality. The accuracy of historical contents and graphic reconstructions have been an important topic: from the beginning the work has been conducted in synergy with the Department of Archaeology of the University of Salento, that has supervised the accuracy in terms of historical data.

After a preliminary context study, we passed to design a technological system for presenting archaeological sites in an interesting and appealing way. For this reason, we developed a system that allows users to visualize the 3D reconstruction of the archaeological buildings and artefacts that no longer exist. This is made possible by the augmented reality technology. The user can view historical contents (3D models, texts, audio) related to the area simply by framing a planimetric map of the location of interest with a mobile device/smartphone. Obviously, visitors can be provided with this map in the archaeological site museum, for an indoor fruition.

This version was developed by using the *Vuforia* framework [10] integrated in the *Unity 3D* environment [11]. *Vuforia* allows to associate virtual objects with 2D or 3D targets, including image targets, object target and multi target. It provides an online Target Management System to process the image target (the aerial photographs of the archaeological sites) and detect the features which will be used for the image recognition (Fig. 1).

Figure 2 shows a view of the AR section of the app: after selecting the historical period of interest, the user can view the 3D reconstructions related to the respective historical phase directly superimposed on the orthophoto. Furthermore, some information buttons provide a description for each part of the building to allow a full understanding of the history of the ancient structure. We studied also a smart interaction mode that allows the user to access to contextual specific information by simply getting closer to some point of interest (POI) with the camera of the smartphone, without tapping any button.

Through this first version of the app, the user is provided with a useful tool to contextualize the museum collection in relation to the archaeological site of provenance. Since the application has been developed for an indoor fruition, now the need is to provide an extension that allows a direct fruition on the archaeological site. It must be taken into account that not all the archaeological sites are available to be directly visited: in this case a remote fruition is recommended.

However, when the site is open to public visits, it is important to encourage the fruition directly in situ: this is one of the goal we are addressing with this new project. Finally, a tool able to make up a network of multiple archaeological contexts open to the public would be even more helpful, by defining thematic paths based on georeferenced data. This is the starting point for the proposal based on virtual portals, which we are going to describe in this paper.

Fig. 1. Feature recognition of the image target

2 Related Works

The development of an augmented reality experience directly on site requires facing some issues related to outdoor scenarios, which are still a topic of intense research. The goal of an outdoor augmented reality system is to allow users to

Fig. 2. Augmented visualization of 3D model on the image target

move freely and without constraints in the environment, as well as to visualize and interact in real time with geo-referenced data by means of his smartphone. In particular, most of the AR outdoor applications aim at extending the visible by means of virtual reconstructions superimposed on the ruins in the real scene. This requires a great effort in terms of tracking techniques, accuracy and virtual content registration. The trend in outdoor scenarios is to exploit hybrid tracking. A fusion of various classic tracking methods can yield better results than their separate use. One of the most important and most cited instances in the archaeological domain that use hybrid tracking is ARCHEOGUIDE [12] (Augmented Reality-based Cultural Heritage Onsite GUIDE), which combines markerless tracking and Global Positioning System (GPS) to determine the viewpoint pose. It aims to develop interactive methods for accessing outdoor cultural heritage information, by providing an opportunity to visualize the 3D reconstructed damaged site.

More experimental solutions for the archaeological field have been proposed in the last years. In [13] the authors developed a location-based AR mobile application, supporting the exploration of a given archaeological site from an aerial perspective, by exploiting unmanned aerial vehicle (UAV) and augmented reality.

There are several critical issues emerging in this field, well summarized in [14], that include the need to address problems related to user mobility and the impossibility to use a marker-based tracking. The outdoor scenario is a highly dynamic and unprepared environment and generally there is the need to adopt an approach based on hybrid tracking. These critical issues have widened the gap between the accuracy of an indoor application and an outdoor one.

A different possibility for outdoor archaeological fruition is offered by the so-called virtual portals [15], which propose a jump between AR and VR. Although it does not provide a precise one-to-one correspondence between the virtual reconstruction and the archaeological remains, it leads to the possibility of opening a virtual gate connecting the present and the past.

In [16], the authors define virtual portals as three-dimensional doorways that connect one virtual location to another and can be entered by users in order to move to and come back from that place. In particular, the former is a faithful replica of the real environment that the user sees before wearing the VR viewer, while the latter is an imaginary environment.

Nevertheless, to the best of our knowledge, the use of virtual portals in the augmented reality field and, in particular, as a tool for the enhancement of archaeological sites, has not yet produced meaningful case studies in literature. By exploiting the development potential of ARkit and ARcore libraries, it becomes much easier to implement "AR Portals", allowing users to physically cross a virtual portal in the AR browser and experience different spatiotemporal realities. This makes users able to create, experience and share content-rich stories in AR. Some experimentation in this field has been conducted by Nedd [17], a French company that develops immersive experiences for brands and organizations. A first prototype combining AR, geoinformation and social networks is proposed, in order to create a highly interactive and personalized storytelling experience.

3   Work Hypothesis

As already mentioned, in this paper we want to define the hypothesis for an extension of the previous work in order to transcend traditional communication methods and propose an innovative and alternative tool for the outdoor and on-site fruition of the archaeological area. Inspired by many series or movies and 3D first-person action video games, we propose an alternative solution, as a plausible "shortcut" that performs a "time travel" by using the virtual portals concept, in order to bring the user from the real environment to the virtual one.

According to the work hypothesis, the aim is not to offer (like in the most of the outdoor AR cases) a one-to-one correspondence between the virtual reconstruction and the archaeological remains: on the contrary, we consider a simplification of the problem that only marginally deals with the tracking issues discussed above. Following our design idea, those virtual portals will be accessible through a smartphone or a mobile device: a POIs network will be provided to the user in order to discover an itinerary of geolocated virtual portals. Through these virtual gates we want to provide a "time travel", a transition between the real (the present) and the virtual (the past) environment.

4   Switching from AR to VR: Virtual Portals

In more concrete terms, the real visit experience of an archaeological site sheds light on the visit issues. Considerable difficulties arise in understanding the chronological sequence of the site itself when the user looks at the ruins, characterized by the overlapping of different aged structures. In view of the above, virtual portals can represent a useful solution to these issues, in order to increase the understanding of these scenarios, without losing sight of the reality.

Firstly, portals are virtual objects that are properly registered in the real scenario framed by the device camera. Their alignment with the surrounding environment should give the illusion of being physically in front of a door. By means of mixed reality techniques, in which virtual objects have a spatial awareness, soil surfaces are detected to position the portal gate properly. A light estimation module will offer a higher level of realism, by setting the lighting of objects according to external lighting conditions.

Once gates are placed in the environment, users will be able to reach the virtual space by simply passing through portals. The related virtual scenario will represent a historical reconstruction of the site. The virtual space where the user will move with his/her device will be an augmented virtuality space: the portal will represent the point of return to the starting real space (consisting of the camera images).

In such approach, the portal is a door that allows a transition from the AR scenario (i.e. reality plus a portal on the virtual past) to the Augmented Virtuality (AV) scenario (i.e. the virtual world plus a portal on the real present) (Fig. 3).

We considered two archaeological areas as case studies, both excavations sites of the University of Salento, as an instance of a wider network we aim to create. The opportunity to put onto the web various archaeological contexts is granted by the use of geo-referenced maps. They exploit the device's on board sensors and GPS received to locate the user and the various portals, positioned according to the points of interest spread over the territory. In the style of a PokmonGO-like [18] location-based mobile game, the user will be provided with a map in order to navigate around the archaeological site and to explore it. When the user approaches one of the virtual portals he/she can tap on the button on the phone's screen and enable the AR visualisation. In that way, the portal will be properly placed on the AR screen as if it were in front of the user. He/she will be able to get inside the portal in order to experience a full virtual immersion into the ancient environment, although still present into the real world.

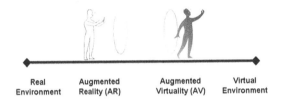

Fig. 3. Virtual portals in the virtual continuum

4.1   Implementation

While the earlier version of the application exploited an image target based tracking and used *Vuforia*, the new one needs other kind of technologies in terms of software development kit (SDK), even though the development core is still based on the *Unity 3D* environment.

The need for an AR framework that offers the possibility of developing a realistic first person game experience leads to the adoption of tools such as *ARCore* [19] for Android systems and *ARKit* [20] for iOS systems. Both the frameworks combine information coming from the device's motion sensing hardware with computer vision analysis of the scene visible to the device's camera. They can compare that information with motion sensing data and track differences in the positions of those features across video frames. The result is a high-precision model of the device's position and motion. Both the frameworks detect flat surfaces in the camera image and report their position and sizes to place virtual portals in the scene and pass through them.

5   Expected Results and Further Work

This article discussed the feasibility of a project aimed at extending outdoor the fruition of archaeological sites, by exploiting mixed reality technology.

Virtual portals are media that allow the transition from reality (present) to virtuality (past) and vice versa. They are located outdoor at some points of interest, so that the user has the perception that the travel passes through different space-temporal dimensions, and this is actually what we want to evoke as user personal experience. By identifying some paths among these portals, we aim at creating a territorial network of "smart" archaeological sites: this network would inspire and stimulate visitors in keeping this engaging experience. The main goal is to provide an interactive learning scenario in which visitors play an active role during the exploration.

The idea for the future development is to provide the user with a mobile application showing a map of the archaeological site, which keeps track of his position and displays where the virtual portals are located. These will be displayed in augmented reality when the user taps on them. For these purposes it needs to choose the tracking methodology and, consequently, the augmented reality framework for GPS support and 3D model registration in outdoor scenario; the realization of the virtual portal effect must be taken into account.

The *ARCore* framework enables devices to sense the environment, understand the world and interact with virtual content as seen through the camera. The development of 3D portals as access to a navigable virtual environment can be carried out thanks to the integration of ARcore in the Unity environment. Thanks to motion tracking, ARCore identifies some features through the phone's camera which, combined with phone's inertial sensors, determines the position and orientation of the phone. Besides key points, ARCore places the portal on top of the identified planes, by detecting the surfaces and understanding the world around them.

Another important topic concerns the 3D reconstruction of the ancient contexts: in the future development it will be necessary to evaluate the 3D modeling software as well as the reconstruction criteria and requirements for partially or totally lost contexts (through a cross-check with historical-artistic documentation).

Furthermore, after the import of 3D models, textures and lights in *Unity 3D*, we will need to properly associate 3D objects with the virtual portal, making them visible only from inside. This means that the shaders' model needs to be manipulated in order to achieve the portal effect. We need also an appropriate plugin for the design of an interactive map which displays the geolocalized points of interest. For instance, *Go Map* is suitable [21] for developing location based games and applications with Unity 3D. It is directly customizable from the Unity inspector and provides support for different types of map providers like *OpenStreetMap*, *Mapzen* and *Mapbox*.

We gave some hints about a possible design for this idea and a subsequent development of the project, as result of multidisciplinary studies. It will be also necessary to test the output, by means of some tasks provided to the user, in order to evaluate the effectiveness and usability of the product.

References

1. Francesco Cervellini, D.R.: Comunicare emozionando. L'edutainment per la comunicazione intorno al patrimonio culturale. DISEGNARECON **4**(8), 48–55 (2011)
2. Bekele, M., Pierdicca, R., Frontoni, E., Malinverni, E., Gain, J.: A survey of augmented, virtual, and mixed reality for cultural heritage. J. Comput. Cult. Heritage **11**(2), 7 (2018)
3. De Paolis, L.T., Aloisio, G., Celentano, M.G., Oliva, L., Vecchio, P.: Experiencing a town of the middle ages: an application for the edutainment in cultural heritage. In: IEEE 3rd International Conference on Communication Software and Networks, pp. 169–174 (2011)
4. De Paolis, L.T., Aloisio, G., Celentano, M.G., Oliva, L., Vecchio, P.: A game-based 3D simulation of Otranto in the middle ages. In: Third International Conference on Advances in Computer-Human Interactions, pp. 130–133 (2010)
5. De Paolis, L.T., Aloisio, G., Celentano, M.G., Oliva, L., Vecchio, P.: MediaEvo project: a serious game for the edutainment. In: 3rd International Conference on Computer Research and Development, vol. 4, pp. 524–529 (2011)
6. Paolis, L.T.: Walking in a virtual town to understand and learning about the life in the middle ages. In: Murgante, B., et al. (eds.) ICCSA 2013. LNCS, vol. 7971, pp. 632–645. Springer, Heidelberg (2013). https://doi.org/10.1007/978-3-642-39637-3_50
7. Cisternino, D., Gatto, C., De Paolis, L.T.: Augmented reality for the enhancement of apulian archaeological areas. In: De Paolis, L.T., Bourdot, P. (eds.) AVR 2018. LNCS, vol. 10851, pp. 370–382. Springer, Cham (2018). https://doi.org/10.1007/978-3-319-95282-6_27
8. D'Andria, F.: Gli insediamenti arcaici della Messapia. Ricerca evalorizzazione di una ricchezza. L'area archeologica di Località Castello aSan Vito dei Normanni - La ricerca come risorsa, Brindisi, pp. 2–8 (1998)

9. D'Andria, F., Mastronuzzi, G., Melissano, V.: La chiesa e la necropoli paleocristiana di Vaste nel Salento. Rivista di Archeologia Cristiana **82**, 231–322 (2006)

10. Vuforia (2018). https://developer.vuforia.com

11. Unity 3D (2018). https://docs.unity3D.com

12. Gleue, T., Dähne, P.: Design and implementation of a mobile device for outdoor augmented reality in the archeoguide project. In: Proceedings of the 2001 Conference on Virtual Reality, Archeology, and Cultural Heritage, pp. 161–168. VAST 2001. ACM, New York (2001)

13. Botrugno, M.C., D'Errico, G., De Paolis, L.T.: Augmented reality and UAVs in archaeology: development of a location-based AR application. In: De Paolis, L.T., Bourdot, P., Mongelli, A. (eds.) AVR 2017. LNCS, vol. 10325, pp. 261–270. Springer, Cham (2017). https://doi.org/10.1007/978-3-319-60928-7_23

14. Zendjebil, I., et al.: Outdoor augmented reality: state of the art and issues. In: 10th ACM/IEEE Virtual Reality International Conference (VRIC 2008), Laval, France, pp. 177–187, April 2008

15. Liu, C., Fuhrmann, S.: Enriching the GIScience research agenda: fusing augmented reality and location-based social networks. Trans. GIS **22**(3), 775–788 (2018)

16. Bruder, G., Steinicke, F., Hinrichs, K.: Arch-explore: a natural user interface for immersive architectural walkthroughs, pp. 75–82 (2009)

17. Nedd: Augmented Reality Portal Demo (2018). http://nedd.me/en

18. Niantic: PokemonGO (2018). https://pokemongolive.com/it

19. ARCore (2018). https://developers.google.com/ar

20. ARKit (2018). https://developer.apple.com/arkit

21. Go Map (2018). https://gomap-asset.com

Education

Intercultural Communication Research Based on CVR: An Empirical Study of Chinese Users of CVR About Japanese Shrine Culture

Ying Li, YanXiang Zhang$^{(\boxtimes)}$, and MeiTing Chin$^{(\boxtimes)}$

Department of Communication of Science and Technology,
University of Science and Technology of China, Hefei, Anhui, China
ly306@mail.ustc.edu.cn, {petrel,mtching}@ustc.edu.cn

Abstract. Previous studies have shown that CVR can promote users' knowledge acquisition, meaning understanding and emotional experience. This study demonstrates the feasibility and advantages of CVR in intercultural communication through empirical research. A virtual reality video about Japanese shrine culture was produced by shooting and editing, and a Chinese user was selected for a controlled experiment. It turns out that CVR can create an immersive cultural environment for users in intercultural communication and bring a good cultural experience, but it also has some shortcomings that need to be continuously improved through technical means.

Keywords: CVR · Intercultural communication · Japanese shrine culture

1 Introduction

CVR (Cinematic Virtual Reality) is a broad term that could be considered to include a growing range of concepts, from passive 360° videos to interactive narrative videos that allow the viewer to affect the story [1]. With the continuous development of CVR, its technical effects are quickly noticed by other industries and applied to different fields such as education, medical care and media. CVR is applied to intercultural communication and the research results are still relatively few. But the research results in other fields provide support for its feasibility study, which mainly reflected in three aspects: knowledge acquisition, meaning understanding and emotional experience. These three dimensions are also the three aspects of user satisfaction in the process of intercultural communication. This study verifies its communication effect through empirical research, and study its problems in the process of communication.

1.1 Knowledge Acquisition

Users could acquire knowledge from text annotations, pictures and sounds through watching video. Both CVR and ordinary video can present this there factors, but the effects are different. Alan Cheng et al. found that VR technology can take advantage of culturally relevant physical interactions to enhance language learning [2]. Passig et al. found that personal experience in the VR environment can promote parents' awareness

L. T. De Paolis and P. Bourdot (Eds.): AVR 2019, LNCS 11614, pp. 277–285, 2019.
https://doi.org/10.1007/978-3-030-25999-0_24

of children with dyslexia [3]. This way of acquiring knowledge is to achieve visual representation of cognition through experience.

At the same time, the 3D technology is considered to be able to enhance users' attention and memory. Terlutter et al. found that 3D movies can significantly enhance attention and memory of prominent brands compared with traditional movie [4]. Cognitive theory believes that human cognition is limited. Therefore, when users are exposed to 3D video, limited information processing capabilities will be shunted. In this case, it may cause users to pay more attention to the key information and ignore non-key information.

1.2   Meaning Understanding

In traditional cognitive theory, Craik believes that meaning understanding is more difficult than sensory and surface feature analysis [5]. Therefore, in the process of CVR communication, reaching the user's meaning understanding is more profound than knowledge learning. Sato et al. found that watching 360-degree video can effectively improve students' reflective ability and strengthen students' understanding of related learning [6].

In addition to deepening the understanding of knowledge, the significance of understanding is also reflected in the improvement of their own learning skills. In the field of skills training, especially in the field of medical training, CVR is widely used. Through research, Yoganathan found that when training medical stuff on knotting skills, medical stuff who learned through CVR scored higher in the final evaluation of knotting techniques than those who learned through traditional 2D video [7]. CVR can provide users with a more comprehensive perspective and face-to-face teaching style, and promote users' self-think after acquiring knowledge, thus showing better learning results.

1.3   Emotional Experience

Compared with traditional video, the three-dimensional panoramic effect highly restores the real effect, and users can obtain a stronger sense of immersion and experience, which is particularly effective in emotional experience. CVR can induce users to generate more intense emotional experience and physiological reflection than traditional video [8]. Keen believes that empathy can be used for rhetorical purposes and to reinforce the user's pre-existing emotions [9]. Therefore, CVR can also enhance users' emotions by arousing their empathy based on emotional experience.

On the one hand, this emotional experience of CVR is used for the treatment of many emotional comforts. On the other hand, it promotes communication and understanding among different groups. By allowing Aboriginal people to experience the social and emotional experiences of new immigrants, Passig found that VR can promote aboriginal's understanding of new immigrants and enhance their social awareness [10]. This is actually a way of intercultural communication approach that achieves mutual understanding by simulating the lifestyle and emotional experiences of different cultural groups.

2 Method and Implementation

The participants were divided into two groups, the experimental group and the control group. Each participant in the experimental group watched a CVR of 3 min and 52 s. Each participant in the control group watched a normal video of 3 min and 21 s. The CVR material and the clip are all completed by the researcher. The ordinary video is found from the YouTube and has the same core content with CVR. Then the predictive tests and interviews were conducted, and a questionnaire model was designed based on the results. According to the dimensions of the questionnaire model, a specific evaluation questionnaire was designed and a control experiment was conducted.

2.1 Preliminary Preparation

Both the CVR and the normal video screens are a day of worship scenes of the Fudatenjin Shrine in Tokyo, Japan. The specific content is the worship process and the culture of Shichi-Go-San Festival.

Use the insta360one camera to shoot CVR footage. The video pixel is 3840 * 1920. Using the method of flat angle shooting from different directions and fixed and moving shots.

Based on the narrative order of the visiting shrine, the CVR used Adobe Premiere for editing. The narrative is arranged by interspersing individual short videos in a long shot. In addition, visual guidance symbols has been added to help users to focus on key information. Firstly, use text annotation to add subtitle prompts next to important information points. Secondly, using graphic annotation, such as arrow which points to other angles to ask users to rotate picture (Fig. 1).

Fig. 1. Graphic annotation in CVR

2.2 Evaluation Process

Through the pre-test experiments conducted by 4 volunteers, the editing and post-production of CVR was improved. Then through 21 interviews, the questionnaire dimensions and theoretical models were determined after the trial. The integration model based on the UTAUT model and the 4C theory is established through research.

UTAUT (Unified Theory of Acceptance and Use of Technology) has four core dimensions: effort expectancy, performance expectations, social impact and facilitating conditions.

4C marketing theory is based on consumer demand and have four basic elements of the marketing mix: customer, cost, convenience, and communication.

2.2.1 Pre-test Experiment

Four volunteers conducted pre-test experiments, watching the video of the experimental group and the video of the control group, and answering the questionnaire. According to the opinions of the volunteers, researcher optimized editing and added visual guidance and background sounds.

On the other hand, the dimension of the questionnaire has been improved. First, the economic factor is added, that is, whether the cost will affect the user's willingness to use CVR for cultural exchange. Second, the description of the problem in the questionnaire is changed more specifically.

2.2.2 Hypothesis

Two assumptions are made through pre-test feedback.

H1: The application of CVR in the intercultural communication field is better than ordinary video, that is, participants who watch CVR have higher performance expectations and effort expectancy than those who watch ordinary video.

H2: Convenient conditions and perceived cost are important factors for the acceptability of CVR.

2.2.3 Interview Feedback

Through the interviews with 21 volunteers on the problem, it is found that most users have certain expectations for CVR, and CVR is superior to ordinary video in intercultural communication. This is mainly reflected in the degree of information presentation and the advantage of new technology.

Compared with ordinary video, CVR is richer in information presentation. So users can selectively receive information according to their own needs. At the same time, the rich information can deepen the impression of the subject. The most important is the three-dimensional space created by the panorama, which enables users to grasp the information based on the spatial level, making it more intuitive and more acceptable.

CVR is more advanced than ordinary video. User's interest can be enhanced by using new technology to generate the points of interest and memory points for the content. Visual impact and immersive feelings make the spread of culture more direct and arouse more emotional resonance. In particular, it was found that users between the ages of 20 and 60 are more interested in and more receptive to CVR. A male user of age 80 showed a negative attitude when watching CVR.

For the immersion of CVR, most users think that it is related to CVR's own resolution, editing mode, picture stability and lens speed. A 22-year-old male user suggested that "a long-lens can be used" to avoid the picture jumping caused by the clip.

For visual guidance symbols, users generally agree that they do not affect immersion, but feedback on the degree of guidance is mixed. Some users think that visual guidance symbols should be added to highlight key content. However, some users believe that it does not have to be too much because the user's interest points are different. Compared with the degree of guidance, the annotations and explanations should be added.

2.2.4   Questionnaire Model

The research object is the feasibility of CVR in intercultural communication, which is to measure the acceptance of CVR in intercultural communication. Therefore, it involves not only the user's acceptance of new technologies, but also whether CVR has value in the market. Therefore, based on the UTAUT model, 4C theoretical elements are added to the core dimension and appropriate modifications are made according to the requirements. Finally, a user acceptance evaluation model for CVR application to intercultural communication is established.

The core variables of the integrated UTAUT and 4C theoretical models are defined as follows:

- Performance expectation: after users experience CVR, it is considered to play a role in intercultural communication.
- Effort expectation: how much effort is required to obtain a better experience when the users experience CVR.
- Facilitating conditions: the convenience of the channel will affect the user's willingness, behavior and satisfaction.
- Perceived cost: the cost perceived by the user before the experience, and the cost will also affect the willingness and behavior.

According to the integrated UTAUT and 4C theoretical model, four basic elements are used as the evaluation dimension (Fig. 2). The specific indicators are shown in the Fig. 2.

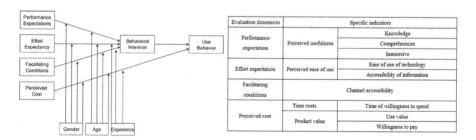

Fig. 2. Integrate UTAUT and 4C theoretical models and evaluation dimension

2.2.5   Experiment

The experiment was divided into experimental group and control group, and each group had 20 participants. In experimental group, there were 14 women (70%) and 6 men. The average age is 27.5 years old. In control group, there were 15 women (75%) and 5 men. The average age is 27 years old. Each person can only watch the video once and answer the questionnaire within 15 min. The questionnaire includes two aspects of answering questions and evaluation.

2.2.6   Data Collection and Analysis

A total of 40 questionnaires were collected. From the results of the answer, the accuracy of the knowledge is higher in control group than the experimental group. But the correctness rate of the experimental group is higher. In the knowledge problem, the correct rate of the experimental group is 66.25%. The correct rate of the control group is 76%. In the comprehension question, the correct rate of the experimental group is 87.5%. The correct rate of the control group is 65%. From the perspective of objective answering effects, ordinary video is better at transmitting knowledge information, while CVR is better at transmitting conceptual information (Fig. 4).

Fig. 3. Perceived usefulness of video: cognition and comprehension

Fig. 4. Perceived usefulness to video: immersion

When users watch CVR, the perceived usefulness is significantly greater than that when they watch ordinary video. The members of experimental group felt that self-cognition level and understanding degree after the experiment was higher than that of the members of the control group (Fig. 3). Although after the objective evaluation, the accuracy of the answer to the understanding level is different but not very large, but for the user's own perception, the feeling of acquisition through CVR is more intense. The members of the experimental group were more immersed in it than those of the control group, and the sense of presence, participation and movement were far greater. It can be seen that in the intercultural communication of CVR, users have higher performance expectations.

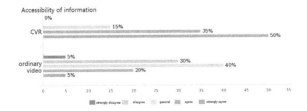

Fig. 5. Perceived ease of use: ease of access to information

Since the average age of users is 27.25 years old, younger and more technically acceptable, the difference in technical operation difficulty between CVR and ordinary video does not cause any trouble to users. In the terms of information acquisition, the experimental group is obviously easier to obtain the information of their own needs than the control group (Fig. 5), especially in the aspects of detail presentation, CVR has stronger information capacity in the process of communication to present more specific and richer information. Overall, users watch CVR whose effort expectation in the intercultural communication process are expected to be higher.

67.5% of users think that CVR is not easy to obtain, and 90% of users think that the ease of access will affect the intention to use. Convenient conditions are one of the obstacles that prevent users from receiving intercultural information when using CVR.

After questionnaires, it was found that 25% of users can spend less than 5 min on CVR, and 55% of users can spend more than 5 min on CVR. Others think that the time they accept on CVR depends on the content. 82.5% of users believe that CVR can promote cultural understanding and communication if it is valuable. If CVR is applied to intercultural communication and is more mature, 75% of users are willing to pay for it. In intercultural communication, time cost and economic cost are important factors affecting the acceptance of CVR, but when the development is relatively mature, most users have the willingness to pay.

3   Conclusion and Discussion

Through experiments, interviews and questionnaires, the different effects after users using CVR and ordinary video were compared. It is found that the application of CVR in intercultural communication is feasible and has academic significance and application value.

3.1   Conclusion

Combined with the experimental results and the interview contents, the paper discusses the communication effect and influencing factors of CVR's application in intercultural communication. It is considered that CVR has several characteristics compared with ordinary video.

3.1.1   Creating a Cultural Environment

CVR can create a cultural environment for users, which is reflected in both space and atmosphere. In the physical space, CVR can fully display the spatial layout, architectural features, cultural rituals, and character activities, making the user's sense of time and space stronger. In interviews, some users mentioned that "the map is formed in the mind", and this three-dimensional sense of space spreads the spatial layout knowledge that is difficult to transmit in ordinary video. CVR can express the physical space and the cultural factors through the panoramic stereo. The combination of these elements forms the cultural environment which in turn forms the cultural atmosphere. This atmosphere also helps users understand the cultural environment. Due to the technical characteristics of VR technology, users can be immersed in it and experience the cultural environment of other cultures. Through the data, we can understand that when users experience CVR, their awareness, comprehension and immersion are higher than ordinary video. In fact, in the cultural environment created by CVR, users actually gain the recognition and understanding of other cultures.

3.1.2   Create a Cultural Experience

CVR creates a cultural experience for users, which is embodied in the event experience and emotional experience. The participation of users watching CVR is far higher than that obtained by watching ordinary videos. We can experience different cultural rituals such as drawing, worship, etc. through the immersive effects in CVR. In this process, users will develop awareness and understanding of culture.

In previous studies, it was found that CVR has the effect of empathy. In a panoramic environment, user's feelings and emotions are more complicated. This research has also proved this point. In CVR, users' emotions are more intense than that of ordinary videos, that is, they can feel the happiness or serious emotions of people around them. In the cultural experience, this empathy effect deepens the user's experience and has a very important positive impact on the user's understanding of the culture. In the interview, some users mentioned their feelings and atmosphere, which they felt helped him understand the cultural connotations of the Japanese shrine and the festival.

3.1.3   Need Technical Means to Avoid Deficiencies

Use visual guidance. Due to the panoramic effect, CVR can provide a lot of information, but also cause information redundancy. Everyone who conducts interviews feels that there is a sense of excess information, and proper visual guidance can help users focus on key information. This visual guidance can be used to express key information from the main perspective, or by graphical means such as arrows. Through the interviews, it is found that this kind of annotation does not arouse the users' antipathy, and does not weaken its immersion. However, in the process of use, it is necessary to cooperate with the explanatory text labeling, especially the fit of time and direction is very important.

The speed is slowing down. In the process of video shooting and editing, you need to pay attention to the rhythm, because CVR is a panoramic effect in each frame, that is, each frame is rich in information much larger than ordinary video. If the rhythm is too fast, it will cause the user's dizziness, and the user will spend more energy to eliminate the vertigo, instead of understanding or memorizing the content.

3.2   Research Value

This study confirms the feasibility of CVR in intercultural communication, which can promote user's cultural cognition and understanding more than ordinary video. As CVR technology brings strong immersion to users, when applied to intercultural communication, it can fully display the characteristics of space architecture, character activities and cultural festivals, thus rendering a cultural atmosphere. In human civilization, basic emotions are always common. The empathy effect brought by CVR can also promote mutual understanding and communication among different cultures through basic emotional perception.

References

1. MacQuarrie, A., Steed, A.: Cinematic virtual reality: evaluating the effect of display type on the viewing experience for panoramic video. In: 2017 IEEE Virtual Reality (VR), pp. 45–54. IEEE, March 2017
2. Cheng, A., Yang, L., Andersen, E.: Teaching language and culture with a virtual reality game. In: Proceedings of the 2017 CHI Conference on Human Factors in Computing Systems, pp. 541–549. ACM, May 2017
3. Passig, D., Eden, S., Rosenbaum, V.: The impact of virtual reality on parents' awareness of cognitive perceptions of a dyslectic child. Educ. Inf. Technol. 13(4), 329 (2008)
4. Terlutter, R., Diehl, S., Koinig, I., Waiguny, M.: Who gains, who loses? Recall and recognition of brand placements in 2D, 3D and 4D movies (2013)
5. Craik, F.I.: Levels of processing: past, present… and future? Memory 10(5–6), 305–318 (2002)
6. Sato, S., Kageto, M.: The use of 360-degree movies to facilitate students' reflection on learning experiences. In: 2018 International Symposium on Educational Technology (ISET), pp. 266–267. IEEE, July 2018
7. Yoganathan, S., Finch, D.A., Parkin, E., Pollard, J.: 360 virtual reality video for the acquisition of knot tying skills: a randomised controlled trial. Int. J. Surg. 54, 24–27 (2018)
8. Ding, N., Zhou, W., Fung, A.Y.: Emotional effect of cinematic VR compared with traditional 2D film. Telematics Inform. 35(6), 1572–1579 (2018)
9. Keen, S.: A theory of narrative empathy. Narrative 14(3), 207–236 (2006)
10. Passig, D., Eden, S., Heled, M.: The impact of Virtual Reality on the awareness of teenagers to social and emotional experiences of immigrant classmates. Educ. Inf. Technol. 12(4), 267–280 (2007)

Augmented Reality to Engage Preschool Children in Foreign Language Learning

Elif Topsakal[1] and Oguzhan Topsakal[2(✉)]

[1] Uludag University, Bursa, Turkey
[2] Florida Polytechnic University, Lakeland, FL 33805, USA
otopsakal@floridapoly.edu

Abstract. The goal of this study is to see if Augmented Reality (AR) helps to motivate preschool kids and engage them in foreign language learning activities. The study conducts an experimental study to compare traditional approaches with an AR app for teaching a foreign language to preschool kids. Data were collected from experiments performed at three daycares on students at three age groups; 4, 5, and 6 years old. Surveys were also conducted to receive input from teachers and kids. The results show that there is significant engagement increase in learning a foreign language when AR apps utilized in the classroom.

Keywords: Augmented reality · Foreign language learning · Preschool · Early education · Engagement · English learning

1 Introduction

Learning a foreign language has become essential and increasing number of parents would like their kids to start learning a foreign language early. Kids have a better capability to learn a foreign language at early ages. However, engaging little kids with learning activities is challenging. To teach the foreign language effectively, it is important to motivate the learner, increase their interest and encourage them through engaging activities [1–4].

As futurist and writer Arthur C Clarke stated, "Any sufficiently advanced technology is indistinguishable from magic". Augmented reality is the technology that is used to enhance/augment our view of the physical world by placing computer-generated (virtual) information or objects [5]. Augmented reality (AR) is like magic for children [6–8]. AR is an excellent tool for children that fascinates, surprises, grabs attention, entertains and makes them engaged. This establishes a perfect environment for foreign language learning.

Recent technological improvements in mobile devices and AR software development kits enabled entrepreneurs to realize their AR product ideas. These products, including the wildly popular and successful apps like 'Pokemon Go' made AR a well-known concept among people. As AR becomes a familiar concept, it becomes easier to introduce them to teachers and come to a consensus that AR apps for kids have huge potential to attract kids' attention due to its engaging, surprising and fun features.

© Springer Nature Switzerland AG 2019
L. T. De Paolis and P. Bourdot (Eds.): AVR 2019, LNCS 11614, pp. 286–294, 2019.
https://doi.org/10.1007/978-3-030-25999-0_25

We have developed an educational set that utilizes AR to start teaching English to pre-school kids. The app aims to teach animal names and related action words. The main components of the educational set include 40 image cards (images of animals), 60 word cards (written text of animal names and action words such as walk, run, jump, etc.) and the AR app for Android and iOS devices. The set is developed through a grant from TUBITAK (The Scientific and Technological Research Council of Turkey) and teaches English for Turkish native speakers. AR app provides an interactive and engaging environment through games and fun interactions. We have conducted an experimental research study on kids between ages 4 and 6 at daycares and identified that kids learn a foreign language significantly better using the mobile AR apps when compared to conventional methods. Figure 1 below shows the content of the educational set and kids playing with it at a daycare center.

Fig. 1. On the left, contents of the 'Magical Animals' English learning set and on the right, kids playing with the educational set at a daycare center.

2   Related Work

The research on Augmented Reality gained momentum in recent years. Researchers have been exploring the utilization of the augmented reality technology in many fields such as tourism, advertisement, training, military, medicine, and education. Education is one of the fields that promise significant improvements and changes due to the utilization of the augmented reality [9, 10].

There have been studies to see the effect of augmented reality to teach a foreign language to students [1–4, 6–8, 11–13]. The studies about teaching a foreign language were mostly performed on students who are at elementary [3, 7, 12], secondary [1, 11] and college levels [14]. To the best of our knowledge, there was only one study [6] to test the effectiveness of AR on pre-school kids. However, this study only involved a very limited number of pre-school kids and it was based on observations [6].

In education, one of the most important factors to facilitate learning is motivation. This is true especially in foreign language learning [1, 6, 11–13]. The early education of kids at pre-school ages can only be motivated through games and surprise factors. AR is very surprising to kids and it is like a magic as it brings virtual objects to existence. In fact, studies conducted on early age school children indicated that many children described the AR apps as magic [6–8]. AR app development for education is challenging because the design of the educational experience needs to be different and user-centric when AR is utilized [3, 11]. For a successful AR app that provides an effective learning environment, certain design patterns need to be followed [15, 16].

3   System Design and Architecture

To teach a foreign language (English) for kids whose native language is Turkish, we developed an AR app that we call 'Magical Animals'. The app teaches 40 animal names along with 20 action words such as run, walk, dance, sit, jump, etc. The AR app was developed utilizing the Unity3D development environment and Vuforia AR libraries. The AR app utilizes image, text and voice recognition to provide an engaging environment for kids. The app works both on iOS and Android devices both on phone and tablets. The Unity3d development environment was chosen because once the app is developed in the Unity3D environment using C# programming language, the app can be exported to multiple platforms including iOS and Android which helps us save development time and cost. This also helps us to preserve the same unified design over multiple platforms.

Fig. 2. Kids are interacting with an animal 3D model via the 'Magical Animals' App

The interaction between the AR app and the user is done as follows: After the installed AR app started, the user can show an animal's image card or an animal's name text card to the camera of the mobile device (phone or tablet) to pop up animal's 3D model over the image or the text. As the user sees the 3D model, he or she can move the card to experience the animal from different angles on the mobile device's screen. The user can also show a card to the camera that has an action word written on it (for example a card with the 'dance' or 'walk' word on it) and the displayed 3D model of the animal starts animating (for example starts dancing or walking). Figure 2 shows that kids are interacting with the AR app.

4   Methodology

The study has been conducted in three daycares in Turkey on pre-school kids at ages 4, 5 and 6. Total of 87 kids participated to the study; 21 of these kids were in the fours years old group, 25 of these kids were in the five years old group and 42 kids were in the six years old group.

We utilized the experimental research study with pre-test and post-test applied to experiment and control groups to assess the effectiveness of utilizing the AR app for teaching English to pre-school kids. While the control group learned English with conventional flash cards, the experimental group learned English with the AR app. We distributed kids at each age group into experimental and control groups randomly.

First, we met with the teachers at daycares to share the details of our research study. We introduced them the AR concepts as well as the AR app that would be used during the study. We described the lesson plans and the activities in details. We also let them know that we would be conducting surveys to teachers and pre-tests and post-tests to kids. During our initial meetings, we also asked the following questions and received the answers listed below:

- How many of the students have used a tablet before? About 80% of the students already used tablets.
- What methods and tools did they use to teach English to kids? Some of the tools they used are flash cards, computers, presentation slides, books, and videos.
- How do they assess kids' level of English? In one of the daycares, they do not assess how much the kids learned English. However, in the other daycares, they ask questions to get a feel about how much they learned.
- Did the kids learn animal names in English before? One of the daycares claimed that kids learned about 15 animal names already, and the other two claimed they learned about 5 animal names.
- How is the attitude of teachers and parents to use tablets and technological tools for teaching? The attitude is positive since these tools attract kids' attention.

Before starting the teaching sessions, experimental and control groups took the pre-test to assess their English level at the beginning of the study. At the end of teaching sessions by using the AR app and by using the conventional flashcards, we applied the same test used in the pre-test as the post-test to both experimental and control groups to assess how much each group improved their English. During the teaching sessions, both control and experimental groups both spend 3 h to learn the English animal names. We prepared the lesson plans for the teaching sessions so that the same teaching style is followed at all three daycares.

For 4, 5, and 6 years old experiment and control groups, 15, 20, and 25 animal names were taught respectively. With the experimental group, the teacher first introduced the AR app to the kids and showed how to play with it. Then, in small groups, kids started to play with the AR app that was installed on an Android tablet. When an animal picture is shown to the camera of the Android tablet, the App pops up the 3D model of the animal, and then the audio is played pronouncing the name of the animal in English. With the control group, the teacher first showed pictures of each animal and repeated its name three times. Then, the teacher let each student say the animal names three times. Later, teacher and kids started to play a game where the teacher says an animal name and kids find the picture of the animal.

5   Results

During the pre-test and post-test, pictures of the animals were shown to the kids and their English names were asked. Table 1 shows the percentage of correct answers.

Table 1. Percentage of correct answers

		Control		Experiment	
		Pre-test	Post-test	Pre-test	Post-test
4 years olds	Daycare A	6.67%	13.33%	0.00%	33.33%
	Daycare B	2.86%	3.81%	4.76%	12.38%
	Daycare C	26.67%	30.00%	42.22%	55.56%
	All Daycares	8.00%	10.00%	14.55%	26.06%
5 years olds	Daycare A	2.50%	7.50%	0.00%	27.50%
	Daycare B	26.43%	30.00%	29.38%	35.63%
	Daycare C	23.33%	26.67%	43.33%	55.00%
	All Daycares	21.67%	25.42%	28.08%	38.85%
6 years olds	Daycare A	3.50%	14.00%	5.00%	36.50%
	Daycare B	28.00%	29.14%	28.00%	34.29%
	Daycare C	49.33%	55.33%	48.67%	64.00%
	All Daycares	24.76%	30.86%	25.14%	43.62%

Having the results in percentages helps us more conveniently compare changes between pre-test and post-test and between control and experiment groups as well. We computed the percentage as follows: For example, 6 years old control group were asked 25 animal names during their pre-test and post-test. We have 21 students in the control group and 21 students in the experiment group. If the group had identified all animal names correctly during a test, they would have given 525 (21 × 25) correct answers. The sum of the correct answers given by all students in the experimental group during the pre-test and post-test were 132 and 229 respectively. These number of correct answers corresponds to 25.14% (132/525) and 43.62% (229/525) as listed in Table 1 above.

Figure 3 below shows how much increase was achieved between pre-test and post-test for age levels 4, 5, and 6. As it can be noticed, the increase is always higher in the experimental group.

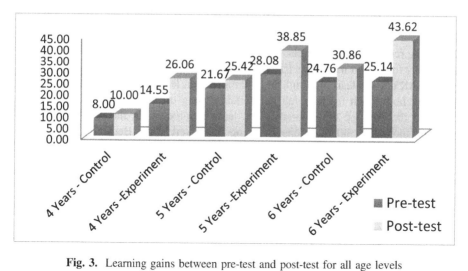

Fig. 3. Learning gains between pre-test and post-test for all age levels

Figure 4 below shows how many percent control and experiment groups increased the number of their correct answers for age levels 4, 5, and 6. For example, the control group of 6 years olds increased their total number correct answers by 24.62% while the experimental group of 6 years olds increased their total number of correct answers by 73.48%. Figure 4 shows that the experimental group increased their vocabulary significantly better than the control group at all age levels.

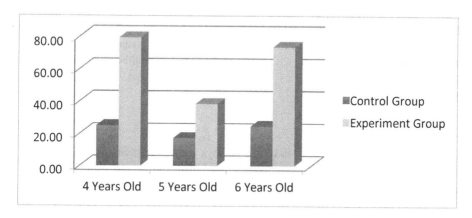

Fig. 4. Percentage of improvements in the correct answers given compared for control and experiment groups at 4, 5 and 6 years old levels.

6 Teacher Comments and Observations

During the research study, we also gathered data via surveys to learn how to improve the AR app to make it more user-friendly and effective, how much did the kids enjoyed utilizing AR app for learning English, how much their level of attention and focus was improved, how much teachers felt comfortable using the AR app for teaching, how much did the AR app contributed to the learning process. In the surveys, teachers answered more than 40 questions by using the Likert scale. Besides, they had the opportunity to provide explanations for their answers.

Teachers expressed that kids' attention span was prolonged while studying with the AR app and they were more focused on the learning process. Since kids found it fun to play with the app, it would be possible for them to spend more time with the app. It was also observed that kids wanted to play with the AR app over and over again. Teachers also suggested that classroom management was easier when AR app was used because kids were very much engaged with the learning process. The animations of the animals' 3D models captured kids' attention and improved communication between the kids since they often tried to mimic the animations and showed it to their friends. However, sometimes AR app caused sharing problems since most of the kids wanted to interact more with the app instead of letting his/her friends take the turn to play. Teachers also commented that since the AR app provides both visual and audio feedback to the kids, the retention level of the learned words would be higher. Teachers praised the AR app for using real pictures of animals instead of cartoon animal characters because the kids would have difficulty identifying the animals from their pictures if they initially learned their name from cartoon characters. Teachers also expressed that when they teach using conventional books or flash cards, they pronounce the English words with an accent however with the AR app, they can play the audio recorded by native English speakers. This enables kids to learn the correct pronunciation of the words. It is also noted that having the 3D model pop up from the text of the word also gives the kids an opportunity to match the written name with its meaning even before they start learning how to write and read.

7 Conclusion

The results of this experimental study and surveys were consistent with the results of the current literature because all the studies concluded with positive outcomes about utilizing AR in education and in foreign language learning. We have seen that kids learn English more effectively and achieve much better scores in the post-tests using the AR app when compared to the post-test scores of kids studied English using the conventional methods (flashcards). While the kids learning through the conventional methods increased their vocabulary 25%, 17%, and 24% for 4, 5, and 6 age levels respectively, kids utilizing the AR app increased their vocabulary 78%, 38%, and 73%.

Based on the feedback received from the teachers through surveys and based on our observations, experiment group showed improved attention span, increased engagement with the subject, and higher motivation to learn. Moreover, the teachers reported that it was easier to manage the classroom. It was observed that utilizing audio of native speakers for the pronunciation of words resulted in increased correctness of the kids' pronunciation of the new vocabulary.

The dramatic increase in learning outcomes encourages us to continue exploring the potential of AR for teaching a foreign language to early age learners. We explain how we plan to continue this research in the future directions section below.

8 Future Directions

We have seen a great potential in AR for engaging kids for foreign language learning. We plan to continue with the following research studies:

- Perform a similar experimental study to teach Spanish to kids that have English as the native language. That would allow us to compare the effectiveness of AR in different cultures and languages,
- Perform an experimental study to teach different subjects such as foods, clothes, numbers, etc. to compare if the subject was a differentiator factor for the engagement,
- Perform an experimental study utilizing the audio recordings of words and voice recognition functionality of the AR app to compare the correctness of pronunciations learned through the AR app and the conventional methods,
- Enhance the current AR application to utilize gamification techniques and variable rewards to test the effectiveness of gamification and rewards,
- Enhance the current AR application to add voice-bots that would guide kids in the learning process and perform the experimental study to test the effectiveness of voice-bots.

Acknowledgments. We would like to thank the management and teachers of the three daycares that provided the environment to perform this study. The daycares were in Bursa, Turkey and their names are Ilkem Kres Gunduz Bakimevi, Deniz Yildizi, Gelisim Atolyesi. We also would like to thank Aslinur Kilinc for her help performing the study in the daycares.

References

1. Gundogmu, N., Orhan, G.: Foreign language teaching with augmented reality application. Eurasia Proc. Educ. Soc. Sci. **4**, 309–312 (2016)
2. Vate-U-Lan, P.: An augmented reality 3D pop-up book: the development of a multimedia project for English language teaching. In: Proceedings - IEEE International Conference on Multimedia and Expo, pp. 890–895 (2012)
3. Chang, Y.J., Chen, C.H., Huang, W.T., Huang, W.S.: Investigating students' perceived satisfaction, behavioral intention, and effectiveness of English learning using augmented reality. In: Proceedings - IEEE International Conference on Multimedia and Expo (2011)
4. Scrivner, O., Madewell, J., Buckley, C., Perez, N.: Augmented reality digital technologies (ARDT) for foreign language teaching and learning. In: FTC 2016 - Proceedings of Future Technologies Conference, pp. 395–398. Institute of Electrical and Electronics Engineers Inc. (2017)

5. Carmigniani, J., Furht, B.: Augmented reality: an overview. In: Furht, B. (ed.) Handbook of Augmented Reality, pp. 3–46. Springer, New York (2011). https://doi.org/10.1007/978-1-4614-0064-6_1
6. Dalim, C.S.C., Piumsomboon, T., Dey, A., Billinghurst, M., Sunar, S.: TeachAR: an interactive augmented reality tool for teaching basic English to non-native children. In: Adjunct Proceedings of the 2016 IEEE International Symposium on Mixed and Augmented Reality, ISMAR-Adjunct 2016, pp. 344–345. Institute of Electrical and Electronics Engineers Inc. (2017)
7. Barreira, J. et al.: MOW: augmented reality game to learn words in different languages case study: learning english names of animals in elementary school. In: Rocha, A., Calvo Manzano, J.A., Reis, L.P., Cota, M.P. (eds.) Information Systems and Technologies. IEEE (2012)
8. Yilmaz, R.M.: Educational magic toys developed with augmented reality technology for early childhood education. Comput. Hum. Behav. **54**, 240–248 (2016)
9. Saidin, N.F., Halim, N.D.A., Yahaya, N.: A review of research on augmented reality in education: advantages and applications. Int. Educ. Stud. **8**, 1–8 (2015)
10. Pellas, N., Fotaris, P., Kazanidis, I., Wells, D.: Augmenting the learning experience in primary and secondary school education: a systematic review of recent trends in augmented reality game-based learning. Virtual Reality, pp. 1–18 (2018)
11. Küçük, S., Yilmaz, R.M., Göktaş, Y.: Augmented reality for learning English: achievement, attitude and cognitive load levels of students. Egitim ve Bilim **39**, 393–404 (2014)
12. Solak, E., Cakır, R.: Investigating the role of augmented reality technology in the language classroom. Croatian J. Educ. **18**, 1067–1085 (2016)
13. Salmon, J., Nyhan, J.: Augmented reality potential and hype: towards an evaluative framework in foreign language teaching. J. Lang. Teach. Learn. **3**, 54–68 (2013)
14. Ibrahim, A., et al.: ARbis Pictus: a study of vocabulary learning with augmented reality. IEEE Trans. Visual Comput. Graphics **24**, 2867–2874 (2018)
15. Antonaci, A., Klemke, R., Specht, M.: Towards Design Patterns for Augmented Reality Serious Games (2017)
16. Emmerich, F., Klemke, R., Hummes, T.: Design patterns for augmented reality learning games. In: Dias, J., Santos, P.A., Veltkamp, R.C. (eds.) GALA 2017. LNCS, vol. 10653, pp. 161–172. Springer, Cham (2017). https://doi.org/10.1007/978-3-319-71940-5_15

A Study on Female Students' Attitude Towards the Use of Augmented Reality to Learn Atoms and Molecules Reactions in Palestinian Schools

Ahmed Ewais[1,2]([⊠]) [iD], Olga De Troyer[2] [iD], Mumen Abu Arra[1], and Mohammed Romi[1]

[1] Computer Science Department, Arab American University, Jenin, Palestine
ahmad.ewais@aaup.edu
[2] WISE, Computer Science Department, Vrije Universiteit Brussel, Brussels, Belgium

Abstract. The chemical reaction between different molecules is an important learning subject in a chemistry course. Especially for elementary school students, this can be an abstract concept and therefore difficult to understand. One way to facilitate this learning process is to use Augmented Reality (AR) technology, which is considered as an added value compared to classical learning materials such as textbooks, 2D images, video, etc. Among the different advantages of using AR technology in the context of educational domain is the fact that 3D technology is offering a safe environment especially if the students have to perform critical tasks such as simulating chemical reactions. This work investigates the students' attitude towards the use of a mobile AR application for learning atoms and molecules reactions. In particular, we focused on female students as in general female students show less interest in science and technology than male students. We were keen to investigate whether the use of AR technology could change this attitude. The students are able to interact with the different AR components to explore the possible reactions in a 3D interface, to see the structure and shape of atoms and molecules, and to view related descriptions in their native language, i.e. the Arabic language. The mobile AR application was evaluated by a class of 12–13 years old (7th grade) students in a Palestinian primary school. The number of participants was 50, all female students. After analyzing the results, we can conclude that the female students had a positive attitude toward the use of this AR application in their learning process.

Keywords: Augmented Reality · Attitudes · Chemical reactions · Female students

1 Introduction

Since the last two decades, Augmented Reality (AR) has been increasingly getting researchers' interest. The first use of AR applications was in the 1990s. It is believed that the term Augmented Reality was coined by the Tom Caudell in 1990 [1]. AR is defined as a technology that allows to give an overlay of computer generated data on

© Springer Nature Switzerland AG 2019
L. T. De Paolis and P. Bourdot (Eds.): AVR 2019, LNCS 11614, pp. 295–309, 2019.
https://doi.org/10.1007/978-3-030-25999-0_26

the user's view of the real environment [2]. AR uses virtual reality combined with video, images, audio, etc. [3].

Researchers proposed the use of AR in teaching and learning because of the main advantage of AR over traditional media. AR is visually and interactively richer than classical learning materials [1]. Accordingly, it is considered as more attractive and motivating than traditional learning materials [2]. For instance, AR can integrate different formats such as text, audiovisual materials, and 3D models into a real environment. Some researchers [4, 5] already reported comparative studies between traditional classes and classes that use AR technology, which confirmed that AR improves students' learning.

There is also a large number of research studies that are related to the possible use of AR in educational contexts [6]. These studies report advantages, effectiveness, and users' attitudes toward the use of AR in educational settings. According to [7], in most cases when using AR for educational purposes, it is used in higher educational settings. Furthermore, among the different domains (social science, engineering, health, etc.) considered for the use AR, the science domain is most common. This is related to the fact that AR is considered as an effective tool for learning concepts and topics that cannot easily be conducted in real world, for which there is a lack of special devices and hardware, and because of the ability of visualizing abstract or complex concepts. AR also offers better ways of learning by providing physical and virtual objects in a rich sensory context [8].

With the increase spread of smartphones and tablets, researchers start investigating the use of such devices in different educational contexts [9, 10]. Furthermore, the recent hardware technologies along with different developed software applications make it possible to run mobile-based AR applications.

This article presents a study about female students' attitudes toward the use of AR for learning chemistry at the primary school level. In particular, the used AR application demonstrates the concepts of atoms and molecules structure and molecules reactions as given in the chemistry course. The focus is on female students because in general female students show less interest in science and technology than male students. In addition, a recent world bank report [11] emphasizes the need to encourage female students to complete their 12 years of education as this can increase economic benefits. Furthermore, it is known that male students, in general, have a good experience with 3D games, Virtual Reality and Augmented Reality [1, 12, 13]. In general, female students are less familiar with 3D games, Virtual Reality and Augmented Reality. Therefore, it is quite important to specifically investigate female students' attitudes and motivation toward the use of AR technology in the learning process as a negative attitude may have a direct impact on their learning achievements.

There are different arguments for integrating new methods and techniques to learn and teach chemistry. First, for many primary school students, chemistry is not an easy course. Some concepts such as atoms, molecules and chemical interaction are tremendous difficult for those students. Furthermore, it is difficult for them to understand how molecules can interact with each other and produce new chemical materials.

Another important aspect is related to the fact that chemical resources are limited in Palestinian schools. Another important reason is the safety issue which is related to doing experiments with young students in the lab. AR can play an important to role to avoid dangerous situations for students. Furthermore, AR can be used to represent molecules for which real counterparts are unavailable or too expensive.

This article is structured as follows. The next section presents prior work related to investigating the use of AR in educations contexts. Section 3 introduces our research method and the setup of the conducted evaluation. Section 4 presents evaluation results followed by discussion and recommendation in Sect. 5. Finally, Sect. 6 concludes this article with future work.

2 Literature Review

In the educational context, motivation is defined as the student's desire to engage in a learning environment [14]. Since researchers started to investigate the use of AR technology in educational settings, a number of studies investigated student's motivation toward the use of AR technology in learning and students' learning achievements [10, 15]. For instance, researchers [16] observed that the students' reaction regarding reading an AR textbook by displaying related text, an avatar, sound, etc. was positive; a high motivation was one of the important results of the conducted evaluation. Another recent work [17] shows that using an AR-based multidimensional concept map with 65 students with average age of 11 years in elementary school could improve the students' learning achievements.

In general, learning experience is directly related to usability [18]. Therefore, different researchers reported on usability studies related to AR applications. For instance, the study in [19] investigated usability aspects of an AR book which allows children to learn about fish conservation. The results showed that learners were interested in using AR technology because of the good usability results of the developed system [10]. showed that an important challenge is the difficulty that users may experience to handle and interact with an AR interface. Also the researchers in [15, 20] showed that a usable interface is important to avoid technical problems while the students are using an AR application. Moreover, researchers [21] suggested that AR technologies need to optimize the use of the device's memory, and need to be fast enough to smoothly display images, 3D models, audio, video, etc.

Furthermore, AR technology has been explored in different domains. For instance, games and entertainment use AR technology, and [22] mentioned that AR games can be considered as a fun, motivating, and engaging environments for students to learn different topics. Other research work [5] indicated that using a SMART AR application is a good tool especially for weak students to gain knowledge about different learning concepts. The students were able to explore transportation, animals, etc. using 3D models inside an AR desktop-based application.

Moreover, in [23] researchers showed that secondary school students are able to play collaboratively an inquiry-based mobile AR game to show related information and to solve different detective cases. Another interesting work is presented in [4] and showed a comparative study with 10 year old children between using a mirror AR interface and traditional teaching methods to learn about the sun and earth. Unexpectedly, the results showed that the students who used AR technology were less engaged than those who used traditional learning resources. Therefore, a number of design requirements have been suggested such as enabling the teacher to add learning content to the AR application. Also, AR has been used in mathematics courses [24]. The work in [25] investigates the students' laboratory skills and attitudes toward science laboratories. The conclusion was that AR technology helped students to build a positive attitude towards physics laboratories. Furthermore, the students' laboratory skills were improved by the use of different AR components such as animation, video, images, and 3D models. However, the proposed application is a desktop-based application.

3 Research Method

This section presents the research goal and the research question of our work. Next, a description about the developed AR mobile application is presented. Finally, the performed usability and acceptance evaluation is explained.

3.1 Research Goal and Question

The main goal of this study was to investigate female students' attitude toward the use of an AR mobile application in a chemistry course. Therefore, the research question was formulated as follows: "what is the student's attitude towards using a mobile AR application in a chemistry course for female students of a Palestinian primary school?"

3.2 AR Application

To answer the research question, a suitable mobile AR application was needed. For this purpose, an interview with the class teacher has been done to identify some of concepts that the students experience as difficult to learn. Molecules reaction was selected to be simulated using a mobile AR application. Therefore a mobile AR application was developed to enable students to explore the different reactions between a number of atoms and molecules. It is important to mention that this AR application was only created for the purpose of evaluating both its usability and student's acceptance for using AR applications in their chemistry course. This implies that the functionality of the application is limited and does not cover all possible atoms, molecules and interactions. The selected atoms and reactions were based on the advice of the chemistry teacher.

The 3D representation for each atom and molecule enable students to visualize and master the structure of the different components. For instance, the student is able to visualize atoms of Oxygen, Hydrogen and water in a 3D representation. They are able to see that a water molecule is composed of one Oxygen atom and two Hydrogen atoms.

The AR mobile application was developed as a marker-based Android application using Unity3D and Vuforia SDK. The AR mobile application is able to read special cards, called "markers", using the mobile camera. These markers are used to represent the following atoms: Oxygen, Zinc, Hydrogen, Sodium, Chlorine, and Copper and two molecules Ammonia and Hydrochloric. Each marker contains the name of the atom, its associated number from the periodic table, and a symbol or icon to represent it.

Once the marker is placed in the scope of the smartphone's camera and according to the used marker, a 3D object is displayed to represent the associated atom. Furthermore, an Arabic explanation is displayed to the students about the current atom. Figure 1 shows a marker and the 3D object representing Cooper "Cu".

Fig. 1. AR mobile application

The application is able to display the possible reaction between the two atoms and will present the formed molecule as a 3D object. This is achieved by placing the two markers close to each other in the scope of the smartphone's camera. Figure 2 displays such a situation for the atoms Na and Cl presented as 3D objects. Furthermore, a description about the possible interaction between both atoms is displayed in Arabic at the bottom of the smartphone screen. If there is no possible reaction between the two atoms, a message will be displayed saying that there is no interaction possible.

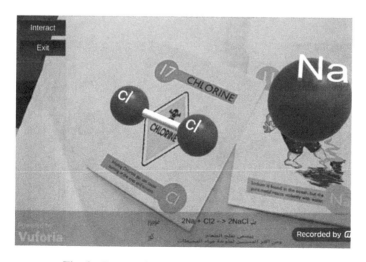

Fig. 2. Two markers used for two different atoms

Once the student presses the interact-button, a 3D representation of the formed molecule will be displayed together with an icon to symbolize the molecule. Figure 3 shows the 3D structure of the salt molecule formed from Na and Cl and it shows the icon for salt. Moreover, an Arabic explanation about salt is shown on the screen.

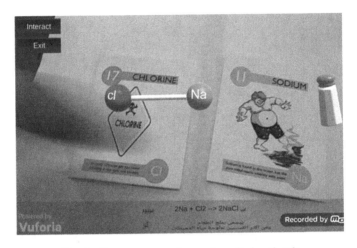

Fig. 3. Two atoms reaction and resulted molecule

The application also enables the students to visualize molecules by using specific markers. For instance, Fig. 4 shows two molecules Ammonia and Hydrochloric which form a new component called ammonium chloride. The predefined reactions are the following: $H + O_2 = H_2O$ forming Water, $Cu + O_2 = Cu_2O$ forming Black Copper Oxide, $2Na + Cl = 2NaCl$ forming Salt, $NH_3 + HCl = NH_4Cl$ forming Ammonium Chloride.

Fig. 4. Two molecules and possible interaction

3.3   Evaluation

The evaluation mainly focused on usability and acceptance of using an AR-based learning application in a chemistry course. The evaluation was conducted in an academic setting inside a lab in a Palestinian school.

The evaluation was carried out with 50 female students of one Palestinian primary school in Jenin. The students were 12–13 years old and in a class of the 7th grade. They were all following a chemistry course.

Both qualitative and quantitative evaluations were adopted in this research work. This hybrid approach was used to allow digging deeper into the students' opinions and experiences.

The evaluation took place during a one-hour session and consisted of three steps. The first step was an introductory session, which lasted for 10 min, about the AR application where one of the researchers presented the AR Molecules Reaction application to the students. He explained the markers and showed how they can be used to view the structure and possible interactions. Moreover, he showed how the possible interaction between two atoms is presented: the name of molecule resulted from the interaction, the Arabic description about the atoms, the status of resulted molecules, the Arabic and English name of molecules, and a 3D representation of the resulted molecule. Figure 5 shows a picture of this step.

In the second step, the students were divided into groups of 5 students. A 10 min slot was given to the students to try out the application. Eight markers were given to each group and a mobile device containing the application. The students were asked to explore the application and try to find out how they could interact with markers using the AR application. Each group of students could interact, visualize, discuss, and learn without a teacher's direct instructions. They were able to see which atoms can interact and their result.

Fig. 5. Application demonstration

Finally, in the third step, paper-based questionnaires were distributed to all students and they were asked to fill them out within 15 min. Students were able to contact their teacher in case they needed clarifications about a question. Afterwards, volunteer students were interviewed during another 15 min.

The paper-based questionnaire was composed of questions that are related to the System Usability Scale (SUS) [26]. In particular, there were 8 questions on usability, 2 on perceived usefulness, 4 on attitude, 3 on intention to use, and 3 on perceived ease of use. Each question was evaluated using a Likert scale (from 1 to 5). General questions were used to gather information about the participants, such as their background in Virtual Reality, AR, the use of 3D games, and the use of AR technology. This was complemented with a qualitative questionnaire to delve deeper into the students' opinions and experiences. The questions of the qualitative questionnaire were selected and reviewed by an HCI expert and the course teacher to ensure their validity. The questions were mainly used to encourage the students to discuss about their opinions and feelings regarding the mobile AR application. Finally, it is important to mention that all questions were in Arabic language to avoid misunderstanding due to the language used.

In the paper-based questionnaire, all statements were mandatory and all answers had a scale from 1 (Strongly Disagree) to 5 (Strongly Agree). For each individual evaluation feedback on a positive question, a score of 3 or higher is considered as "good" (positive feedback). Each individual evaluation feedback on a negative formulated question is considered "good" when the score is 3 or lower (positive feedback).

As we were looking for a critical and objective evaluation, we adopted a number of scientific steps to avoid bias. First, the students were encouraged to be critical and objective. Next, to try to exclude bias, negative and positive formulated questions were used in the questionnaire. Any contradictory results between these questions led to disregarding that evaluation form. Moreover, the students were informed that the evaluation results would only be used to evaluate the provided AR application and to reveal advantages and disadvantages of such an application. They were also informed that the evaluation would be done anonymously so that there was no need to indicate their names or IDs on the questionnaire.

4 Results

This section presents the results concerning the general information about the participants, the SUS questionnaire (usability and attitudes), and the interview.

4.1 Demographic Data

The evaluation was conducted with a group of 50 volunteer students with almost homogeneous background about AR applications. The results show that 68% (34) of the students already knew Virtual Reality (VR). However, the majority of the participants (30) had no or very little experience with Augmented Reality (AR). Such result is expected as AR technology is a relatively new trend. 60% of the participants were not at all familiar with AR technology and more than 70% of participants had never used AR technology or used it only once.

4.2   Usability and Attitude

In general, the obtained results were good. Table 1 provides an overview of the number of questions that were rated with good, neutral, and poor for the different categories. As shown in the provided table, no questions were rated as poor. Two questions for Usability, one for Attitude and one for Intention to use received a neutral rating. All other questions, 6 for Usability, 2 for Perceived usefulness, 3 for Attitude, 2 for Intention to use, and 3 for Perceived ease of use, received a good evaluation.

Table 1. Summary of the quantitative evaluation

Category	Score			
	Good	Neutral	Poor	Total
Usability	6	2	0	8
Perceived usefulness	2	0	0	2
Attitude	3	1	0	4
Intension to use	2	1	0	3
Perceived ease of use	3	0	0	3
Total number of all questions				20

This result provides a first indication that such an AR application can be used in a chemistry course for elementary schools' students in Palestine.

Next, we provide the details for each part of the questionnaire. To make it easy for the readers to distinguish between the positive formulated questions and negative formulated questions, we added at the beginning of each negative formulated question a star symbol (*) and put them in italic.

A number of findings can be revealed from Table 2. For instance, question 'USA5' ("I would imagine that most people would learn to use this AR mobile application very quickly") gained the highest average with second lowest standard deviation. This indicates that the AR mobile application can be learned quickly.

The rest of the questions that received good average scores were 'USA3' ("I found that the various functions in this AR mobile application are well integrated") and 'USA7' ("I felt very confident using the AR mobile application") for positive formulated questions and 'USA1' ("I found the AR mobile application unnecessarily complex"), 'USA4' ("I thought there is too much inconsistency in this AR mobile application"), and 'USA6' ("I found the AR mobile application very cumbersome to use") for negative formulated questions. Such results indicated that the AR application in general has good usability.

A number of questions (2) were given a neutral score, i.e. 'USA2' ("I think that I would need the support of a technical person") and 'USA8' ("I needed to learn a lot of things before I could get going with this AR mobile application"). The standard deviations are relatively high for both questions. This result is expected as most of the participants had not yet used an AR application. However, this issue can be solved by

Table 2. Usability evaluation results

Code	Questions	Min	Max	Avg.	Std.
USA1	*1. *I found the AR mobile application unnecessarily complex*	1	4	1.72	0.68
USA2	* 2. *I think that I would need the support of a technical person to be able to use this AR mobile application*	1	5	3.38	0.73
USA3	3. I found the various functions in this AR mobile application are well integrated	3	5	4.1	0.28
USA4	*4. *I thought there is too much inconsistency in this AR mobile application*	1	4	1.7	0.59
USA5	5. I would imagine that most people would learn to use this AR mobile application very quickly	2	5	4.32	0.36
USA6	*6. *I found the AR mobile application very cumbersome to use*	1	4	1.68	0.68
USA7	7. I felt very confident using the AR mobile application	2	5	4.18	0.37
USA8	*8. *I needed to learn a lot of things before I could get going with this AR mobile application*	1	5	3.38	0.66

providing a proper introductory session for the students in the chemistry class or providing them with a manual on how to use the AR mobile application.

Concerning the questions related to the perceived usefulness, the results (see Table 3) show an increase in perceived attention (PU1) and the students considered the class as interesting and useful (PU2). This result corresponds with results obtained in previous research such as [27].

Table 3. Perceived usefulness evaluation results

Code	Questions	Min	Max	Avg.	Std.
PU1	9-I paid more attention in this class than any others	1	5	3.76	0.37
PU2	10-This class has been useful and interesting	0	5	4.24	0.451

Concerning students' attitudes, the students mentioned that they behaved better in this class and they would like to attend more classes that use AR technology. The majority of the students found that the use of AR technology could help them to pass their exams. However, this finding needs further investigation by comparing students' results in exams with and without using AR technology during the course. Table 4 shows the results of the questions related to the attitude aspect. The rating of question ATT2 ("It is easier to follow the teacher's explanation in this type of classes") was neutral. This could be due to the fact that some of the students were too curious to experiment with the tool instead of focusing on the teacher's instructions.

Table 4. Students' attitude evaluation result

Code	Question	Category	Min	Max	Avg	Std.
ATT1	11- I would like to take more classes like this	Attitude	1	5	4.22	0.4
ATT2	12- It is easier to follow the teacher's explanation in this type of classes	Attitude	1	5	3.28	0.69
ATT3	13- I behaved better in today's class than I did in other classes	Attitude	0	5	3.96	0.38
ATT4	14- I believe this AR material will help me pass the exam	Attitude	0	5	4.32	0.42

These findings echo those reported in the literature which also stated that using AR technology in education has the potential to improve students' attitudes and increase their motivation [15, 17, 28]. However, some researchers [10, 29] attribute this effect to the novelty effect and this means that the effect would only be temporally. In other words, once the student gets familiar with the use of AR technology, the effect on his attitude and motivation could diminish.

With regards to students' intention to use AR technology, the results showed that the majority of the students like to use the mobile AR application for learning chemistry. As evidenced by the results on question ITU1 and ITU3, students are willing to use the AR application at home and on a frequent basis. Table 5 shows the results of the questions that are related to the intention to use part of the questionnaire.

Table 5. Intension to use evaluation result

Code	Question	Min	Max	Avg.	Std.
ITU1	15- I would like to use this AR material at home	1	5	3.64	0.73
ITU2	*16- I prefer the classic book over the new materials	1	5	2.38	0.97
ITU3	17-I think that I would like to use this AR technology frequently	1	5	3.82	0.39

Concerning the perceived ease of use aspect, high average scores were obtained. This is a good indication that the students found the mobile AR application easy to use and learn (Table 6).

Table 6. Perceived ease of use evaluation result

Code	Question	Min	Max	Avg.	Std.
PEOU1	*18- It was not easy for me to use the AR markers	1	5	1.98	0.78
PEOU2	19- This AR material has been easy to learn and use	0	5	4.18	0.37
PEOU3	20. I think the AR mobile application is easy to use	2	5	4.38	0.31

4.3 Interview

As mentioned earlier, after filling out the questionnaire, 10 volunteered students were selected for an interview. The results for each question were as follow:

All 10 students answered the question "What do you think about using an AR tool for learning molecules interaction?" with positive feedback and with some notable appreciation for the use of AR in the chemistry course. Among the different comments, one student stated: "It was an interesting experience for me to use AR application in this activity".

The answers on the question "Why do you think this tool is helpful? In what other classes could the tool help you?" indicate also a great appreciation for the value that AR adds to the learning activities in a chemistry course. A number of participants (5) were convinced that AR makes it easy, not only to understand the structure of atoms and molecules, but also to see the result of the interaction between atoms or molecules, by using markers and 3D representations.

Answers concerning the question "Do you wish to use mobile AR application to learn chemistry in the future? Why?", all participants (10) were positive. Most of the participants (6) mentioned that the application should be available on the smartphone of the students so that they can discuss and learn as a group even after class time. Others considered AR technology as an interesting and entertaining experience. Similar students answers are also found in the literature, see for instance [10].

In reply to the question "Do you prefer to use an AR application on a smartphone or on a personal computer using a mouse to interact with the application?", 8 participants indicated a desire to use AR technology on a smartphone rather than on a personal computer, as a smartphone can be use everywhere and anytime.

Furthermore, participants were encouraged to state what they did not like about the application. The question was formulated as follows: "Do you think that AR mobile application has disadvantages? Which ones?". Some participants (4) thought that there is a need to add more details about atoms, molecules, and interactions. The level of detail given should depend on the level of the student (beginner, intermediate or advanced student).

Finally, students were asked the question "Can you offer some advice for improving this AR learning tool?" and two students indicated that the use of the periodic table as a marker could be better than having a marker for each atom. Four other participants suggested to create more markers so that they could explore all atoms in the periodic table.

5 Discussion

The overall results of the conducted evaluation were quite positive. Usability was rated well despite the fact that most of the participants lacked a minimum experience with AR applications and there was no true training period foreseen to enable students to get acquainted with the use of AR. Furthermore, students had a positive attitude towards the use of AR for learning chemistry.

In general, the results from the qualitative evaluation are consistent with the results of the quantitative evaluation. The following points show that the consistency and validity of the conducted evaluation is good:

- In the quantitative evaluation, the participants perceived the AR application as useful and easy to use. This is also confirmed in the qualitative feedback by the fact that the interviewees were considering AR mobile applications as interesting, easy to use and entertaining. Furthermore, other comments mentioned the usefulness of 3D visualization and the interaction with 3D models.
- Good appreciation results for both attitudes and intention to use were confirmed in the qualitative evaluation by the fact that the interviewees were considering AR as a good tool for understanding atoms, molecules and interactions. Furthermore, its use was suggested for different other courses such as math, physics, etc.
- There was a high appreciation for using AR over traditional materials. This is also confirmed by participants in the qualitative evaluation. They mentioned that it would be beneficial to provide them with mobile AR applications for different courses.

6 Conclusion

This research allows to conclude that AR technology was well received by our participants (all females) and provides several advantages over classical learning material. This study is also useful for policy makers in education in Palestine. Our research work provides a first evidence for the potential of AR technology in Palestinian elementary schools, and more in particular for chemistry courses, and to encourage female students to complete their 12 years of education. This specific AR application can help 12–13 years old (7th grade) students to understand the structure of the atoms, molecules and possible interactions between different atoms and molecules.

As mentioned earlier, the intention of measuring the influence of the mobile AR application on the learning outcome was not in the scope of this study. For this it will be needed to examine the effectiveness of the use of AR technology in respect with knowledge retention, acquired skills, etc. This can be done by means of a pre-test and post-test or by comparing an experimental group with a control group. This is planned for future work.

We also intend to conduct a study with school teachers to allow them to give their feedback on their perceived effectiveness of using AR technology as a teaching tool inside chemistry courses in Palestinian universities. We know relatively little about the attitude and acceptance of primary school teachers for using AR technology into their lessons.

Acknowledgment. The first author gratefully acknowledges the financial support by Zamalah seventh scholarship in undertaking the research to publish the result in cooperation with WISE research lab at the Vrije Universiteit Brussel-Belgium.

References

1. Lee, K.: Augmented reality in education and training. Link. Res. Pract. Improv. Learn. **56**(2), 13–21 (2012)
2. Azuma, R.T.: A survey of augmented reality. Presence: Teleoperators Virtual Environ. **6**(4), 355–385 (1997)
3. Parker, J.R.: Algorithms for Image Processing and Computer Vision. Wiley, Hoboken (2011)
4. Kerawalla, L., Luckin, R., Seljeflot, S., Woolard, A.: 'Making it real': exploring the potential of augmented reality for teaching primary school science. Virtual Reality **10**(3–4), 163–174 (2006)
5. Freitas, R., Campos, P.: SMART : a SysteM of augmented reality for teaching 2nd grade students. In: Proceedings of the 22nd British HCI Group Annual Conference on People and Computers: Culture, Creativity, Interaction, vol. 2, pp. 27–30, April 2008
6. Wu, H.K., Lee, S.W.Y., Chang, H.Y., Liang, J.C.: Current status, opportunities and challenges of augmented reality in education. Comput. Educ. **62**, 41–49 (2013)
7. Chen, P., Liu, X., Cheng, W., Huang, R.: A review of using Augmented Reality in education from 2011 to 2016. Innovations in Smart Learning. LNET, pp. 13–18. Springer, Singapore (2017). https://doi.org/10.1007/978-981-10-2419-1_2
8. Dunleavy, M., Dede, C., Mitchell, R.: Affordances and limitations of immersive participatory augmented reality simulations for teaching and learning. J. Sci. Educ. Technol. **18**(1), 7–22 (2009)
9. Farley, H., et al.: How do students use their mobile devices to support learning? A case study from an Australian Regional University. J. Interact. Media Educ. **1**(14), 1–13 (2015)
10. Akçayır, M., Akçayır, G.: Advantages and challenges associated with augmented reality for education: a systematic review of the literature. Educ. Res. Rev. **20**, 1–11 (2017)
11. The World Bank: Not Educating Girls Costs Countries Trillions of Dollars, Says New World Bank Report. Washington (2018)
12. Johnson, L., Smith, R., Willis, H., Levine, A., Haywood, K.: The 2011 Horizon Report (2011)
13. Cai, S., Wang, X., Chiang, F.K.: A case study of Augmented Reality simulation system application in a chemistry course. Comput. Human Behav. **37**, 31–40 (2014)
14. Keller, J.M., Litchfield, B.C.: Motivation and Performance. In: Reis, R.A. (ed.) Trends and Issues in Instructional Design and Technology. Merill Prentice Hall, New Jersey (2002)
15. Lin, H.C.K., Chen, M.C., Chang, C.K.: Assessing the effectiveness of learning solid geometry by using an Augmented Reality-assisted learning system. Interact. Learn. Environ. **23**(6), 799–810 (2015)
16. Dünser, A., Hornecker, E.: Lessons from an AR book study. In: Proceedings of the 1st International Conference on Tangible and Embedded Interaction, TEI 2007 (2007)
17. Chen, C.H., Huang, C.Y., Chou, Y.Y.: Effects of augmented reality-based multidimensional concept maps on students' learning achievement, motivation and acceptance. Univers. Access Inf. Soc. **18**, 1–12 (2017)
18. Cuendet, S., Bonnard, Q., Do-Lenh, S., Dillenbourg, P.: Designing Augmented Reality for the classroom. Comput. Educ. **68**, 557–569 (2013)
19. Lin, H.C.K., Hsieh, M.C., Wang, C.H., Sie, Z.Y., Chang, S.H.: Establishment and usability evaluation of an interactive AR learning system on conservation of fish. Turkish Online J. Educ. Technol. **10**(4), 181–187 (2011)
20. Squire, K., Klopfer, E.: Augmented reality simulations on handheld computers. J. Learn. Sci. **16**(3), 371–413 (2007)

21. Yu, D., Jin, J.S., Luo, S., Lai, W., Huang, Q.: A useful visualization technique: a literature review for augmented reality and its application, limitation & future direction. In: Huang, M., Nguyen, Q., Zhang, K. (eds.) Visual Information Communication, pp. 311–337. Springer, Boston (2009). https://doi.org/10.1007/978-1-4419-0312-9_21

22. Schrier, K.: Using augmented reality games to teach 21st century skills. In: ACM SIGGRAPH 2006 Educators program on - SIGGRAPH (2006)

23. Bressler, D.M., Bodzin, A.M.: A mixed methods assessment of students' flow experiences during a mobile augmented reality science game. J. Comput. Assist. Learn. 29(6), 505–517 (2013)

24. Bujak, K.R., Radu, I., Catrambone, R., MacIntyre, B., Zheng, R., Golubski, G.: A psychological perspective on augmented reality in the mathematics classroom. Comput. Educ. 68, 536–544 (2013)

25. Akçayir, M., Akçayir, G., Pektaş, H.M., Ocak, M.A.: Augmented reality in science laboratories: the effects of augmented reality on university students' laboratory skills and attitudes toward science laboratories. Comput. Human Behav. 75(1), 334–342 (2016)

26. Brooke, J.: SUS - A quick and dirty usability scale. Usability Eval. Ind. 30(9), 189–194 (1996)

27. Wojciechowski, R., Cellary, W.: Evaluation of learners' attitude toward learning in ARIES augmented reality environments. Comput. Educ. 68, 570–585 (2013)

28. Hwang, G.J., Wu, C.H., Kuo, F.R.: Effects of touch technology-based concept mapping on students' learning attitudes and perceptions. Educ. Technol. Soc. 16(3), 274–285 (2013)

29. El Sayed, N.A.M., Zayed, H.H., Sharawy, M.I.: ARSC: augmented reality student card an augmented reality solution for the education field. Comput. Educ. 56(4), 1045–1061 (2011)

Intelligent System for the Learning of Sign Language Based on Artificial Neural Networks

D. Rivas[1(✉)], Marcelo Alvarez V.[1(✉)], J. Guanoluisa[1(✉)],
M. Zapata[1(✉)], E. Garcés[2(✉)], M. Balseca[2(✉)], J. Perez[1(✉)],
and R. Granizo[1(✉)]

[1] Universidad de las Fuerzas Armadas ESPE, Sangolquí, Ecuador
{drrivas,rmalvarez,jfguanoluisal,mbzapata,japerez9,
ragranizo}@espe.edu.ec
[2] Pontificia Universidad Católica del Ecuador sede Ambato, Ambato, Ecuador
{egarces,jbalseca}@pucesa.edu.ec

Abstract. Sign language has become a form of communication for people who have some type of hearing impairment, in order to relate to the world around him and perform daily activities including education, work performance and personal development. In the knowledge society, the development of new Technologies for Information and Communication (TIC), they constitute a teaching tool that changes the conditions for learning and communication as the world is digitized. Therefore, the development of an Ecuadorian sign language learning system based on the use of a gesture sensor and an artificial neural network multilayer feed-forward, implementing the Backpropagation algorithm, which takes advantage of the parallel property of decreasing the time required by a processor to distinguish the existing relationship between given patterns, represents a solution for improving the teaching-learning process.

The system consists of three main layers: The first is responsible for the acquisition of data through a gestural device; the second uses the library of the gesture sensor to perform the acquisition and storage of information of the hand and the position of the fingers. Finally, the output layer, once the training of the neuron is analyzed, generates the real-time recognition of sign language.

Keywords: CSCL · Artificial neural networks · Pattern recognition · Leap Motion

1 Introduction

The World Health Organization (WHO) reports that more than 5% of the world population suffers from some type of hearing disability, among them, 466 million are adults and 34 million are children [1]. In Latin America, the presence of hearing problems is 1.6% in children under 14 years of age; in people older than 15 years it is 8%; and 38.62% corresponds to people over 65 years of age. In Ecuador, according to the National Council for Disability Equality (CONADIS) there are 437,268 people with some degree of disability; Of these, 14.17% have hearing impairment, subdividing this percentage of the population as follows: 8.13% are children under 12 years, 13.13%

© Springer Nature Switzerland AG 2019
L. T. De Paolis and P. Bourdot (Eds.): AVR 2019, LNCS 11614, pp. 310–318, 2019.
https://doi.org/10.1007/978-3-030-25999-0_27

people between 13 and 65 years, and 19.05% people over 65 years. Of a total of 58,161 people with occupationally active disabilities, 16.03% are people with hearing disabilities, of this percentage 16.24% are people whose age is between 18 to 29 years, 78.49% people between 30 to 64 years and 5.28% they are people over 65 years [2].

Traditional education is mostly focused on people without any type of disability, which makes it difficult for people with disabilities to have access to a quality education that has the improvements and technological advances developed throughout the world [3]. Therefore, promoting the inclusion of people with disabilities and developing new teaching tools is considered an appropriate option in the field of education, new methodologies such as Computer-Supported Collaborative Learning (CSCL) are presented as an emerging area of the learning sciences regarding how people can interact together with the help of computers [4].

There is an interest in projects dealing with artificial neural networks, or gesture sensors, focused on the teaching of sign language, as can be indicated in the following works; for example, in the article developed by [5], consists of the use of Leap Motion, a computerized pattern classification module and a graphical user interface for the teaching of Ecuadorian sign language.

The proposal made by [6], consists of a Sign Language Teaching Model (SLTM) designed to develop different Communication Skills (CS) in deaf children within a Collaborative Learning Environment with Mixed-Reality (CLEMR). In the work developed in [7], he proposes the use of a new information technology to model sign language based on the human computer interaction model. In addition, in [8] the recognition of patterns through artificial neural networks is proposed.

Based on the above, a learning system is presented that uses as input device a gesture sensor, which acquires the complete data of the hand such as: size of the fingers, center of the palm, orientation with respect to the axes; performs the recognition and classification of the signs through the development of a neural network [9], on the other hand, it contains three environments created especially to improve the learning experience of sign language, to contribute in the process of educational and professional training of people with some degree of hearing disability.

This work is divided into six sections, including the introduction. The second section deals with the training of the artificial neural network, in Sect. 3 the development of the system interface is exposed; Sect. 4 contemplates the tests performed on the system on users with and without disabilities, the presentation of the results analysis is found in Sect. 5 and finally the conclusions are located in Sect. 6.

2 Neural Network Classifier

In the recognition phase, a multilayer feed-forward Neural Network is used, with the Backpropagation algorithm; which takes advantage of the property of reduction of time required by a computer processing unit; to distinguish the existing relationship between given patterns [10]. The neural network has three layers: input, hidden and output; the first is constituted by m entries, the hidden layer is composed of n neurons and a logarithmic sigmoid activation function f^1, the output layer is composed of six neurons with a logarithmic sigmoid activation function f^2; this structure can be seen in the Fig. 1.

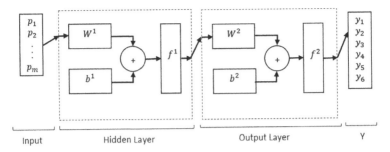

Fig. 1. Architecture of a three-layer artificial neural network feed-forward.

The output given by the supervisor is responsible for establishing the response that forms the network from a specific input, in this context we have the binary combination 000001 which is equal to 1, up to 100111 equivalents to 39; the same that represents the letters A to Z and to the numbers from 1 to 9 respectively.

The vector Y of the Neural Network, for an input vector P = (P1; ...; Pm), is given by the following expression:

$$Y = f^2 \left[W^2 f^1 \left(W^1 P + b^1 \right) + b^2 \right] \tag{1}$$

When the output Y, express a probability that P belongs to a letter of the alphabet, W1 and W2; they express the matrices of the weights of the first and second layers respectively, b1 and b2 determine the level of threshold or activation of the neuron of the input and hidden layer.

The activation function used in the Artificial Neural Network is a logarithmic sigmoid function because the output vector has values of 0 and 1, $f1 = f2 = 1 = (1 + e^{(-z)})$.

To perform the training of the neuron, a database was created in Amazon Web Server (AWS) with values of the angle between the normal vector and the direction of the palm of the hand, angles of the palm with respect to the x, y, z; the angles generated by the direction vector of the following phalanx, except the distal phalanx in the thumb. If the vector of entries belongs to a letter, the output will give us an approximate vector, for example, if letter is A, the output gives us a vector equal to [0 0 0 0 0 1], from number 9 the output is [100111].

3   Development of the Neural Network

The implemented network is a multilayer perceptron, structured by an input layer that has a vector with 21 features, a hidden layer with 25 neurons and an output layer with 6 neurons. Weights and tracks were initialized with randomic values between −0.5 and 0.5. The output layer and the hidden layer have a logarithmic sigmoidal activation function that limits the value of the output between 0 and 1, adopting values very close to 0 when the input is small, but as the value increases on the abscissa axis the function happens to be growing.

The weights and the values of the roads were calculated by the method of back propagation, which starts from the last layer to the first, which causes the increase in error; to find the gradient of the error totalized in each neuron, the new weights and paths are recalculated to the inverse direction together with the value of Alpha, which for this network has a value of 0.5, which represents the normal learning rate avoiding jumps in said learning, Fig. 2.

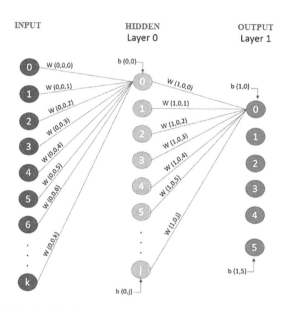

Fig. 2. Graphic description of the developed neuronal network

4 Structure of the System

The architecture of the system consists of three main layers: Inputs, Script and Outputs Fig. 3. The input layer is responsible for the acquisition of data by means of a gestural device; the script layer uses the library of Leap Motion, to perform the acquisition of the position of the fingers when forming a signal, the storage of the information of the hand is done in MySQL, for the effect the Amazon Web Server platform also communicates the Matlab and MySQL programing interfaces for data processing and neuron training. Finally, the output layer generates the recognition in real time of Sign Language, by creating a teaching-learning system made in Unity 3D.

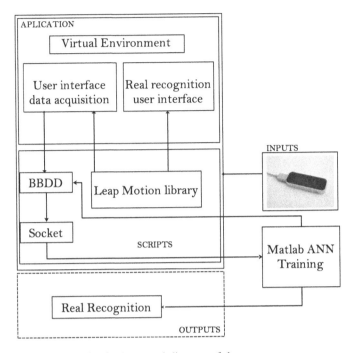

Fig. 3. Structural diagram of the system.

4.1 Description of the Graphical User Interface (GUI)

The Fig. 4a, presents the window of the environment menu with the following options: "Instrucciones", pressing this button gives access to a manual of necessary instructions for the manipulation of the system, "Aprendizaje", access directly to a window where the created learning environments are located, "Salir" closes the whole application.

Fig. 4. (a) Selection menu window. (b) Learning environment window

The Fig. 4b, presents the learning environment window, which consists of three environments: ordered letters, disordered letters and letter leakage.

The environment ordered letters, allows a game that offers the recognition of letters ascending form the A–Z, like numbers 1–9; this environment serves for user training and learning, there is no type of limit of use, nor time; if the user performs the indicated signal the learning system continues until analyzing the configurations and numbers presented Fig. 5.

Fig. 5. Ordered letters interface

The disordered letters environment, allows access to a new window where a set of 10 letters is presented in different order, in this environment the user can corroborate the level of learning.

5   Tests Performed

In this section the teaching-learning capacity of the sign recognition system is evaluated, which will be based on three aspects: response of the neural network recognition, learning tests and finally the usability of the system.

5.1   Response of the Neural Network Recognition

To verify the performance of the neural network, the analysis of the area under the curve (AUC) is used, using the receiver operating characteristic curves (ROC), which is a graphic representation between the success rate (probability of recognizing the sign done) against the false alarm rate (probability of not recognizing a signal) for tasks that have only two possible outcomes (yes/no, acknowledges/does not recognize).

Figure 6 shows the results obtained from the Total AUC of the static configurations, obtaining a maximum value of 0.9779, and of the dynamic configurations with a maximum value of 0.9700, therefore, it is considered as a system perfect classifier since it exceeds 0.5.

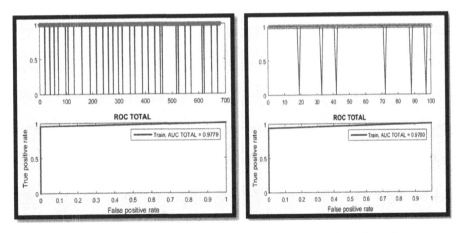

Fig. 6. (a) AUC - Static configuration, (b) AUC - Dynamic configuration

The Table 1 shows the results obtained from the area under the ROC curve, obtained from each static and dynamic configuration that forms the Sign Language.

Table 1. Area under the curve (static and dynamic configurations)

Nº.	Letters	AUC	Nº.	Letters	AUC	Nº.	Letters	AUC
1	A	0,9750	11	J	0,9750	21	R	0,9742
2	B	0,9742	12	K	0,9750	22	RR	0,9750
3	C	0,9742	13	L	0,9727	23	S	0,9742
4	CH	0,9750	14	LL	0,9750	24	T	0,9742
5	D	0,9750	15	M	0,9750	25	U	0,9750
6	E	0,9742	16	N	0,9742	26	V	0,9742
7	F	0,9712	17	Ñ	0,9750	27	W	0,9735
8	G	0,9750	18	O	0,9735	28	X	0,9750
9	H	0,9992	19	P	0,9750	29	Y	0,9992
10	I	0,9750	20	Q	0,9985	30	Z	0,9437

5.2 Learning Tests

The tests of the learning system are carried out with a total of 10 users, of which 4 are children who have hearing impairment and 6 children without hearing impairment.

The learning tests starts with a stage of recognition of the system by the user, then the three learning modes must be completed: Sorted letters mode, disordered letters mode and learning mode or letters leakage; in the first, the whole alphabet of the sign language is ordered in an orderly manner AZ, the second mode shows 10 letters of the alphabet interchangeably, and the third consists in making the greatest number of letters of a word in a time of 30 s at the end, the time is analyzed and the score obtained is indicated.

The learning curve on the part of the users who have auditory disability is appreciated, it can be observed that the performance and the level of learning increases in the 4th and 5th tests. The performance and level of learning of users who do not present any type of hearing impairment where it is observed that the learning curve increases exponentially.

Figure 7, shows the average learning curve obtained in children with hearing impairment versus children without hearing impairment, graphically you can see the similarity of the curves, which show an exponential growth, which allows to deduce that the system fulfilled the teaching objective.

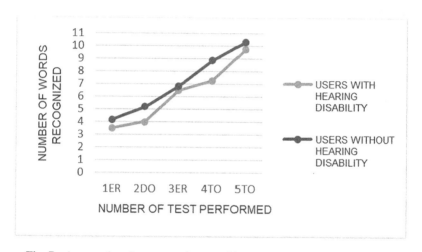

Fig. 7. Average learning curve of users with and without hearing impairment

6 System Usability

The SUS questionnaire allows evaluating the usability of the system once the learning is analyzed, a test is performed, collecting the data for analysis and study; The results obtained by the users are presented to each question posed [11].

For the system to be acceptable in terms of usability, the global average must exceed 68% according to the SUS questionnaire standard.

The results obtained in the questionnaire exceed the standard value by 80.25%. Therefore, it can be considered that the learning system developed is comfortable in its use, it does not present complexity in the manipulation and, above all, the results indicate that It is useful for learning sign language.

7 Analysis of Results

The system based on artificial neural network presented a better response in comparison to the statistical classifier, since it was obtained more than 97% of correct answers compared to 87%, derived by the use of the statistical classifier.

The multifaceted neural network was adapted correctly to the data entered by the user, allowing to maintain an interaction between the student and the system, which is evidenced by an improvement in learning.

The use of assistance systems for teaching in cases of people with special abilities, allows the student to progress in their learning as it does not depend on the presence of an instructor to practice the different signs.

The teacher can integrate computer tools in their teaching process, including people with disabilities, allowing them to contribute to the development of their daily activities.

8   Conclusions

The development of the Ecuadorian sign language learning system through neural networks is a tool that has allowed the inclusion of new technologies in the learning of basic sign language for children with and without deterioration of the auditory system.

The responses of the average learning curves performed in the 10 users with and without auditory disability show an exponential growth, which indicates that the higher the number of training sessions the learning level increases.

References

1. Organización Mundial de la Salud, "Sordera y pérdida de la audición," March 2018
2. Consejo Nacional para la Igualdad de Discapacidades, "Estadísticas de Discapacidad," Consejo Nacional para la Igualdad de Discapacidades, Ecuador, Estadístico (2017)
3. Palomares Ruiz, A.: Liderazgo y empoderamiento docente, nuevos retos de la educación inclusiva en la sociedad del conocimiento (2016)
4. Järvelä, S., et al.: Enhancing socially shared regulation in collaborative learning groups: designing for CSCL regulation tools. Educ. Technol. Res. Dev. **63**(1), 125–142 (2015)
5. Rivas, D., et al.: LeSigLa_EC: learning sign language of Ecuador. In: Huang, T.-C., Lau, R., Huang, Y.-M., Spaniol, M., Yuen, C.-H. (eds.) SETE 2017. LNCS, vol. 10676, pp. 170–179. Springer, Cham (2017). https://doi.org/10.1007/978-3-319-71084-6_19
6. Cadeñanes Garnica, J.J., Arrieta, M.A.G.: Augmented reality sign language teaching model for deaf children. In: Omatu, S., Bersini, H., Corchado, Juan M., Rodríguez, S., Pawlewski, P., Bucciarelli, E. (eds.) Distributed Computing and Artificial Intelligence, 11th International Conference. AISC, vol. 290, pp. 351–358. Springer, Cham (2014). https://doi.org/10.1007/978-3-319-07593-8_41
7. Rautaray, S.S., Agrawal, A.: Vision based hand gesture recognition for human computer interaction: a survey. Artif. Intell. Rev. **43**(1), 1–54 (2015)
8. Pigou, L., Dieleman, S., Kindermans, P.-J., Schrauwen, B.: Sign language recognition using convolutional neural networks. In: Agapito, L., Bronstein, M.M., Rother, C. (eds.) ECCV 2014. LNCS, vol. 8925, pp. 572–578. Springer, Cham (2015). https://doi.org/10.1007/978-3-319-16178-5_40
9. Mendoza, G., Fernanda, J., Zapata Sarzosa, M.B.: Desarrollo de una red neuronal para la clasificación y reconocimiento del lenguaje de signos ecuatoriano (2018)
10. Liu, W., Wang, Z., Liu, X., Zeng, N., Liu, Y., Alsaadi, F.E.: A survey of deep neural network architectures and their applications. Neurocomputing **234**, 11–26 (2017)
11. Lewis, J.R., Utesch, B.S., Maher, D.E.: Measuring perceived usability: the SUS, UMUX-LITE, and AltUsability. Int. J. Hum.-Comput. Interact. **31**(8), 496–505 (2015)

Measuring and Assessing Augmented Reality Potential for Educational Purposes: SmartMarca Project

Emanuele Frontoni[1]([✉]), Marina Paolanti[1], Mariapaola Puggioni[2], Roberto Pierdicca[2], and Michele Sasso[3]

[1] Department of Information Engineering, Universitá Politecnica delle Marche, Via Brecce Bianche 12, 60131 Ancona, Italy
e.frontoni@staff.univpm.it
[2] Dipartimento di Ingegneria Civile, Edile e dell'Architettura, Universitá Politecnica delle Marche, 60100 Ancona, Italy
r.pierdicca@staff.univpm.it
[3] UBISIVE s.r.l., Via Dell'Universitá 13, 63900 Fermo, FM, Italy
info@ubisive.it

Abstract. Augmented and Virtual reality proved to be valuable solutions to convey contents in a more appealing and interactive way. Their use is nearly embracing several domains like medicine, geospatial applications, industry, tourism and so on. But among the others, the one that might benefit the most by their use is the Cultural Heritage. In fact, given the improvement of mobile and smart devices in terms of both usability and computational power, contents can be easily conveyed with a realism level never reached in the past. However, despite the tremendous number of researches related with the presentation of new fascinating applications of ancient goods and artifacts augmentation, few papers are focusing on the real effect that these tools have on learning. In fact, whether a disposable use of such tools seems to have a great benefit in terms of visual impact for the users, the same cannot be said about the long-term effect they have on the users, especially for education purposes. Within the framework of SmartMarca project, that will be briefly explained in these pages, this paper focuses on assessing the potential of AR applications specifically designed for Cultural Heritage. More specifically, tests have been conducted on an Augmented Reality experience upon different paintings. For evaluating the benefits of such technology in terms of learning, we have performed our experiment on classrooms of teenagers. By testing different learning approaches, we were able to evaluate and assess the effectiveness of using these technologies for the education process. The paper will even argue on the necessity of developing new tools to enable users to become producers of contents of AR/VR experiences, since up to now there no exists a platform specifically designed for an agile creation, even for not skilled programmers.

© Springer Nature Switzerland AG 2019
L. T. De Paolis and P. Bourdot (Eds.): AVR 2019, LNCS 11614, pp. 319–334, 2019.
https://doi.org/10.1007/978-3-030-25999-0_28

1   Introduction

Augmented Reality is now expanding in many sectors and fields, not only those specifically dedicated to digital technology, but equally in those of everyday life. Even in educational environments, the use of new technologies that develop Augmented Reality proposals is spreading [1–3]. Great development has taken place in the Primary School thanks to the didactic paths that are closer to the learning transmitted through the game, which is the preferential model of development of Augmented Reality applications. Reconstructing a 3D image by framing a page of a book or increasing the didactic content of a text thanks to the use of a device makes it possible to involve students' attention. Some disciplines lend themselves particularly to the use of these new didactic forms, such as the sciences [4]. The scientific disciplines generally have a greater field of application in this new learning methodology so that these increased spaces, presenting themselves through mobile devices widely used by the younger generations, create greater familiarity and therefore relaxation during the learning process [5]. The often difficult and complex aspect of scientific disciplines, such as mathematics, take on the form of play or a enjoyable quiz. The topics covered in a lecture are transformed into survey data in which they often involve themselves in competitive ways. A survey is then carried out through interactive processes that involve students and stimulate their learning. New terms are coined as "learning by searching" [6], "Inquiry based science education" [7] which translates the different processes that are implemented through the construction of a knowledge that is based on researches, surveys and modeling construction. The possibility of being completely immersed in the artifact, in the case of monuments, or deepening its contents and meanings, in the case of paintings or sculptures, without losing contact with the real environment that surrounds them, is essential to enjoy immersive, experiential and interactive of the work of art itself [8, 9]. This perspective presents innumerable advantages in terms of dissemination of cultural heritage by creating direct interactions between the reality of the artwork and its knowledge mediated by the digital tool. The technology facilitates in this sense the empathetic approach with the good that, favored by the enhanced use of the sense of sight, produces a greater response of attention from the student. The process of learning, however, requires other factors that produce in the student, beyond the immediate and emotional response to the proposal of content, even a permanence in time of them. This condition favors a correct re-elaboration and use of information in different situations and contexts. The so-called skills must, in other words, produce a capacity for understanding and reworking the acquired data to produce cross and multidisciplinary skills. It is therefore essential to use the technologies without impoverishing the student's ability to create his own patrimony of skills that is created in a personal way, refining the cognitive techniques through the study and personal re-elaboration. In the didactic activity, normally modulated on the class and on the single students, the use of the technologies could lead to a method of work that is not flexible and can be modeled on the variable context of the students. The learning experience is modified by technological means [10] so it is important to study the effects of such technologies on learning itself, in particular on the student's ability to reuse in different fields, as learned during the lesson. The goal of creating a cognitive tool, which should be Augmented Reality, which facilitates participative and metacognitive learning processes through an active observation as suggested by Dunleavy and Dede [11], must be supported by a thorough development of the terms of verification of the valence long-term teaching of the contents transmitted. It is Dede himself who claims that technology has advantages that affect the way the content is designed. The variables that are used depending on the type of technology could lead to results that should be interpreted as a guide. A certain result is the motivational one of the student who, through technological skills, manages to build paths that are more familiar and participated to him. A real judgment on learning requires a more careful path to the relapse over time of the educational activities carried out with Augmented Reality.

Given the above, the purpose of this work is to verify the real potential of the digital tool within a didactic environment that includes all the processes related to teaching and learning, possibly with long-term effects on the student's cultural and personal training. This process should also be supported by an ability to re-elaborate what has been learned, its decoding and subsequent implementation in other professional and cultural fields. The work carried out with this research therefore aims to offer a contribution to the studies on the use of Augmented Reality environments and applications in the educational field, verifying how the contents and methods of these new paths can promote real learning and how through an effective didactic action we can direct the student to an effective cognitive process and to re-elaborate knowledge, which will last over time. Tests have been performed within the framework of SmartMarca project, that will be briefly explained in Sect. 3. During the project duration, several outputs have been produced in terms of multimedia experience; in fact the platform underling the project was specifically designed to manage AR/VR contents related to the Cultural Heritage displaced among the south part of Marche Region, in Italy. The advantage of such approach is that, with a cloud based architecture, contents distributed in a quite wide territory can be exploited by different users in every location they are, with the benefits of spreading the knowledge thanks to the use of new technologies. Among these users, students and teenagers are well oriented toward the use of new media. The outputs of the project have been thus exploited and tested among this kind of users, with specific focus in assessing AR potential for learning in the field of Cultural Heritage.

2   Related Work

In the literature there are few data and in-depth research that allow to support the actual educational value of Augmented Reality that does not go beyond motivational development, the creation of a relaxed learning atmosphere, a collaborative participation and a communication facilitation. It is necessary to measure the effect of educational content. There are still a few jobs that are based on the theory of learning and its validation. The main reason lies in a generalized lack of significant research projects oriented for a real involvement of a wide public. Tourism oriented services might act as a flywheel toward this direction. In the literature however some projects moves towards this direction. It has been proved in fact, that the use of ICT tools and specifically AR can contribute in the process of helping visitors to enhance their knowledge [12]. Beside improving the quality of experience for the visitors [13], AR can be adopted in tourism related project.

From the e-learning standing point, Mobile Augmented Reality (MAR) has proved to be a winning solution [15]. In Garau and Ilardi [14] for instance, a specific application was designed allowing people to download contents related to the Cultural Heritage (CH) area they were discovering. Given the huge disposal of CH related artifact produced in the latest years, contents such as paintings of buildings have been augmented, especially in archaeological site. A good example can be found [14]. And more, with the specific purpose of testing such contents for learning purposes, in [15] similar solutions have been tested. By the way, there is neither a clear direction among researchers in the way AR/VR can be evaluated for learning purposes, nor well structured methodologies to quantify their benefits.

In [16] the research objectives are highlighted: *As a first goal, this paper aims to measure the effect of AR educational content to show whether or not it is useful. Although there are many educational AR prototypes in the current literature, only a few are developed by interdisciplinary groups and base their work on learning theory. Even if the current state-of-the-art execution of AR educational content is effective, it can only be replicated to other contexts if a guideline exists for applying AR for education. As a second goal, we provide a guideline for effective AR content by first summarizing the state-of-the-art implementation and evaluation of AR prototypes.*

The use of Augmented Reality as a learning technology had already been included in the Horizon Report 2011, highlighting the great advantages offered by learning in the use of digital data superimposed on the real world [17]. In Horizon Report Preview 2019 Higher Education Edition developed by Educase published in February 2019 aims to develop within a year or less mobile learning that is based mainly on learning no longer focused on the App but on synchronized access from different devices everywhere and at any time[1]. This is why the use of Augmented Reality makes learning more active and collaborative, although this is still an early experience. The educational experiences promoted in the educational field with possible applications of Augmented Reality are many and have different connotations between the different levels of education and the different disciplines [18]. Currently, both at an international level and in Italy there are several experiences in the educational field that in some way put attention to the possible applications of Augmented Reality in the teaching-learning processes but still do not have significant data about their concrete consequences. Certainly the thrusts determined by the technological evolution

---

[1] https://library.educause.edu/-/media/files/library/2019/2/
2019horizonreportpreview.

and by the introduction of digital tools on the market that increasingly use this technology with particular reference to mobile devices, which are no longer simple cell phones, impose rapid response times and fine-tuning research programs that are able to give the appropriate indications to face the inevitable changes also in the educational field [17]. These important changes in the forms of teaching must therefore be monitored, studied and evaluated to understand the real learning levels of the students that go beyond the simple involvement or participation of the students. All studies agree that students show interest and consensus on the use of immersive and interactive technologies [19], but the results on the actual learning of content remain an open question. Some research even lead to a significant reduction in the ability to build learning outcomes due to the overload of information due to immersive technology that distracts the learner from a correct reprocessing of data [20].

3 Brief Description of SmartMarca Project

The SmartMarca project[2] aims to enhance the cultural and tourist heritage of the local area (specifically "Fermano"), introducing innovative and digital systems to support managers and users. These tools are currently still under used and need a cultural leap in the territories with respect to the Digitization of Cultural Heritage and Tourism in general. The project explores the relationships between the user, the space and technologies, looking with a new key to understanding this synergistic relationship of interaction that links the material space to the digital space. SmartMarca intends to enhance the cultural offerings of the territory, through the use of advanced technologies such as Augmented and Virtual Reality, beacons and geolocalization systems to provide users with contextual services when exploring the territory. The main innovation brought by the project is the management platform, which has been conceived to manage different contents and output from a single cloud based service. Then, the micro-services structure enable the managers of the platform to exploit different contents with different output. A general overview of the ICT architecture of the project can be found in Fig. 1. The common denominator of the project is the concept of Senseable Space used to define a new scenario in which the user is provided with contextual services, but at the same time is able to measure actions, analyze them and react accordingly, creating an exchange of seamless information [21]. Not being the main purpose of this paper, we will not describe in deep the SmartMarca project. However, to acquaint the reader about the complexity of the project, it is important to highlight that the platform is able to manage heterogeneous kind of multimedia (e.g. 360° images, video, virtual tours, AR/VR, 3D models) every of which can be used with different devices, on-site and on the web. An example of 3D models exploitation can be seen in Fig. 2. However, for the matter of this article, it is fair to say that, to test multimedia for learning purposes, only the AR application have been used. To achieve these goals, contents can be exploited by the users thanks to the use of a mobile

---

[2] http://www.marcafermana.it/it/SmartMarca/.

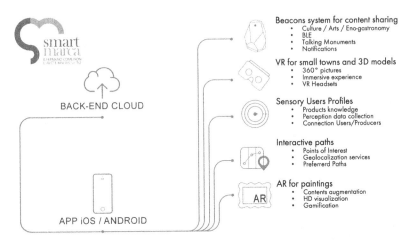

Fig. 1. General overview of the project architecture. The system allows to manage different multimedia contents in an all-in-one solution.

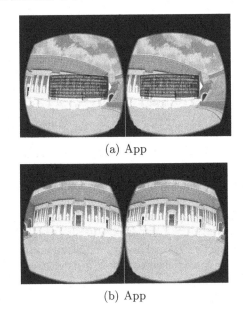

(a) App

(b) App

Fig. 2. Virtual Reality visualization of Falerone amphitheater 3D model. The picture represent the visualization when the users interact with VR headsets.

application, specifically designed for iOS[3] and Android[4]. Some screenshots of the developed service can be found in Fig. 3.

---

[3] https://itunes.apple.com/it/app/smartmarca/id1404807790.

[4] https://play.google.com/store/apps/details?id=it.ubisive.smartmarca&hl=it.

For the specific case of this paper, we will focus on the AR services specifically designed to augment two of the most important paintings of the "Fermano", to improve accessibility and learning of the painting itself. Through Augmented Reality it is possible to recognize a painting simply by framing it with the camera of your mobile phone to access increased contents, as for example (Fig. 4):

Fig. 3. Screen shots of the app running. From left to right, the home page, the geolocalized POIs, a detail of a POI, a possible itinerary.

- highlighting certain areas of the painting to provide accurate information on a specific feature;
- reproduction of video superimposed to the painting;
- display of images and texts in overlay.

In addition to this feature we want to obtain an ultra-high definition image to allow, through the use of the application, the display of details of the paintings that would be impossible to see in the original.

4 Methodology

Within the SmartMarca project some experiences of Augmented Reality have been developed, related to the interpretation of artworks present in the territory of Fermo: the "Adorazione dei pastori" of P. P. Rubens preserved in Palazzo dei Priori in Fermo and "Paesaggio" of Osvaldo Licini, present in the artist's house (museum in Monte Vidon Corrado). The app allows the reading of the artwork through tags that, deepening the critical contents of the paintings, offer a more attentive view of the details and the entire work. The product is mainly aimed at tourists, but in the drafting of the texts it was also thought of a possible didactic use that allows to start a first proposal of knowledge of the artist and of the work in its essential contents. These tools have been used to undertake an educational validation path of AR.

Some screenshots of the application running can be found in Fig. 4(a) and (b).

The research activity has been structured in different stages and modalities, summarized in the following Table 1. This is just the first step of the research, whilst the second stage will be conduct as a future work with VR, as described in Sect. 6.

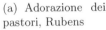

(a) Adorazione dei pastori, Rubens

(b) Paesaggio, Licini

Fig. 4. AR application in front of the paintings.

Table 1. First Step of the research path: description of the methodology used in the learning-teaching process, using AR application from SmartMarca project.

First step				
With device	Explanation of the artwork by the teacher	Check online with the *Socrative* support	Data collection and statistical definition	Analysis, data comparison and conclusions
	Without explanation of the artwork	Check online with the *Socrative* support		
Without device	Explanation of the artwork by the teacher	Check online with the *Socrative* support		

In the secondary school curriculum, the programs have provided a first approach to the history of art through interdisciplinary links that link various disciplines (Italian, History, Drawing). The students are then introduced to a comparative reading of History and History of Art which includes, although simplified, a synchronic and diachronic method of study. The use of pictorial works also favors the use of the image as a tool facilitating the vision of the object of study. The students, coming from the two-year period and the three-year period of a technical institute and a high school, were introduced to the reading of the works - original or copied - through different methodologies, which provided a brief or thorough introduction by the guide and/or the teacher. At the end of the explanation the students were invited, through their smartphones, to access the dedicated app within the SmartMarca application. Framing the painting they had access to Augmented Reality content through overlay texts and the visualization of details, to which the markers were linked (see Fig. 5).

Fig. 5. Students using the application before the learning test.

This information was subsequently verified through a series of multiple-choice questions included within *Socrative*[5] - an online application that allows verification and collection of results, data and statistics related to student learning. The questions were structured to verify the effective comprehension of the content of the texts included in the App, expanding the request to a reprocessing of the explanations. In addition to notional requests, which can be assimilated without understanding their meaning and without reworking their content, requests were made regarding critical interpretations of the symbolic meanings of the elements present in the paintings. The questions included in the *Socrative* program and explained with the help of images, were structured as in Tables 4 and 5, which can be found in the Appendix section of this manuscript.

Questions have been structured with keywords that matches with the overlay contents of the AR application. This was done to facilitate the link between the

---

[5] https://socrative.com.

image and its related comment. All questions are with multiple choice (four options with only one correct) in order to have homogeneous data to be later elaborated with statistical meaning. However, for some questions keywords have been removed to catch the effective learning effect on the student. Questions Q5, Q6, Q7, Q21 in Table 4 are related to the main theme of the hands of the subjects within the artwork scene. This theme was choose for its evocative value. Students have been asked to re-elaborate what was visualized in the app, to stress test their meta-cognitive learning. In the questionnaire related to Table 5 questions to check the ability to re-elaborate contents and concepts are Q8, Q9, Q10, Q11, since they attempt to deepen some details of the painting like tree and house, the main theme of the whole painting.

For the sake of completeness, we report in Figs. 6 and 7 the questionnaire results as it appear in the application used for this test, namely Socrative.

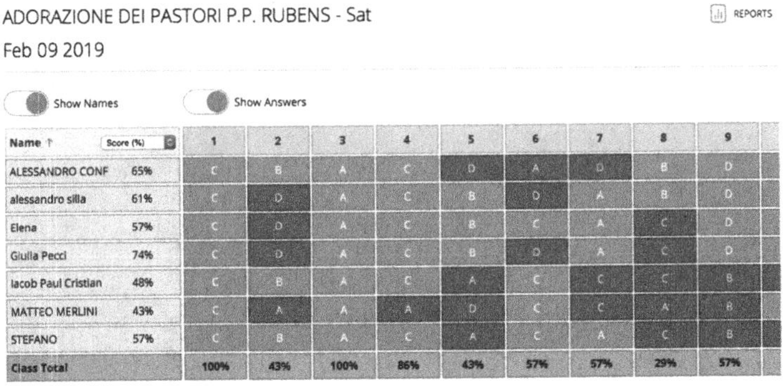

ADORAZIONE DEI PASTORI P.P. RUBENS - Sat Feb 09 2019   REPORTS

Show Names   Show Answers

Name ↑	Score (%)	1	2	3	4	5	6	7	8	9
ALESSANDRO CONF	65%	C	B	A	C	D	A	D	B	D
alessandro silla	61%	C	D	A	C	B	D	A	B	D
Elena	57%	C	D	A	C	B	C	A	C	D
Giulia Pecci	74%	C	D	A	C	B	D	A	C	D
Iacob Paul Cristian	48%	C	B	A	C	A	C	C	C	B
MATTEO MERLINI	43%	C	A	A	A	D	C	C	A	B
STEFANO	57%	C	B	A	C	A	C	A	C	B
Class Total		100%	43%	100%	86%	43%	57%	57%	29%	57%

Fig. 6. Questionnaire data report "Adorazione dei pastori" of Socrative program

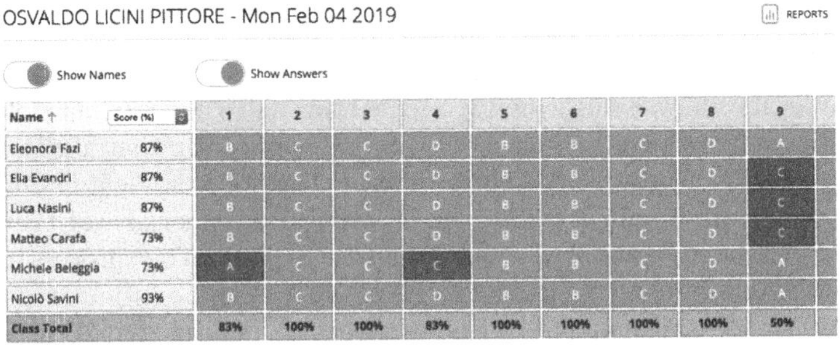

OSVALDO LICINI PITTORE - Mon Feb 04 2019   REPORTS

Show Names   Show Answers

Name ↑	Score (%)	1	2	3	4	5	6	7	8	9
Eleonora Fazi	87%	B	C	C	D	B	B	C	D	A
Ella Evandri	87%	B	C	C	D	B	B	C	D	C
Luca Nasini	87%	B	C	C	D	B	B	C	D	C
Matteo Carafa	73%	B	C	C	D	B	B	C	D	C
Michele Beleggia	73%	A	C	C	C	B	B	C	D	A
Nicolò Savini	93%	B	C	C	D	B	B	C	D	A
Class Total		83%	100%	100%	83%	100%	100%	100%	100%	50%

Fig. 7. Questionnaire data report "Paesaggio" of Socrative program

5   Results and Discussion

In Fig. 8 the comparison between different testing method are reported. Percentages are related to the kind of learning process by the students, with or without the support of AR. It is very interesting to note that Q8, Q9, Q10, Q11 have an higher percentage of positive answers in the combined modality lesson with AR (85,25%). The only use of the app is not sufficient to reach a higher degree of in depth knowledge, hence the student is not able to elaborate meta-cognitive processes (less that 50%). The classical lecture confirms an average of 65% of right answers. Thus, the use of AR increases the learning process, even if its use alone is not enough. The overall statistics collected during the tests can be found in Table 2 and Fig. 9.

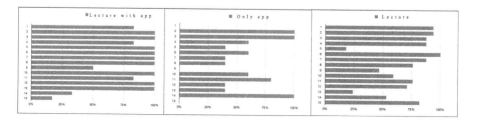

Fig. 8. Comparison of different learning methods. Left chart reports the number of right answers after the combination of classical lecture and AR. Central charts are the answers after the sole use of AR. Right chart the correct answer after the lecture without AR.

Table 2. Data final report. Comparison of all data collected with the different didactic approaches: classical lecture, only app and app plus lecture.

	Licini	Rubens	Average
Lectures	70.11%	46.25%	58.18%
App	50.82%	57.33%	54.07%
Lectures+App	83.33%	57.85%	70.59%

The results obtained in this first stage of the research path gave the following results:

1. The only frontal lesson is still a good means of transmitting contents and skills, valid for a satisfactory average learning response (58.18%).
2. The only use of the app is not a valid means of learning even if it is motivating and innovative (54.07%).
3. The synergy between the frontal lesson and the use of the app is the educational tool with the greatest results in terms of learning, with results much above average, an element that validates the value in terms of learning (70.59%).

The classical lecture still keep maintained its own value to share and spread the knowledge, since involves a bidirectional involvement between students and

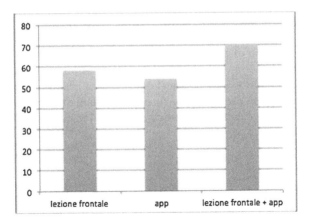

Fig. 9. Data final report

teacher. The dynamics of teaching-learning involves various factors: emotional, experiential, relational, didactic, communicative, psychological. For this reason it is not a simple passage of contents but a construction of links between the learner and the student. It is also true that learning also takes place through gestures and interaction with the environment and with the surrounding reality. In the particular case of art being able to visit a museum, to know an archaeological site, to admire a monument constitutes an added value to the knowledge of the work. The technology favors this type of process by providing tools that involve the user, increasing the ability to obtain information, details, that otherwise would be difficult to find and consult in real time. The teaching experience with the contribution of Augmented Reality contributes to improve the effectiveness of the learning process by producing scenarios enriched with content that can be easily found and used in flexible and interconnected ways. The information provided by the teacher is then enhanced through a psychological and gestural involvement that can not be transmitted by technology. On the other hand, the wealth of interrelated information is mediated by versatile devices that create experiences in which the students find a response to methods, that, nowadays, they are familiar with. The first data provided by this research leads to the conclusion of a mediated use between new technologies and traditional teaching. One cannot undertake innovative educational programs without validating their long-term effectiveness. It is therefore necessary to verify another aspect, namely that of experiential learning, that is, "learning by doing". From the data collected so far it is clear that the didactic action has greater value through the intervention of the teacher who, through an interrelationship with the students, succeeds in transmitting the contents in a more incisive manner. The contribution of Augmented Reality is edifying for a greater involvement of the students in the training activity, and demonstrates that increases the learning ability of the students, as demonstrated by the increasing number of correct answers.

6   Conclusion and Future Works

In this work, a comparative test to assess the benefits of using AR for learning processes has been presented. To achieve this result, we made use of an existing project, namely SmartMarca. The project aims to propose the implementation of omnichannel strategies targeted for each main territorial feature through the use of innovative technologies that can be inserted into paths for the enhancement and dissemination of the landscape/cultural heritage in order to expand (in terms of innovation and quality) proposed offers, promote them on a global scale to finally generate economic value. The same contents conveyed through the project have been used as a flywheel to test students when using these multimedia solutions. As a first output of this study, we can state that AR can contribute in the learning process, even if it cannot be used to completely replace the teacher, which still remains the main driver of the knowledge for the students. This research is a preliminary, but promising test, that will be enforced in the future work. First of all, we will evaluate the long-term effectiveness of AR in the learning process, by providing the same test at different times. Beside this, it will be necessary to consider the use of more intellectual AR systems where context to the students is brought through gamification scenarios and bidirectional interaction. Moreover, we will expand the research even for Virtual Reality (Table 3). In fact, the course of study and research then provides a second verification phase aimed at the creation of a 3D model of an archaeological site (see Fig. 2) by the students, with the aid of the modelling tools. In the second part of the research we will examine the value of teaching impact of new technologies through a path that sees the creation of an app by the students within an educational activity that includes the development of the teacher's frontal lesson, the direct participation of the student in the transmission of the contents of the lesson itself. A monument or part of an archaeological site will be surveyed in order to return it with a computer graphics program. Realized the three-dimensional model of the artifact will apply information overlay. The results in terms of learning will be a further element of knowledge of the potential of new technologies in the field of education and training. What

Table 3. Second step of the research path: learning effect of VR by creating 3D models by the students.

Second step				
Realization of a 3D model of an archaeological site and/or a monument (with the use of autoCAD and SketchUp)	Tagging the 3D model	Check online with the *Socrative* support	Data collection and statistical definition	Analysis, data comparison and conclusions
Creating an Aura with HP Reveal (Aurasma)	Application to the monument or to the archaeological site			

will emerge from the study will contribute to enrich the pedagogical research aimed at the introduction of Augmented and Virtual Reality in the educational learning scenario.

A Appendix 1

Table 4. Survey administered to students upon completion of the teaching experience on "Adorazione dei Pastori" di P. P. Rubens

Adorazione dei Pastori		
Code	Question	Answer
Q1	What are the elements present in the mantle of the Madonna that recall the late-ancient Christian tradition?	4 options only 1 right
Q2	In representing the face of the Madonna, Rubens was inspired by...	
Q3	The sense of vitality of the Madonna's face comes...	"
Q4	The open mouth of the Madonna alludes...	"
Q5	The hands of the Virgin are...	"
Q6	The hands of the Virgin move for...	"
Q7	The hands of the Madonna...	"
Q8	The Child is the protagonist of the painting. The painter paints him...	"
Q9	The light that comes from the Baby Jesus illuminates the face of the Virgin and of the other characters...	"
Q10	The straw on which the Baby Jesus is laid, full of light, seems to be burning while...	"
Q11	The figure of St. Joseph is confused with the colors of the background of the painting, because...	"
Q12	The landscape on the background of the painting and the figure of Saint Joseph...	"
Q13	The shepherd with sheepskin is poorly dressed, leaning on a stick and has the face of an old man. It is confused with the bottom of the painting and is placed in the extreme part of the painting. The Rubens wanted...	"
Q14	The elderly pastor puts his hand on his forehead...	"
Q15	The red of the young kneeling shepherd's tunic represents...	"
Q16	The pose of the kneeling shepherd recalls for the Rubens...	"
Q17	The old woman with raised hands represents...	"
Q18	In realizing the character of the old woman, Rubens imitates a great master of painting...	"
Q19	The four angels that accompany the shepherds are positioned high up in the canvas and dominate the scene. The composition allows to appreciate...	"
Q20	The composition of the angels made by Rubens testifies to the passion of Rubens for...	"
Q21	The hands of the different characters in the painting represent...	"
Q22	Among the different characters which seems to be a stranger to the composition...	"
Q23	The painting was intended for the Oratory of the Church of...	"

Table 5. Survey administered to students upon completion of the teaching experience on Osvaldo Licini

Paesaggio		
Code	Question	Answer
Q1	The "Paesaggio" by Osvaldo Licini is donated to the Municipality of Monte Vidon Corrado in 2015 in memory...	4 options only 1 right
Q2	The painting the "Paesaggio" represents a view...	
Q3	The birthplace of Osvaldo Licini is...	"
Q4	The artist usually paints "en plein air", that is...	"
Q5	The rich colors of the painting recalls the works of...	"
Q6	The clouds seem to transform...	"
Q7	The hills and the sky converse with each other...	"
Q8	The leafy tree in the foreground...	"
Q9	The small farmhouse...	"
Q10	The sign on the canvas...	"
Q11	The line has the purpose...	"
Q12	The painting the "Paesaggio" is made by the artist...	"

References

1. Pierdicca, R., Paolanti, M., Frontoni, E.: eTourism: ICT and its role for tourism management. J. Hospit. Tour. Technol. **10**(1), 90–106 (2019)
2. Yuen, S.C.Y., Yaoyuneyong, G., Johnson, E.: Augmented reality: an overview and five directions for AR in education. J. Educ. Technol. Dev. Exch. (JETDE) **4**(1), 11 (2011)
3. Wu, H.K., Lee, S.W.Y., Chang, H.Y., Liang, J.C.: Current status, opportunities and challenges of augmented reality in education. Comput. Educ. **62**, 41–49 (2013)
4. Chen, P., Liu, X., Cheng, W., Huang, R.: A review of using augmented reality in education from 2011 to 2016. Innovations in Smart Learning. LNET, pp. 13–18. Springer, Singapore (2017). https://doi.org/10.1007/978-981-10-2419-1_2
5. Pierdicca, R., Frontoni, E., Pollini, R., Trani, M., Verdini, L.: The use of augmented reality glasses for the application in Industry 4.0. In: De Paolis, L.T., Bourdot, P., Mongelli, A. (eds.) AVR 2017. LNCS, vol. 10324, pp. 389–401. Springer, Cham (2017). https://doi.org/10.1007/978-3-319-60922-5_30
6. Yin, C., Sung, H.-Y., Hwang, G.J., Hirokawa, S., Chu, H.-C., Flanagan, B., Tabata, Y.: Learning by searching: a learning environment that provides searching and analysis facilities for supporting trend analysis activities. J. Educ. Technol. Soc. **16**(3), 286 (2013)
7. Pascucci, A., et al.: Science education research for evidence-based teaching and coherence in learning. In: Proceedings of the ESERA 2013 Conference. ISBN: 978-9963-700-77-6
8. Di Serio, Á., Ibáñez, M.B., Kloos, C.D.: Impact of an augmented reality system on students' motivation for a visual art course. Comput. Educ. **68**, 586–596 (2013)

9. Naspetti, S., Pierdicca, R., Mandolesi, S., Paolanti, M., Frontoni, E., Zanoli, R.: Automatic analysis of eye-tracking data for augmented reality applications: a prospective outlook. In: De Paolis, L.T., Mongelli, A. (eds.) AVR 2016. LNCS, vol. 9769, pp. 217–230. Springer, Cham (2016). https://doi.org/10.1007/978-3-319-40651-0_17

10. Dede, C.: The evolution of distance education: emerging technologies and distributed learning. Am. J. Distance Educ. **10**(2), 4–36 (1996)

11. Dunleavy, M., Dede, C.: Augmented reality teaching and learning. In: Spector, J., Merrill, M., Elen, J., Bishop, M. (eds.) Handbook of Research on Educational Communications and Technology, pp. 735–745. Springer, New York (2014)

12. Kounavis, C.D., Kasimati, A.E., Zamani, E.D.: Enhancing the tourism experience through mobile augmented reality: challenges and prospects. Int. J. Eng. Bus. Manag. **4**, 10 (2012)

13. Pierdicca, R., et al.: Cyberarchaeology: improved way findings for archaeological parks through mobile augmented reality. In: De Paolis, L.T., Mongelli, A. (eds.) AVR 2016. LNCS, vol. 9769, pp. 172–185. Springer, Cham (2016). https://doi.org/10.1007/978-3-319-40651-0_14

14. Garau, C., Ilardi, E.: The "non-places" meet the "places:" virtual tours on smartphones for the enhancement of cultural heritage. J. Urban Technol. **21**(1), 79–91 (2014)

15. Etxeberria, A.I., Asensio, M., Vicent, N., Cuenca, J.M.: Mobile devices: a tool for tourism and learning at archaeological sites. Int. J. Web Based Communities **8**(1), 57–72 (2012)

16. Santos, M.E.C., Chen, A., Taketomi, T., Yamamoto, G., Miyazaki, J., Kato, H.: Augmented reality learning experiences: survey of prototype design and evaluation. IEEE Trans. Learn. Technol. **7**(1), 38–56 (2014)

17. Arduini, G.: La realtà aumentata e nuove prospettive educative. Educ. Sci. Soc. **3**(2), 209–216 (2012)

18. Dunleavy, M., Dede, C., Mitchell, R.: Affordances and limitations of immersive participatory augmented reality simulations for teaching and learning. J. Sci. Educ. Technol. **18**(1), 7–22 (2009)

19. Eastman, J.K., Iyer, R., Eastman, K.L.: Interactive technology in the classroom: an exploratory look at its use and effectiveness. Contemp. Issues Educ. Res. **2**(3), 31–38 (2009)

20. Makransky, G., Terkildsen, T.S., Mayer, R.E.: Adding immersive virtual reality to a science lab simulation causes more presence but less learning. Learn. Instr. **60**, 225–236 (2017)

21. Osaba, E., Pierdicca, R., Malinverni, E., Khromova, A., Álvarez, F., Bahillo, A.: A smartphone-based system for outdoor data gathering using a wireless beacon network and GPS data: from cyber spaces to senseable spaces. ISPRS Int. J. Geo-Inf. **7**(5), 190 (2018)

Augmented Reality in Laboratory's Instruments, Teaching and Interaction Learning

Víctor H. Andaluz$^{(\boxtimes)}$, Jorge Mora-Aguilar$^{(\boxtimes)}$, Darwin S. Sarzosa$^{(\boxtimes)}$, Jaime A. Santana$^{(\boxtimes)}$, Aldrin Acosta$^{(\boxtimes)}$, and Cesar A. Naranjo$^{(\boxtimes)}$

Universidad de las Fuerzas Armadas ESPE, Sangolquí, Ecuador
{vhandaluz1,jlmora2,dssarzosa1,agacosta,
canaranjo}@espe.edu.ec, sistemas@santana.ec

Abstract. This article presents the research carried out to develop an application of augmented reality for Android devices that allows to know and explore an Electrical and Electronic laboratory of the University of the Armed Forces ESPE, as a means of dissemination and learning of equipment as instruments with which they work in this laboratory. Such elements deploy the respective documentation, a virtualized model and in some cases a deployment in its components for the respective maintenance. The results obtained are based on the clear streamlining of the teaching method with the complementation of all the information that is appended to the application. This results in an aggregate use of the laboratory for the diversification and correct use of its elements. The use of the AR application generates an extension of the knowledge already acquired in classes.

Keywords: Augmented reality · Virtual laboratories · Unity · Vuforia

1 Introduction

The search for new knowledge by the human being has been present since the beginning of time. Teaching is considered a fundamental human process for the acquisition of knowledge and is a means to reach different skills and abilities in each individual. The teaching processes have been continuously developed over the years, evolving to continuous improvement, however in recent decades this evolution has been slowing down exponentially due to the limitations of different teaching techniques. The common goal of any educational procedure is to make each and every one of the apprentices acquire the skill or knowledge taught by the teacher in the shortest possible time [1]. In the current era, combined learning is a concept belonging to a teaching methodology whose objective is to link the knowledge acquired in the classroom with experience obtained in the field of work, however, this methodology has limitations due to lack of resources that each teaching institution possesses, however, technological advancement has developed tools capable of immersing the user in a virtual environment in which his perception of the abstract improves considerably, this type of tools has been called immersive technologies.

L. T. De Paolis and P. Bourdot (Eds.): AVR 2019, LNCS 11614, pp. 335–347, 2019.
https://doi.org/10.1007/978-3-030-25999-0_29

Immersive technologies are the application of virtual reality (VR) and augmented reality (AR) in various fields such as education, economy, industry and construction [2]. Since 2016, the media has shown that Virtual Reality can reach homes massively through consumers' electronic devices, such as smartphones, this implies the adoption of this technology in educational environments, providing new learning styles and methods of more efficient teaching [3]. The VR transports the individual to an environment of scenes and objects generated by computer. AR refers to an image processing technology within three-dimensional space in which virtual 3D images created with the computer are combined with real images for interaction [4, 5]. The use of VR and AR in the field of formal education can modify the traditional way of teaching at all levels, making it more agile and interactive [6]. The education of diverse subjects can be complemented with the visualization of objects and processes, which makes knowledge more flexible and dynamic. Currently, tools are created to create virtual educational environments [7–9] and intelligent software tutoring agents [10]. Thanks to the support and reception that this new technology is having, it has allowed the creation of new teaching and learning media, that free us from the limitations of traditional methods, as are the theoretical lessons that do not provide enough knowledge to students or the high cost of acquiring real equipment to perform practices or recognition of technological equipment [11].

The development of Augmented Reality applications has been generated in different areas. In medicine new forms of supervised practices of medical procedures and recognition of various structures of the human body have been provided [6, 12]. Augmented Reality can increase the intraoperative vision of the surgeon by providing a virtual transparency of the patient. While minimally invasive surgery increases operational difficulty since the perception of depth is generally drastically reduced, the field of vision is limited and the instrument transmits the sense of touch [13]. An AR intravenous injection simulator has also been developed to train veterinary and preveterinary students to perform canine blood sampling [14]. In museums the addition of this resource improves the visit with the additional sample of information of the exhibitions because they cannot provide background information of the collections, such as lost landscapes, collapsed buildings or objects that are too fragile due to their great value and physical limitations. In education in general, different paths have been provided to expedite teaching [10, 15]. The AR creates possibilities for collaborative learning around virtual content and non-traditional environments, which facilitates meaningful experiences [16]. Multiple AR systems have been developed and tested for learning through hands-on studies that are often conducted in laboratory settings [17]. These systems and the learning activities they provide are designed in conjunction with teachers for their own classrooms [18]. An AR system is useful when teaching language concepts, as was done to educate about the concepts of Arabic vocabulary to children in kindergarten [19]. The AR also offers great opportunities for teaching science and engineering [8], since these disciplines emphasize practical training. Virtual laboratories are proposed to extend the experience [9], where the student is allowed to perceive sensations and explore learning exercises that, in some cases, may exceed those offered by traditional laboratory classes [20, 21].

The purpose of this document is the development of an application through the use of AR that focuses on two parts. (i) PAPER AR is a sheet that can be printed on paper so that the student through a mobile device, regardless of whether it is in the laboratory or not, can visualize the 3D models of the equipment and instruments that exist in the laboratory possess the documentation of the team, such as the deployment of the elements that compose it, this will improve their understanding and at the same time ensure a correct handling of the laboratory instruments. (ii) LAB AR is nothing more than the interaction of the student with the laboratory and the various equipment and instruments that compose it, in which the application will display all the attached documentation when it focuses on the desired element with the camera of the mobile device. This facilitates learning, accompanies the teaching methodology making it more agile and flexible.

In this document, it is divided into chapters. In Sect. 1 is the introduction. Section 2 details how the proposal intervenes in education. In Sect. 3 specifies the structure of the system, the operation at the level of the modules. In Sect. 4 is the development of the application. Section 5 presents the experimental results and discusses what has been obtained.

2 Formulation of the Problem

The current education system aims for future professionals to integrate their theoretical knowledge, hours of field practice, so that students are able to solve problems effectively in the area in which they are developed.

Academically, a large part of higher education institutions does not have the means to complement what they have learned in classrooms, the theoretical part, with time for field work, the practical part, in this way developing a deficient learning process in students. The absence of workshops and laboratories in which you can apply the knowledge acquired in the theoretical part, represent a disadvantage in teaching, because the practical part can fill the gaps or doubts that would have been in what was learned in the classroom. It is essential for all professionals to be able to perform in their area of work, be familiar with the process that is being carried out and the instruments that this entails.

The constant training of future workers is vital in the professional field, thanks to the companies that have a highly trained staff can adapt to the constant changes that exist in the market. This ability that companies have to adapt to the market depends on the capacity of each one of the operators of the same, thus generating proposals and innovative solutions to different situations that arise.

Another obstacle that is present in the current education systems is the use of "obsolete" laboratories; The lack of proper maintenance or care that these have, leads to the deterioration of the instruments or stations, in addition to the absence of financing by the institutions prevents acquisition of improvements in the workshops that may allow learning along with the technological advance in industrial processes. This is the point at which virtualization comes into action, in which all the problems founded can be resolved with a virtual laboratory with an open structure that allows the manipulation of all the devices, knowing their characteristics and becoming more familiar with them. You can use the most modern modules that are present in industrial processes, developing an interactive learning tool for future professionals.

The relationship of students with different instruments and learning modules of a laboratory in a virtual environment can be individual or multi user, allowing in this way a better interaction between teacher-student, student-student, thus generating an optimization of resources. In this context, it is proposed to develop an augmented reality application in which by means of a common device such as a smartphone, information can be obtained from the different equipment that is present inside a laboratory, its functioning parts and the possibility to carry out a preventive maintenance to the instruments of the same one, in addition to this thanks to specially designed sheets, the laboratory instruments can be generated in a virtual way in the screens of the smartphones, for the interaction of the user with the instrument facilitating the learning and generating a more interactive environment in the classrooms as illustrated in Fig. 1.

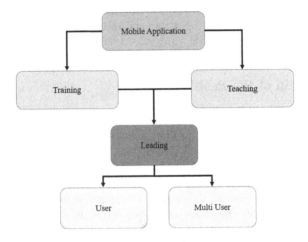

Fig. 1. Multi-user training structure

The development of an application of augmented reality as a learning tool aims to improve the teaching-learning methods of higher level institutions by optimizing resources and direct interaction with the instrument or equipment with which is working, In addition to this, you can get a guide to the use of the equipment and the possibility of the user performing the maintenance without having a trained person to be guided at all times, thus improving the self-learning of the user by generating skills that will help the time of your professional development.

3   Problem's Structure

In the training of a professional it is of the utmost importance to have quality training in the handling of different instruments and industrial equipment, this need calls for the use of different methods for the development of training processes through new technologies, in the which the user can obtain relevant information about the equipment and instruments with which he will work, combining the data obtained through augmented reality with real data, thus obtaining an innovative and intuitive experience within the reach of his mobile device; In addition to this, it is necessary to take into account the possibility of generating a virtual instrument in the comfort of the classroom in which the learning team can be better appreciated. In the field of engineering, a didactic, intuitive and interactive training is promoted for the student.

This article deals with the implementation of an augmented reality application divided into two fundamental parts for a teaching tool: *(i) PAPEL AR,* which deals mainly with the generation of virtual instruments and equipment on the screen of the user's smartphone by means of recognition of images and icons printed on a sheet of paper. *(ii) LAB AR* this part of the application grants the user training and assistance for the handling of industrial equipment and instruments by means of the visual estimation of the information of the objects focused with the camera of the mobile device.

As shown in Fig. 2; In the Vuforia block, the images will be uploaded to the Vuforia server, which will later be used for the identification and generation of virtual instruments, this server will examine the image, obtaining the characteristic points of each of these and determining their level of acceptance or recognition, generating a specific target for each image, to which a specific function will be assigned; it must be borne in mind that the images to be treated must be clear and have a considerable number of characteristics to increase the level of recognition.

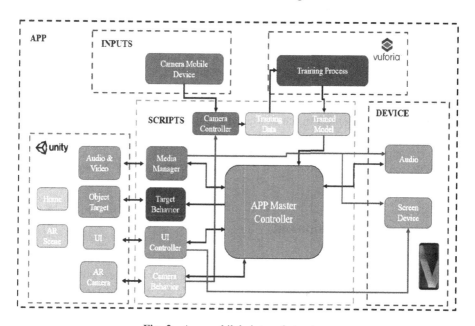

Fig. 2. App mobile's internal structure

In the Scripts block of the augmented reality application, a sub-block of APP Master Controller is implemented, this is in charge of processing the information obtained from the camera Controller which in turn is in charge of processing the images that is obtained from the camera back of the mobile device; the APP Master Controller

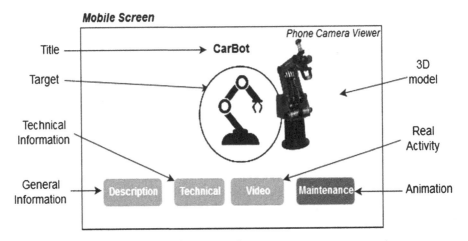

Fig. 3. Sketch of the elements that intervene in the augmented experience

is also responsible for managing the targets obtained from a local server that data that is downloaded from the Vuforia server, and also has main management with the Media Manager who is responsible for the audio management and screen of the device and the UI manager which communicate with Unity.

The Unity block is in charge of the development of the application in which, as can be seen in Fig. 2, it has a main relationship with the multimedia control of the input device that is the telephone, through the target obtained by Vuforia, Unity can apply to it the functionality it wants, such as a graphical interface, UI, or an augmented reality environment that is displayed on the device's screen, such as multimedia content characteristic of each object in the downloaded database.

Last, but not least, the Device block that works with the smartphone on which the developed application of augmented reality is installed, the operating system of the mobile device must be Android and it must have a back camera to be able to use the application.

4   Application Development

This section describes the process carried out for the implementation of the application, the partition for planning and the outline of the elements that are involved in the creation of each experience. on the screen of the mobile device on the viewfinder of the video camera, when one of the registered targets is identified.

The target is the main element within the scene, since it is the starting point to load the elements that make up the augmented experience, all the physical elements of the laboratory are related to an objective, it is important that all the objectives have the same results characteristic, so that the recognition is fast and consistent at the moment of exploring each 3D model; Then in Fig. 4 we can see how Vuforia highlights the characteristic points of a target with yellow color, and in this way we determine if an objective is valid for the application.

Fig. 4. Characteristic points of the target of a robotic arm (Color figure online)

The next element that contributes the most to the augmented experience are the 3D models that are placed on the identified targets, these 3D models are built in CAD applications and optimized at the moment of importing them into Unity so that the experience is fluid.

The following elements that complement the experience increase are the title that appears in the upper part of the objective and shows the name of the equipment or the robot that is being analyzed, in addition the user can request general information that is a form in a way of a text box with concise information of the team or robot, in the same way the technical information of the team relevant to the students or specialists can be visualized in an image or a diagram that appears next to the 3D model.

In order for the user to be able to complement their learning or recognition stage, it is important to have an idea of the equipment's real functioning, so the application shows the video with the objective of the task is when the action is required by the user.

Then we will define the elements of the augmented experience, a table will be drawn up in which the elements that are available for each registered team will be defined. As shown in Table 1.

Table 1. Elements of the augmented experience for each instrument

Equipment	Target	3D model	Title	General info	Technical info	Real activity	Animation
Robotic arm	X	X	X	X	X	X	X
UAV drone	X	X	X	X		X	
Wheelchair	X	X	X	X	X	X	X
Humanoid	X	X	X	X	X		
Omnidirectional	X	X	X	X	X	X	X
Manipulator	X	X	X	X		X	X

5 Experimental Results

This section presents the implementation and performance of the augmented reality application that is oriented to the process of learning and training in the handling of laboratory instruments, which consists of two main functions, such as PAPER AR and LAB AR, main tools for serving to facilitate the Pedagogy of future professionals.

5.1   Paper AR

This function of the application is mainly focused as a tool to improve teaching without having to be present in the laboratory, thanks to a carefully designed sheet with the instruments that are inside the laboratory, as shown in Fig. 3.

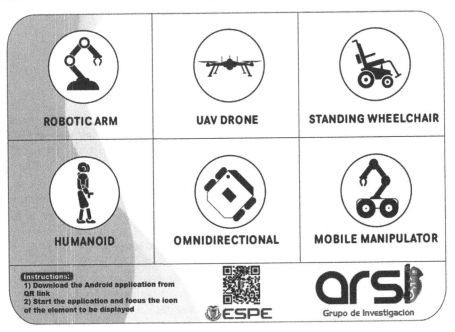

Fig. 5. Paper AR

As can be seen in Fig. 5, the sheet has characteristic icons that, when focused with the camera of the smartphone, generate the 3D model of the equipment in question, in turn the sheet has a series of instructions that help the handling of the user and a QR code that will be used to download the application on your smartphone.

Fig. 6. PAPER AR implementation

In Fig. 6 it is observed how students use the sheet and complement the teacher's explanation. They acquire extra useful information to improve their understanding, once this intuitive tool reinforces their learning process.

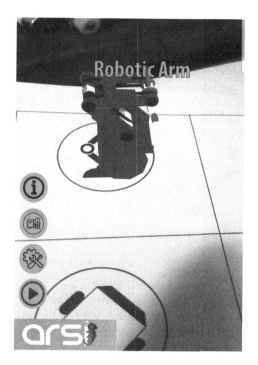

Fig. 7. Available options on a target of the paper AR

In Fig. 7 we can see how the elements of the augmented experience are shown on one of the targets detected in the designed sheet, giving the user the possibility to deepen the learning of the device through the options offered by the application.

In Fig. 8 we can see the two options enabled, a sample of general information of the device and the other option that allows us to see the real operation of the device through a video.

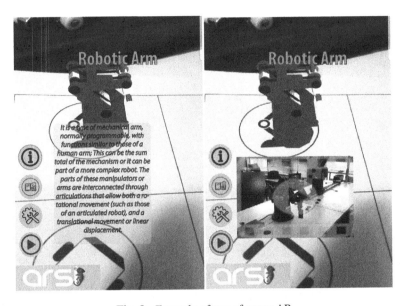

Fig. 8. Example of use of paper AR

5.2   LAB AR

The functionality of LAB AR is essential so that laboratory users can use the equipment without the need for a guide, because they have all the necessary information for their use, characteristics and maintenance within reach of their mobile device. The Fig. 9 Can be seen as the user focuses the robotic arm with his smartphone in order to obtain the data for its correct use.

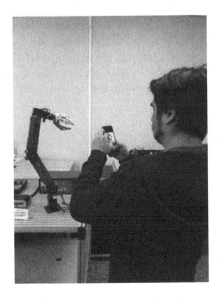

Fig. 9. Person using LAB AR

As has been demonstrated the use of these tools complement the teaching method. With what can be delivered more accessible content for each student. Everyone can have their own disposition.

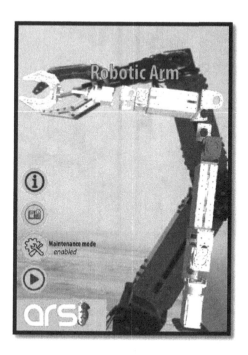

Fig. 10. Maintenance mode of the robotic arm

In Fig. 10, a screenshot of the application can be seen in its working in which it can be seen that the robotic arm, is separated into parts to give a guide to the user how the equipment should be assembled in case of the replacement of an element so that a type of maintenance occurs so that it may be necessary to take into account that there are other functionalities in the application, in the image it is observed that the maintenance mode is active, this active The decomposition by elements of the robotic arm.

In Fig. 11, it can be seen that there is also information about the robotic arm, its functionality, and a brief description of what it does; This type of information serves as a guide for people who are not yet clear about the concept of the equipment they are using; thus fostering a more intuitive learning without needing to interact physically with the team, to avoid any inconvenience due to the inexperience of the user. In turn we can find a video of the operation of the robot focused by the application to have a greater understanding of the functionality of the robot.

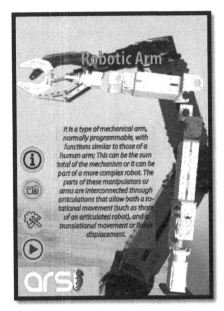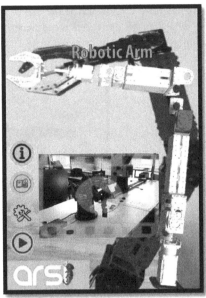

Fig. 11. Robotic arm's description

6   Conclusions

The use of immersive technologies in the field of education contributes exponentially to the training, assistance and management of laboratory equipment and instruments, which will later be reflected in their handling with industrial machinery. One of the main advantages of this application is the reduction of losses that will exist due to poor handling by students in their development stage towards the teams with which they are working, since they do not have a direct interaction with the object real, and avoid bad maneuvers that cause irreparable losses in the instruments. The article shows the development and implementation of the application in an educational environment such as the research laboratory of the university of the Armed Forces ESPE, which has contributed significantly to the students' understanding, in the area of robotics, giving the possibility that they immerse themselves in the subject without a guide.

Acknowledgements. The authors would like to thanks to the Corporación Ecuatoriana para el Desarrollo de la Investigación y Academia–CEDIA for the financing given to research, development, and innovation, through the CEPRA projects, especially the project CEPRA-XI-2017-06; Control Coordinado Multi-operador aplicado a un robot Manipulador Aéreo; also to Universidad de las Fuerzas Armadas ESPE, Universidad Técnica de Ambato, Escuela Superior Politécnica de Chimborazo, Universidad Nacional de Chimborazo, and Grupo de Investigación ARSI, for the support to develop this work.

References

1. Andaluz, V.H., et al.: Multi-user industrial training and education environment. In: De Paolis, L.T., Bourdot, P. (eds.) AVR 2018. LNCS, vol. 10851, pp. 533–546. Springer, Cham (2018). https://doi.org/10.1007/978-3-319-95282-6_38
2. Ortiz, J.S., et al.: Teaching-learning process through VR applied to automotive engineering. In: Proceedings of the 2017 9th International Conference on Education Technology and Computers, pp. 36–40 (2017)
3. Martín-Gutiérrez, J., Mora, C.E., Añorbe-Díaz, B., González-Marrero, A.: Virtual technologies trends in education. EURASIA J. Math. Sci. Technol. Educ. 13(2), 469–486 (2017)
4. Floty, J.: Augmented Reality, Virtual Reality, and Computer Graphics, pp. 589–609 (2018)
5. Kim, T.J., Huh, J.H., Kim, J.M.: Bi-directional education contents using VR equipments and augmented reality. Multimedia Tools Appl. 77(22), 30089–30104 (2018)
6. Kamphuis, C., Barsom, E., Schijven, M., Christoph, N.: Augmented reality in medical education? Perspect. Med. Educ. 3(4), 300–311 (2014)
7. Tuysuz, C.: The effect of the virtual laboratory on students' achievement and attitude in chemistry. Int. Online J. Educ. Sci. 2(1), 37–53 (2010)
8. Villar-Zafra, A., Zarza-Sánchez, S., Lázaro-Villa, J.A., Fernández-Cantí, R.M.: Multiplatform virtual laboratory for engineering education. In: 2012 9th International Conference on Remote Engineering and Virtual Instrumentation, REV 2012 (2012)
9. Loureiro, A., Bettencourt, T.: The use of virtual environments as an extended classroom – a case study with adult learners in tertiary education. Procedia Technol. 13, 97–106 (2014)
10. Lee, K.: Augmented reality in education and training. TechTrends 56(2), 13–21 (2012)
11. Jara, C.A., Candelas, F.A., Torres, F.: Virtual and remote laboratory for robotics e-learning. Comput. Aided Chem. Eng. 25, 1193–1198 (2008)
12. Bichlmeier, C., Sielhorst, T., Heining, S.M., Navab, N.: Improving depth perception in medical AR. In: Bildverarbeitung für die Medizin 2007 (2007)
13. Nicolau, S., Soler, L., Mutter, D., Marescaux, J.: Augmented reality in laparoscopic surgical oncology. Surg. Oncol. 20(3), 189–201 (2011)
14. Lee, S., et al.: Augmented reality intravenous injection simulator based 3D medical imaging for veterinary medicine. Vet. J. 196(2), 197–202 (2013)
15. Kesim, M., Ozarslan, Y.: Augmented reality in education: current technologies and the potential for education. Procedia Soc. Behav. Sci. 47, 297–302 (2012)
16. Bujak, K.R., Radu, I., Catrambone, R., MacIntyre, B., Zheng, R., Golubski, G.: A psychological perspective on augmented reality in the mathematics classroom. Comput. Educ. 68, 536–544 (2013)
17. Akçayır, M., Akçayır, G.: Advantages and challenges associated with augmented reality for education: a systematic review of the literature. Educ. Res. Rev. 20, 1–11 (2017)
18. Cuendet, S., Bonnard, Q., Do-Lenh, S., Dillenbourg, P.: Designing augmented reality for the classroom. Comput. Educ. 68, 557–569 (2013)
19. Hamada, S.: Education and knowledge based augmented reality (AR). Stud. Comput. Intell. 740, 741–759 (2018)
20. Andújar, J.M., Mejías, A., Márquez, M.A.: Augmented reality for the improvement of remote laboratories: an augmented remote laboratory. IEEE Trans. Educ. 54(3), 492–500 (2011)
21. Radu, I.: Augmented reality in education: a meta-review and cross-media analysis. Pers. Ubiquit. Comput. 18(6), 1533–1543 (2014)

Touchless Navigation in a Multimedia Application: The Effects Perceived in an Educational Context

Lucio Tommaso De Paolis$^{(\boxtimes)}$, Valerio De Luca$^{(\boxtimes)}$,
and Giovanna Ilenia Paladini$^{(\boxtimes)}$

Department of Engineering for Innovation, University of Salento, Lecce, Italy
{lucio.depaolis,valerio.deluca,ilenia.paladini}@unisalento.it

Abstract. Natural user interfaces have been widely adopted in several application fields. In particular, they are employed in learning environments to increase the motivation and the engagement of students.

In this paper we evaluate the user experience with a touchless application interface. The reference scenario is an ipermedia learning application based on the Kinect touchless interface: it describes the machines designed by Leonardo Da Vinci, an Italian Renaissance inventor. The user can interact with the graphical interface by means of hand gestures and poses, which allow selecting items from the menus, scrolling text areas, zooming images and so on.

We employed the well-known SUS and USE questionnaires to assess the perceived usability. We prepared also some questions about users' impressions and opinions on the application itself. In this way, we studied how the users were influenced by their experience with the Kinect device in the learning activity.

Keywords: Natural user interaction · Kinect · Gestures ·
User experience · Education · Cultural heritage

1 Introduction

Nowadays, information technology is widely adopted in educational contexts to improve the organization, efficacy and effectiveness of the learning process. Beyond more conservative forms of content delivery, such as Massive Open Online Courses (MOOCs) [1], and despite the benefits deriving from the interactivity of real-time video streaming systems designed for distance learning [2,3], the main need is to enhance the level of attention and engagement of learners. This has led to the rise of edutainment [4], which denotes the use of game oriented environments [5,6] for educational purposes. In such context, the use of engaging multi-channel and multi-sensory platforms [7] can increase the level of attention of learners. In this way, education in cultural heritage often assumes the connotations of informal learning [8], which is not highly structured or classroom-based [9] and is more closely related to practical experience.

© Springer Nature Switzerland AG 2019
L. T. De Paolis and P. Bourdot (Eds.): AVR 2019, LNCS 11614, pp. 348–367, 2019.
https://doi.org/10.1007/978-3-030-25999-0_30

Many game-oriented learning environments [10–12] have been inspired by the ARCS model based on attention, relevance, confidence and satisfaction [13]. In such contexts, natural interaction based on the user movements can play an important role in kinesthetic pedagogical practices [14], which exploit the physical involvement of the human body to enhance the cognitive process. It can be successfully employed to turn students into active learners [15] and improve their spatial visualization skills [16] and mental computation speed [17].

In general, natural interaction enables new communication modalities between people and computers based on gestures [18], voice or even brain waves. It makes computing contextually aware and able to interpret actions that are similar to the ones performed to achieve the same results in the real world. Devices specifically designed to enable this kind of interaction are called Natural User Interfaces (NUIs): they employ various technologies, ranging from infrared cameras to depth sensors, to detect the user movements, which are interpreted as commands. Such innovations have contributed to the development of the concept of User Experience [19], which goes beyond the traditional concept of usability by including emotional and aesthetical factors [20].

The Microsoft Kinect controller is a camera-based device enabling touchless interaction based on the detection of human body movements. It has been widely adopted in several applications for education in cultural heritage, such as the manipulation of 3D replicas of historical relics [21], the creation of augmented artworks [22], interactive presentations of 360 spin images [23], immersive learning museum environments [24] and virtual guides [25].

Other studies have evaluated the usability, effectiveness and efficacy of Kinect-based interfaces in various applications, such as PowerPoint presentations [26] and virtual TV sets [27], where computer-generated objects are combined in real-time with real elements and three types of interaction are provided: virtual object selection, virtual object manipulation and natural gesture interaction.

In this paper we analyze the user impressions after an experience with a Kinect-based multimedia application we developed to describe the machines designed by Leonardo Da Vinci, an Italian Renaissance inventor. The application was firstly presented during a public exhibition and then it was tested with 16 students in a secondary school. We asked users to fill in some questionnaires after their experience with the application. The analysis of the results allowed us to make some considerations on the usability of the Kinect device and on the effects perceived in the considered learning scenario.

The following subsections present and discuss in detail natural touchless interaction and its adoption in learning environments. Then the rest of the paper is structured in this way: Sect. 2 describes the Kinect-based interface we developed for the multimedia application about Leonardo's inventions; Sect. 3 describes the methodology and the questionnaires we adopted to assess the user experience; Sect. 4 presents and discusses the results; Sect. 5 concludes the paper.

1.1   Natural Interaction

Natural interaction is mainly based on gestures [28], which consist in the association of meaningful commands with well-defined positions or movements of some parts of the human body. Unfortunately, gesture-based controls are not always perceived as a natural form of interaction by users, since they have been defined mainly according to what physical devices can recognize. The concept of guessability [29] has been defined as the ability to guess symbolic input without any prior knowledge or learning phase: this new criterion is inspiring a redesign of gesture interfaces through the definition of interactive gesture vocabularies based on users' attitudes and preferences [30].

Besides gestures, other interaction forms concern object selection and manipulation [31]. The former, which usually influences the latter, usually exploits virtual hand or virtual pointing metaphors. According to usability studies, virtual pointing offers a better selection effectiveness, as it allows object selection in a wider area and requires less physical movements [31,32].

The performance of a pointing selection task depends on application-related factors (such as the shape, layout and density of the targets) and device-related factors. Besides noise affecting devices, the absence of a physical support for the user hands during the interaction in the free space makes the selection of small targets less accurate [33]. Such difficulties may increase the cognitive load of users and may require more extensive training phases [31], producing also a sense of frustration that may have a negative influence on the user experience.

Object manipulation can be direct [34] or based on an augmented tangible interface [35]. In the former a virtual object is associated directly with the position of the user hand detected by a sensor device; some complex techniques [34] also reproduce physics-inspired interactions and compute collision and friction forces between real and virtual objects. In the latter a real element manipulated by the user is tracked by an infrared camera and replaced by a virtual object in the same position. According to some experimental tests [27], this seems the most effective and natural interaction modality, since the absence of a physical reference makes the interaction less intuitive.

The application described in this paper exploits the Kinect's hand tracking capabilities to enable touchless interaction based on natural gestures and on object selection and manipulation metaphors.

1.2   Gesture-Based Learning

The adoption of natural interfaces in learning environments can increase intrinsic motivation, interest, engagement and participation [36] especially in younger students [37,38], which feel more attracted by the magic effect and by novelty in general.

Gesture-based learning combines kinaesthetic interactions with auditory and visual information. Kinaesthetic students, whose performance is better when they are physically involved during learning activities, are 15% of the total population [39] and unfortunately they have been mostly ignored by traditional methods. Gesture-based learning is strongly in line with embodied cognition theory, which suggests that manipulating tools can improve the way of thinking, understanding and perceiving the environment [40]. Therefore, the combination of cognitive tasks with physical movements can enhance the efficacy of learning activities, with significant benefits in terms of attention and memory performance [41]. By exploiting human actions, multi-sensory learning increases interactivity, which can be defined as the ability to promptly respond to the user action [42], and is also oriented to constructivism [42], since it fosters the personal creativity of students in an active process of knowledge construction. Moreover, it enables new modalities for children affected by disabilities [43], children with special educational needs [44] and autistic children [45].

Thanks to its multimodality, gesture-based learning can be appropriate not only for kinaesthetic learners, but also for different learning styles and spatial ability levels. It can greatly enrich the user experience and meet the needs of learners with different individual characteristics [46], in line with the theory of multiple intelligences [42] and Kolb's theory of different learning styles [47, 48]. The benefits can be analyzed in terms of affective outcomes, dealing with the perceived learning effectiveness, and cognitive outcomes, dealing with the academic performance [46]: the former seem the most influenced by gesture-based learning, since students think such modality can help understanding contents and discussing ideas.

Unfortunately, some issues can degrade the performance of the motion sensing technology and may partially hamper in particular the adoption of Kinect in school environments.

First of all, there should be at least 2–3 m of empty space between the Kinect and the user for an accurate gesture detection. Also the presence of other people behind the user can partially interfere with the proper functioning of the device. For these reasons, several attempts could be required to find the optimal position of the device in a classroom setting [42,49]. Moreover, the preliminary calibration phase may require time and may disrupt the class [42].

Sometimes users may feel frustrated after a long usage session due to the fatigue and physical effort required to reproduce gestures [50].

The accuracy of Kinect tracking is influenced also by environment conditions and surrounding materials: for instance, an opaque floor seems to allow better performance than a light reflecting floor [51].

From an educational point of view, a drawback can be the difficulty to convey the attention of students back to traditional lessons after the Kinect-based activities [49].

2   The Application Interface

We developed the application described in this paper and an augmented reality mobile application presented in a previous work [52] for an exhibit on Leonardo da Vinci's inventions. Both the applications were designed to describe machines for which physical models were not available starting from the pages of the Atlantic Codex [53].

352 L. T. De Paolis et al.

The user can control the interface elements by means of both gesture interaction and object selection/manipulation based on hand tracking. We employed the second version of Kinect, known as Kinect 2.0 or Kinect One, based on a new architecture providing better performance [54]: it is equipped with a Time-of-Flight camera with a 512×424 resolution, 30 Hz frequency, $70° \times 60°$ field of view, and a color camera able to capture 1080p video at 30 Hz [55].

After the application startup, the user is requested to wave his/her hand to activate the gesture recognition and start the navigation. Then the user can navigate through the graphic interface, where he/she can select an item by holding his/her hand over it for a few seconds. In the same way, the user can select the navigation tab at the bottom of the screen, which will show a dashboard providing Back, Stop and Exit buttons. Figure 1 depicts the typical interface of a menu showing the multimedia content available in one of the application sections. Sheets containing text and images are showed in a scrolling panel. The user can grab and slide the scroll bar on the right or push the up and down arrows on the left to navigate inside the panel. The user can select one of the tab on the top of the panel (*All, Introduction, Images, Video*) to filter the content by category.

Fig. 1. Menu of multimedia content

After the selection of an item from the scrolling panel, the application displays an interface like the one in Fig. 2, which shows the original sketch, a 3D reconstruction and a detailed description of one of Leonardo's invention (the adjustable opening compass). The user can push the right or the left arrow to go to the next or the previous page respectively. Another interaction modality is based on swipe gestures: the user can swipe with right hand from right to left side to go to the next page and swipe with left hand from left to right side to go to the previous page. The interface provides also a combo controller for zooming, made up of a slider and a loupe, on the right side of the panel. The user can grab the slider and move it up and down to zoom in and zoom out respectively. As an alternative, the user can hold his/her hand on the loupe icon below the slider to enable zooming by gestures: he/she can push his/her right hand forward to zoom in and pull it back to zoom out. After the image zooming, the user can perform a grab gesture and move the image by his/her hand.

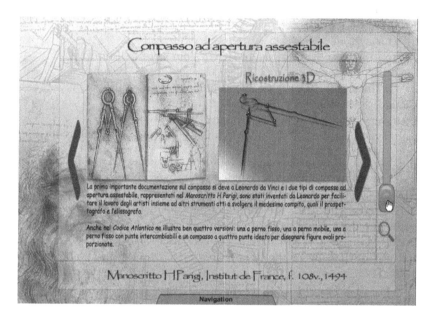

Fig. 2. Description of the adjustable opening compass

3 Assessing the User Experience

Several questionnaires were designed for the evaluation of the user experience [56,57]. We chose two pre-existing questionnaires, the System Usability Scale (SUS) [58,59] and the Usefulness, Satisfaction and Ease of Use (USE) [60] questionnaire, to assess the perceived usability. Moreover, we designed another custom questionnaire to collect users' impressions and opinions on the application and on the efficacy of the designed interface based on the Kinect touchless interaction. For each item of the three questionnaires we adopted a 5-point likert scale ranging from "Strongly Disagree" to "Strongly Agree".

We recruited 16 subjects, aged between 16 and 18, in a secondary school to test our application: after a short explanation about the application features and the touchless control modalities, each subject was asked to test the application by himself/herself and then to fill in the questionnaires.

The following subsections describe the three questionnaires we used to collect the users' impressions.

3.1 System Usability Scale

The System Usability Scale (SUS) [58,59] is made up of a 10 item questionnaire.

1. I think that I would like to use this system frequently.
2. I found the system unnecessarily complex.
3. I thought the system was easy to use.
4. I think that I would need the support of a technical person to be able to use this system.
5. I found the various functions in this system were well integrated.
6. I thought there was too much inconsistency in this system.
7. I would imagine that most people would learn to use this system very quickly.
8. I found the system very cumbersome to use.
9. I felt very confident using the system.
10. I needed to learn a lot of things before I could get going with this system.

The questionnaire can be decomposed into two subscales [61]: *learnability*, which refers to the ability of quickly and independently learning how to use a system, and *usability* in a more strict sense.

The scores of odd items are computed by subtracting 1 from the scale value chosen by the user. The scores of even items are computed by subtracting the scale value chosen by the user from 5. For each user the overall score is computed by multiplying the sum of the item scores by 2.5 to obtain a value between 0 and 100%.

3.2 USE

The Usefulness, Satisfaction and Ease of Use (USE) [60] questionnaire is made up of 30 items divided in four categories. For many applications, *Usefulness* and *Ease of Use* contribute to Usability and are correlated. In the analysis presented in this paper, we considered *Ease of Learning* as a separate component from *Ease of Use*.

We computed the score of each item by subtracting 1 from the scale value chosen by the user. Then we divided the total score by 1.2 to obtain a value between 0 and 100%.

We report below the questionnaire items grouped by each one of these four factors.

– Usefulness

1. It helps me to be more effective.
2. It helps me to be more productive.

3. It is useful.
4. It gives me more control over the activities in my life.
5. It makes the things I want to accomplish easier to get done.
6. It saves me time when I use it.
7. It meets my needs.
8. It does everything I would expect it to do.

- Ease of use

  1. It is easy to use.
  2. It is simple to use.
  3. It is user friendly.
  4. It requires the fewest steps possible to accomplish what I want to do with it.
  5. It is flexible.
  6. Using it is effortless.
  7. I can use it without written instructions.
  8. I don't notice any inconsistencies as I use it.
  9. Both occasional and regular users would like it.
  10. I can recover from mistakes quickly and easily.
  11. I can use it successfully every time.

- Ease of learning

  1. I learned to use it quickly.
  2. I easily remember how to use it.
  3. It is easy to learn to use it.
  4. I quickly became skillful with it.

- Satisfaction

  1. I am satisfied with it.
  2. I would recommend it to a friend.
  3. It is fun to use.
  4. It works the way I want it to work.
  5. It is wonderful.
  6. I feel I need to have it.
  7. It is pleasant to use.

3.3 Custom Application Questionnaire

Besides employing SUS and USE, we prepared also some questions about users' impressions and opinions on the ipermedia application. Beside the effectiveness of the Kinect device when employed in our application (*Kinect effectiveness*) we evaluated the students' opinion on the *content presentation/understandability* and on the system ability to improve *cooperative learning*, increase *interest* and improve *digital competence*.

We computed the score of each item by subtracting 1 from the scale value chosen by the user. Then we converted the total score to a percentage value.

We report below the questionnaire items grouped by each one of these four factors.

– Kinect effectiveness

1. Do you think the Kinect device can facilitate the interaction with the application?
2. What is the naturalness and intuitiveness of Kinect gestures associated with commands?

– Content presentation and understandability

1. Do you think the application content is shown in an effective way?
2. Do you think the application can help in understanding the concepts?
3. How do you evaluate the level of detail of the study achievable through this application?
4. Do you think the information presentation is effective and engaging?
5. How do you evaluate the connections among pieces of information concerning the subject?
6. How do you evaluate the multimedia content organization in order to support the understanding of the subject?

– Cooperative learning improvement

1. Do you think the application could improve the collaboration with the teacher?
2. How do you evaluate the ability to work in group during the use of the application?
3. How do you evaluate the use of ICT to support processes aimed at sharing, collaboration and involvement in a learning content?
4. Do you think the use of such applications could improve the educational processes in a laboratory?

– Increase of interest

1. Do you think the application has enhanced your interest in the subject?
2. Do you think the application has enhanced your interest in the laboratory school activities in general?

– Digital competence improvement

1. Do you think the application has improved your competence in using digital technologies?

4   Test Results

In the following subsections we will evaluate the results collected though the three questionnaires. We will firstly consider the scores of the application questionnaire, the SUS scores and the USE scores individually (Sect. 4.1) and then we will try to detect any correlation among them (Sect. 4.2).

4.1   Main Results

Table 1 reports the mean and the standard deviation of the scores of each component of the application questionnaire. The last row refers to the total score computed over all the questionnaire items.

Content presentation is the component with the highest mean score and the lowest standard deviation: most users expressed a positive opinion on the way multimedia content is organized; moreover, the low standard deviation reveals users mostly agree on the efficacy and effectiveness of this content presentation modality.

On the other hand, the high standard deviation of the *Increase of interest* component reveals users have different opinions about the ability of the Kinect-based approach to enhance the users' interest towards the topics of the multimedia application.

Digital competence is the component with the second highest mean score, but it has also the second highest standard deviation: in general, users think the system could improve their digital competence, but the high variability of the scores reveals users' opinions are discordant on this aspect.

Table 1. Mean and standard deviation of the application questionnaire scores

	Mean	Standard deviation
Kinect effectiveness	64.844	12.670
Content presentation	72.396	11.110
Cooperative learning	67.968	13.599
Increase of interest	64.844	28.027
Digital competence	70.313	25.339
Application total score	69.063	12.061

Table 2 reports the mean and the standard deviation of the SUS scores: besides the total SUS score, we considered also the *usability* and *learnability* SUS components as defined in [61]. The usability in a more strict sense has a higher mean score than the learnability component. However, the high standard deviation on the SUS learnability component reveals users have different opinions about the possibility of quickly and independently learning how to use the

Table 2. Mean and standard deviation of the SUS scores

	Mean	Standard deviation
SUS learnability	61.719	31.396
SUS usability	67.323	14.945
SUS total score	66.250	17.633

Table 3. Mean and standard deviation of the USE scores

	Mean	Standard deviation
Usefulness	68.359	13.344
Ease of use	69.176	15.969
Ease of learning	75.781	15.761
Satisfaction	71.652	14.087
USE total score	70.417	11.851

system. On the other hand, users' opinions are more in accordance on the SUS usability component, as suggested by the lower standard deviation.

Table 3 reports the mean and the standard deviation of the scores of each component of the USE questionnaire. The last row refers to the total score computed over all the questionnaire items.

4.2 Data Analysis

To find out any dependence among the considered variables we performed a principal component analysis of the five factors composing our application questionnaire (Table 4). As highlighted by the barplot in Fig. 3, there is a strong dominance of the first component, which accounts for about 59% of the total variability.

Figure 3 depicts also a biplot diagram, which provides a combined representation of two pieces of information [62]. The left and bottom axes represent the first and the second principal components respectively: in this reference system, dots represent the PCA scores of the samples. At the same time, the top and right axis of the same chart identify another reference system, where loadings, which stand for the influence strength of each considered variable, are depicted as vectors.

Since the first three components account for about 89% of the total variability, we discarded the last two components and applied a varimax rotation [63], which maximizes the sum of the variances of the squared loadings, and obtained the loadings in Table 5. Then we repeated the procedure including also the fourth component, which accounts for about 8% of the total variability, and obtained the loadings in Table 6.

While the results in Table 5 suggest a correlation between *Kinect effectiveness* and *Content presentation*, the results in Table 6 does not reveal any correlation

Table 4. Principal component analysis of the application questionnaire scores

	PC1	PC2	PC3	PC4	PC5
Standard deviation	1.7115557	0.9645988	0.7779061	0.61714066	0.39258880
Proportion of variance	0.5858846	0.1860902	0.1210276	0.07617252	0.03082519
Cumulative proportion	0.5858846	0.7719747	**0.8930023**	**0.96917481**	1.00000000

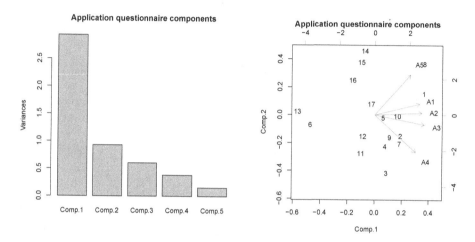

Fig. 3. Barplot and biplot of the variances of the components for the application questionnaire scores. A_1: Kinect effectiveness A_2: Content presentation A_3: Cooperative learning A_4: Increase of interest A_5: Digital competence

Table 5. Loadings of the first three principal components for the application questionnaire

	PC1	PC2	PC3
Kinect effectiveness	−0.105	-	**0.779**
Content presentation	0.103	-	**0.622**
Cooperative learning	**0.654**	0.310	-
Increase of interest	**0.742**	−0.267	-
Digital competence	-	**0.912**	-

between the two components. Both the analyses suggest a correlation between *Cooperative learning* and *Increase of interest*. We performed a more accurate analysis by means of a Pearson's correlation test (Table 7), which confirmed the correlation between *Cooperative learning* and *Increase of interest* and revealed also a correlation between *Kinect effectiveness* and *Content presentation*.

Then we carried out a principal component analysis of the four factors composing the USE metric (Table 8). Figure 4 depicts the barplot and the biplot diagrams. The former highlights a strong dominance of the first component, which

Table 6. Loadings of the first four principal components for the application questionnaire

	PC1	PC2	PC3	PC4
Kinect effectiveness	-	-	**0.964**	-
Content presentation	-	-	-	**−0.978**
Cooperative learning	**0.608**	0.325	−0.207	−0.159
Increase of interest	**0.794**	−0.220	0.151	0.119
Digital competence	-	**0.918**	-	-

Table 7. Pearson's correlation tests for the application questionnaire

	t	p-value	Correlation	Confidence interval
Kinect effectiveness/Content presentation	3.1462	**0.006658**	0.6305222	0.2150498 – 0.8527384
Cooperative learning/Increase of interest	3.7794	**0.001818**	0.6984049	0.3277964 – 0.8827306

accounts for about 61% of the total variability. In a first analysis we applied a varimax rotation to the first two components, which account for about 84% of the total variability, and obtained the loadings in Table 9. Then we repeated the procedure including also the third component, which accounts for about 11% of the total variability, and obtained the loadings in Table 10. Both the analyses suggest a correlation between *Usefulness* and *Satisfaction*. We performed a more accurate analysis by means of a Pearson's correlation test (Table 11) to detect whether *Ease of Use* and *Ease of Learning* are correlated. The test confirmed the correlation between *Usefulness* and *Satisfaction* and revealed also a correlation between *Ease of Use* and *Ease of Learning*.

Table 8. Principal component analysis of the USE metric

	PC1	PC2	PC3	PC4
Standard deviation	1.5608128	0.9585858	0.6666271	0.44786736
Proportion of variance	0.6090341	0.2297217	0.1110979	0.05014629
Cumulative proportion	0.6090341	**0.8387558**	**0.9498537**	1.00000000

Table 12 reveals no evident correlation between the SUS score and each of the five components of the application questionnaire since the p-value is greater than 0.05. For a deeper analysis aimed at detecting any influence of the perceived usability on the other application factors we applied the Pearson's correlation test between each component of the USE metric and each component of the application questionnaire.

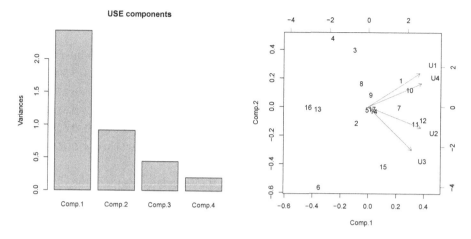

Fig. 4. Barplot and biplot of the variances of the components for the USE metric. U_1: Usefulness U_2: Ease of Use U_3: Ease of Learning U_4: Satisfaction

Table 9. Loadings of the first two principal components for the USE metric

	PC1	PC2
Usefulness	**0.735**	-
Ease of use	0.164	**−0.595**
Ease of learning	−0.144	**−0.798**
Satisfaction	**0.642**	-

Table 10. Loadings of the first three principal components for the USE metric

	PC1	PC2	PC3
Usefulness	**0.616**	0.328	0.334
Ease of use	-	-	**0.913**
Ease of learning	-	**−0.915**	-
Satisfaction	**0.786**	−0.216	−0.219

Table 11. Pearson's correlation tests for the USE questionnaire

	t	p-value	Correlation	Confidence interval
Usefulness/Satisfaction	3.9703	**0.001232**	0.7158212	0.3585311–0.8901907
Ease of use/Ease of learning	2.8528	**0.0121**	0.5930695	0.1572496–0.8355390

Table 13 highlights a possible correlation between the *Usefulness* scores and each of the following components: *Content presentation*, *Cooperative learning* and *Increase of interest*. As highlighted in Table 16, the same three components are correlated also with the *Satisfaction* scores. The Pearson's correlation tests do not reveal any influence of *Ease of Use* and *Ease of Learning* on the application scores (Tables 14 and 15).

Table 12. Pearson's correlation tests between the *SUS* scores and the application scores

	t	p-value	Correlation	Confidence interval
Kinect effectiveness	−0.61339	0.5488	−0.1564261	−0.5925219–0.3505770
Content presentation	−0.036549	0.9713	−0.009436481	−0.4878685–0.4733552
Cooperative learning	0.21838	0.8301	0.05629489	−0.4361512–0.5227940
Increase of interest	0.81889	0.4257	0.2068642	−0.3040075–0.6253332
Digital competence	−0.12045	0.9057	−0.03108467	−0.5041964–0.4563787

Table 13. Pearson's correlation tests between the *Usefulness* scores of the USE questionnaire and the application scores

	t	p-value	Correlation	Confidence interval
Kinect effectiveness	1.964	0.06834	0.45228	−0.03624366–0.76633437
Content presentation	**2.5743**	**0.02115**	**0.5535594**	**0.09934764–0.81686472**
Cooperative learning	**3.7419**	**0.001963**	**0.6948354**	**0.3215927–0.8811900**
Increase of interest	**2.1494**	**0.04833**	**0.5852474**	**0.006002595–0.783220384**
Digital competence	1.6127	0.1277	0.3843955	−0.1180621–0.7301411

Table 14. Pearson's correlation tests between the *Ease of Use* scores of the USE questionnaire and the application scores

	t	p-value	Correlation	Confidence interval
Kinect effectiveness	0.58095	0.5699	0.1483401	−0.3578166–0.5871236
Content presentation	0.28386	0.7804	0.07309591	−0.4223887–0.5349463
Cooperative learning	1.2781	0.2206	0.3133832	−0.1969233–0.6900835
Increase of interest	0.65255	0.5239	0.1661467	−0.3417928–0.5989600
Digital competence	1.1327	0.2751	0.2807131	−0.2311144–0.6708452

Table 15. Pearson's correlation tests between the *Ease of Learning* scores of the USE questionnaire and the application scores

	t	p-value	Correlation	Confidence interval
Kinect effectiveness	−0.14428	0.8872	−0.03722805	−0.5087692–0.4514955
Content presentation	−0.26363	0.7956	−0.06791301	−0.5312178–0.4266588
Cooperative learning	−0.3149	0.7572	−0.08103973	−0.5406264–0.4158010
Increase of interest	−0.32225	0.7517	−0.08291686	−0.5419625–0.4142367
Digital competence	0.14428	0.8872	0.03722805	−0.4514955–0.5087692

Table 16. Pearson's correlation tests between the *Satisfaction* scores of the USE questionnaire and the application scores

	t	p-value	Correlation	Confidence interval
Kinect effectiveness	1.3169	0.2077	0.3219127	−0.1877876–0.6950200
Content presentation	**2.4489**	**0.0271**	**0.5344259**	**0.07237099–0.80761850**
Cooperative learning	**2.4204**	**0.02866**	**0.5299612**	**0.06617196–0.80544144**
Increase of interest	**2.6788**	**0.01717**	**0.5688587**	**0.1214096–0.8241620**
Digital competence	0.081715	0.936	0.02109399	−0.4642578–0.4967028

5   Conclusions and Future Work

In the light of the potential and issues of the Kinect device, we tried to go further by conducting some experimental tests in a real education scenario.

We implemented an ipermedia application on Leonardo Da Vinci's inventions and based the user interactions on gestures detected by the Kinect touchless interface. We analyzed the usability factors of the whole system and their influence on the perceived efficacy and effectiveness of the learning environment. The main results highlighted that the perceived usefulness and the level of satisfaction influence students' opinions about content presentation, the ability of the application to foster cooperative learning and the ability to increase the interest towards the topics. Moreover, results revealed a correlation between the perceived effectiveness of Kinect and the opinions about the content presentation, as well as between the improvement in cooperative learning and the increase in interest.

In future work we will conduct similar experiments also with other devices for gesture detection: we will employ the Leap Motion controller and the new Microsoft Azure Kinect. We will compare the new results with data analyzed in this work in terms of perceived usability and in terms of efficacy and effectiveness in a learning environment.

Also the use of brain computer interfaces [64,65] could give interesting food for thoughts, as data recorded by such devices allow a deeper analysis of the cognitive processes and can be compared with questionnaires' scores to find out any correlation.

References

1. Balaji, B.S., Sekhar, A.C.: The various facets of MOOC. In: IEEE International Conference in MOOC, Innovation and Technology in Education (MITE), pp. 139–142 (2013)
2. Tommasi, F., Melle, C., De Luca, V.: OpenSatRelaying: a hybrid approach to real-time audio-video distribution over the internet. J. Commun. **9**(3), 248–261 (2014)
3. Tommasi, F., De Luca, V., Melle, C.: Are P2P streaming systems ready for interactive e-learning? In: International Conference on Education Technologies and Computers (ICETC), pp. 49–54 (2014)

4. De Paolis, L.T., Aloisio, G., Celentano, M.G., Oliva, L., Vecchio, P.: Experiencing a town of the middle ages: an application for the edutainment in cultural heritage. In: IEEE 3rd International Conference on Communication Software and Networks, pp. 169–174 (2011)
5. De Paolis, L.T., Aloisio, G., Celentano, M.G., Oliva, L., Vecchio, P.: A game-based 3D simulation of otranto in the middle ages. In: Third International Conference on Advances in Computer-Human Interactions, pp. 130–133 (2010)
6. De Paolis, L.T., Aloisio, G., Celentano, M.G., Oliva, L., Vecchio, P.: MediaEvo project: a serious game for the edutainment. In: 3rd International Conference on Computer Research and Development, vol. 4, pp. 524–529 (2011)
7. De Paolis, L.T.: Walking in a virtual town to understand and learning about the life in the middle ages. In: Murgante, B., et al. (eds.) ICCSA 2013. LNCS, vol. 7971, pp. 632–645. Springer, Heidelberg (2013). https://doi.org/10.1007/978-3-642-39637-3_50
8. Lin, A.C., Fernandez, W.D., Gregor, S.: Understanding web enjoyment experiences and informal learning: a study in a museum context. Decis. Support Syst. 53(4), 846–858 (2012)
9. Marsick, V.J., Watkins, K.E.: Informal and Incidental Learning. New Dir. Adult Contin. Educ. 2001(89), 25–34 (2001)
10. Chang, Y.H., Hwang, J.H., Fang, R.J., Lu, Y.T.: A Kinect-and game-based interactive learning system. Eurasia J. Math. Sci. Technol. Educ. 13(8), 4897–4914 (2017)
11. Qian, X.: Construction and application of an educational game based on the ARCS model. World Trans. Eng. Technol. Educ. 12(2), 236–241 (2014)
12. Chang, Y.H., Hwang, J.H., Fang, R.J.: A joyful Kinect-based learning system. In: Proceedings of the 2017 IEEE International Conference on Applied System Innovation: Applied System Innovation for Modern Technology, ICASI 2017 (2017)
13. Keller, J.M.: How to integrate learner motivation planning into lesson planning: the ARCS model approach. In: VII Semenario, Santiago, Cuba (2000)
14. Kosmas, P., Ioannou, A., Retalis, S.: Moving Bodies to Moving Minds: A Study of the Use of Motion-Based Games in Special Education. TechTrends (2018)
15. Kumara, G.W., Wattanachote, K., Battulga, B., Shih, T.K., Hwang, W.Y.: A Kinect-based assessment system for smart classroom. Int. J. Distance Educ. Technol. 13(2), 34–53 (2015)
16. Tsai, C.-H., Yen, J.-C.: Teaching spatial visualization skills using OpenNI and the microsoft Kinect sensor. In: Park, J.J.J.H., Pan, Y., Kim, C.-S., Yang, Y. (eds.) Future Information Technology. LNEE, vol. 309, pp. 617–624. Springer, Heidelberg (2014). https://doi.org/10.1007/978-3-642-55038-6_97
17. Yilmaz, O., Bayraktar, D.M.: Impact of Kinect game on primary school students' mental computation speed. Int. J. Game-Based Learn. 8(4), 50–67 (2018)
18. Anwar, S., Sinha, S.K., Vivek, S., Ashank, V.: Hand gesture recognition: a survey. In: Nath, V., Mandal, J.K. (eds.) Nanoelectronics, Circuits and Communication Systems. LNEE, vol. 511, pp. 365–371. Springer, Singapore (2019). https://doi.org/10.1007/978-981-13-0776-8_33
19. Tullis, T., Albert, B.: Measuring the User Experience: Collecting, Analyzing, and Presenting Usability Metrics, 2nd edn (2013)
20. Bachmann, D., Weichert, F., Rinkenauer, G.: Review of three-dimensional human-computer interaction with focus on the leap motion controller. Sensors 18(7), 2194 (2018)
21. Ramos, E., et al.: Based Kinect application to promote mixtec culture. Procedia Technol. 7, 344–351 (2013)

22. Blanchard, E.G., Zanciu, A.N., Mahmoud, H., Molloy, J.S.: Enhancing in-museum informal learning by augmenting artworks with gesture interactions and AIED paradigms. In: Lane, H.C., Yacef, K., Mostow, J., Pavlik, P. (eds.) AIED 2013. LNCS (LNAI), vol. 7926, pp. 649–652. Springer, Heidelberg (2013). https://doi.org/10.1007/978-3-642-39112-5_80

23. Dondi, P., Lombardi, L., Rocca, I., Malagodi, M., Licchelli, M.: Multimodal workflow for the creation of interactive presentations of 360 spin images of historical violins. Multimedia Tools Appl. **77**(21), 28309–28332 (2018)

24. Yoshida, R., et al.: Novel application of Kinect sensor to support immersive learning within museum for children. In: Proceedings of the International Conference on Sensing Technology, ICST (2016)

25. Weede, O., Muchinenyika, S.H., Muyingi, H.N.: Virtual welcome guide for interactive museums (2014)

26. Cuccurullo, S., Francese, R., Murad, S., Passero, I., Tucci, M.: A gestural approach to presentation exploiting motion capture metaphors (2012)

27. Méndez, R., Flores, J., Castelló, E., Viqueira, J.R.: Natural interaction in virtual TV sets through the synergistic operation of low-cost sensors. Univ. Access Inf. Soc. **18**(1), 17–29 (2019)

28. Garber, L.: Gestural technology: moving interfaces in a new direction [technology news]. Computer **46**(10), 22–25 (2013)

29. Wobbrock, J.O., Aung, H.H., Rothrock, B., Myers, B.A.: Maximizing the guessability of symbolic input (2005)

30. Dong, H., Figueroa, N., El Saddik, A.: An Elicitation Study on Gesture Attitudes and Preferences Towards an Interactive Hand-Gesture Vocabulary (2016)

31. Argelaguet, F., Andujar, C.: A survey of 3D object selection techniques for virtual environments. Comput. Graph. **37**(3), 121–136 (2013)

32. Ruddle, R.: Review: 3D user interfaces: theory and practice Doug A. Bowman, Ernst Kruijff, Joseph J. LaViola Jr., Ivan Poupyrev. Presence Teleoperators Virtual Environ. **14**(1), 117–118 (2006)

33. Herndon, K.P., van Dam, A., Gleicher, M.: The challenges of 3D interaction. ACM SIGCHI Bull. **26**(4), 36–46 (2007)

34. Hilliges, O., Kim, D., Izadi, S., Weiss, M., Wilson, A.: HoloDesk: direct 3D interactions with a situated see-through display. In: Proceedings of the 2012 ACM Annual Conference on Human Factors in Computing Systems - CHI 2012 (2012)

35. Ullmer, B., Ishii, H.: Emerging frameworks for tangible user interfaces. IBM Syst. J. **39**, 915–931 (2010)

36. Hung, C.Y., Lin, Y.R., Huang, K.Y., Yu, P.T., Sun, J.C.Y.: Collaborative game-based learning with motion-sensing technology. Int. J. Online Pedagogy Course Des. **7**(4), 53–64 (2017)

37. Sinha, H., Srivastava, S., Sinha, Y.: Studying the Role of Kinect as a Multi-Sensory Learning Platform for Children (2018)

38. Granja, F.T.M., Escriba, L.R., Lozada, R.M.: MS-Kinect in the development of educational games for preschoolers. Int. J. Learn. Technol. **13**(4), 277–305 (2019)

39. Kennewell, S., Tanner, H., Jones, S., Beauchamp, G.: Analysing the use of interactive technology to implement interactive teaching. J. Comput. Assisted Learn. **24**(1), 61–73 (2008)

40. Chang, C.Y., Chien, Y.T., Chiang, C.Y., Lin, M.C., Lai, H.C.: Embodying gesture-based multimedia to improve learning. Br. J. Educ. Technol. **44**(1), E5–E9 (2013)

41. Chao, K.J., Huang, H.W., Fang, W.C., Chen, N.S.: Embodied play to learn: exploring Kinect-facilitated memory performance. Br. J. Educ. Technol. **44**(5), E151–E155 (2013)

42. Hsu, H.M.J.: The potential of Kinect in education. Int. J. Inf. Educ. Technol. **1**(5), 365 (2013)
43. Too, M.S., Ong, P.T., Lau, S.H., Chang, R.K., Sim, K.S.: Kinect-based framework for enhanced learning of disabled students. In: Proceedings of 2016 International Conference on Robotics, Automation and Sciences, ICORAS 2016 (2017)
44. Ojeda-Castelo, J.J., Piedra-Fernandez, J.A., Iribarne, L., Bernal-Bravo, C.: KiNEEt: application for learning and rehabilitation in special educational needs. Multimedia Tools Appl. **77**(18), 24013–24039 (2018)
45. Mir, H.Y., Khosla, A.K.: Kinect based game for improvement of sensory, motor and learning skills in autistic children. In: Proceedings of the 2nd International Conference on Intelligent Computing and Control Systems, ICICCS 2018 (2019)
46. Shakroum, M.A., Wong, K.W., Fung, L.C.C.: The effectiveness of the gesture-based learning system (GBLS) and its impact on learning experience. J. Inf. Technol. Educ. Res. **15**, 191–210 (2017)
47. Kolb, D.A.: Experiential Learning: Experience as the Source of Learning and Development. Prentice-Hall P T R, Englewood Cliffs (1984)
48. Kolb, D., Rubin, I., McIntyre, J.: Organizational Psychology: Readings on Human Behavior in Organizations. Behavioral Science in Business Series. Prentice-Hall (1984)
49. Angotti, R., Bayo, I.: Making Kinections: Using video game technology to teach math. In: Prato CIRN Community Informatics Conference (2012)
50. Gerling, K., Dergousoff, K., Mandryk, R.: Is movement better? Comparing sedentary and motion-based game controls for older adults. In: Graphics Interface 2013 (2013)
51. Vrellis, I., Moutsioulis, A., Mikropoulos, T.A.: Primary school students' attitude towards gesture based interaction: a comparison between Microsoft Kinect and mouse. In: Proceedings - IEEE 14th International Conference on Advanced Learning Technologies, ICALT 2014 (2014)
52. De Paolis, L.T., De Luca, V., D'Errico, G.: Augmented reality to understand the Leonardo's Machines. In: De Paolis, L.T., Bourdot, P. (eds.) AVR 2018. LNCS, vol. 10851, pp. 320–331. Springer, Cham (2018). https://doi.org/10.1007/978-3-319-95282-6_24
53. da Vinci, L., Marinoni, A.: Il Codice Atlantico della Biblioteca Ambrosiana di Milano. No. v. 1-4 in Grandi opere in facsimile, Giunti (2000)
54. Sarbolandi, H., Lefloch, D., Kolb, A.: Kinect range sensing: structured-light versus time-of-flight Kinect. Comput. Vis. Image Underst. **139**, 1–20 (2015)
55. Lachat, E., Macher, H., Landes, T., Grussenmeyer, P.: Assessment and calibration of a RGB-D camera (Kinect v2 Sensor) towards a potential use for close-range 3D modeling. Remote Sens. **7**(10), 13070–13097 (2015)
56. Assila, A., Marçal De Oliveira, K., Ezzedine, H.: Standardized usability questionnaires: features and quality focus. J. Comput. Sci. Inf. Technol. (eJCSIT) **6**(1), 15–31 (2016)
57. Lewis, J.R.: Measuring perceived usability: the CSUQ, SUS, and UMUX. Int. J. Hum.-Comput. Interact. **34**(12), 1148–1156 (2018)
58. Brooke, J.: SUS - a quick and dirty usability scale. Usability Eval. Ind. **189**(194), 4–7 (1996)
59. Borsci, S., Federici, S., Lauriola, M.: On the dimensionality of the system usability scale: a test of alternative measurement models. Cogn. Process. **10**(3), 193–197 (2009)
60. Lund, B.A.M.: Measuring usability with the USE questionnaire. STC Usability SIG Newsl. **8**(2), 3–6 (2001)

61. Lewis, J.R., Sauro, J.: The factor structure of the system usability scale. In: Kurosu, M. (ed.) HCD 2009. LNCS, vol. 5619, pp. 94–103. Springer, Heidelberg (2009). https://doi.org/10.1007/978-3-642-02806-9_12
62. Gabriel, K.R.: The biplot graphic display of matrices with application to principal component analysis. Biometrika **58**(3), 453–467 (1971)
63. Kaiser, H.F.: The varimax criterion for analytic rotation in factor analysis. Psychometrika **23**(3), 187–200 (1958)
64. Invitto, S., Faggiano, C., Sammarco, S., De Luca, V., De Paolis, L.T.: Interactive Entertainment, Virtual Motion Training and Brain Ergonomy (2015)
65. Invitto, S., Faggiano, C., Sammarco, S., De Luca, V., De Paolis, L.T.: Haptic, virtual interaction and motor imagery: entertainment tools and psychophysiological testing. Sensors **16**(3), 394 (2016)

Industry

Analysis of Fuel Cells Utilizing Mixed Reality and IoT Achievements

Burkhard Hoppenstedt[1]([envelope]), Michael Schmid[2], Klaus Kammerer[1], Joachim Scholta[2], Manfred Reichert[1], and Rüdiger Pryss[1]

[1] Institute of Databases and Information Systems, Ulm University, Ulm, Germany
`burkhard.hoppenstedt@uni-ulm.de`
[2] Zentrum für Sonnenenergie- und Wasserstoff-Forschung Baden-Württemberg, Ulm, Germany

Abstract. Recent advances in the development of smart glasses enable new interaction patterns in an industrial context. In the field of Mixed Reality, in which the real world and virtual objects fuse, new developments allow for advanced procedures of condition monitoring. Hereby, the smart glasses serve as a mobile display and inspection station. In this work, we focus on the applicability of Mixed Reality to monitor data of the spatially resolved current density distribution of a fuel cell. To be more specific, we implemented an IoT approach based on the Message Queuing Telemetry Transport protocol (MQTT) to enable the aforementioned monitoring. The realized solution, in turn, provides a live monitoring as well as an overview feature.

Keywords: Fuel cells · Mixed Reality · IoT · MQTT

1 Introduction

In the context of the *industrial internet of things* (IIoT), also denoted as Industry 4.0 [18], the collection of sensor values becomes more and more crucial. These values are then used, e.g., for *condition monitoring* [23], *process control* [11], or *advanced analytics* (e.g., Predictive Maintenance [14]). The overall goal of a production setting connected through sensors is to increase the production efficiency by (a) reducing downtimes through predictive methods, (b) increasing the production transparency to discover bottlenecks, and (c) enabling data-driven approaches for a self-diagnostics plant. Hereby, machine communication protocols, such as *Open Platform Communications Unified Architecture* (OPC UA) [13] or the *Message Queuing Telemetry Transport* protocol (MQTT) [16], are an essential part to exchange data in the needed distributed architectures [4]. These protocols implement features to ensure the guaranteed delivery of messages and required encryption needs. In this work, an IoT approach based on (1) the MQTT communication protocol and (2) the Microsoft HoloLens smart glass was realized to test its feasibility for the monitoring of current density distribution data of a fuel cell.

© Springer Nature Switzerland AG 2019
L. T. De Paolis and P. Bourdot (Eds.): AVR 2019, LNCS 11614, pp. 371–378, 2019.
https://doi.org/10.1007/978-3-030-25999-0_31

Research into alternative energy sources is particularly important nowadays as the impact of greenhouse gases on the environment through the use of fossil fuels continues to increase as these resources become more and more scarce [3]. One way to overcome these problems could be the use of fuel cells and the expansion of the hydrogen infrastructure. Such energy conversion devices generate electricity using hydrogen and oxygen in an electrochemical process for which water is the only remaining waste product [2]. Therefore, in recent years, a variety of scientific research has been conducted to optimize fuel cells and minimize their manufacturing costs [22].

In our use case, the fuel cell represents a machine that delivers sensor values, whereas a HoloLens is the monitoring application. We connect these two devices via the MQTT protocol for a quick and trustworthy connection. The connection allows to supervise the progress of the fuel cell's sensor values as well as to automatically generated alarms, which, in turn, can be sent to various recipients (e.g., process control staff).

The remainder of the paper is structured as follows: in Sect. 2 related work is discussed, while Sect. 3 introduces the backgrounds on fuel cells, Mixed Reality, and the MQTT protocol. In Sect. 4, the developed prototype is presented, in which the data set, the Graphical User Interface (GUI), and the backend system are presented. Threats to validity are presented in Sect. 5, whereas Sect. 6 concludes the paper with a summary and an outlook.

2   Related Work

The first part of the related work refers to fuel cells. According to the United States Department of Energy (DOE) [7], fuel cells with polymer electrolyte membranes (PEMs) have been developed for use in automobiles since the late 1980's and steady progress has been made to date. Fundamental studies of electrochemical properties are particularly important for improving PEM fuel cells as they can generate current density distribution inhomogeneities due to different reactions and activities in the active cell region. These are also influenced by parameters such as temperature and humidity of the membrane and have a fundamental influence on the life cycle and performance of a fuel cell. By visualizing the current density distribution within the fuel cell as shown in Fig. 1, corresponding information can be obtained [8]. Concerning the second part of related work, augmented reality is used in various use cases to monitor aspects of the real world. In [10], 3D models are compared to real world objections for the purpose of construction supervision. Wireless sensor networks, in turn, are monitored by [9] using an augmented reality interface. The HoloLens, which represents a smart glass of the category Mixed Reality, is often utilized in the medical context (cf. [17] or [15]). However, to the best of our knowledge, a combination of technologies as shown in this work, has not been presented in other works so far.

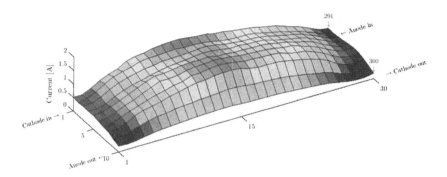

Fig. 1. Visualization of the current density distribution inside a fuel cell

3   Fundamentals

3.1   Fuel Cells

Hydrogen (H_2) is introduced on the anode side and air containing oxygen (O_2) on the cathode side. At the anode, the molecular hydrogen is split into hydrogen nuclei (H^+), also called protons and electrons (e^-), with the help of a catalyst. The protons migrate through the electrolyte membrane - which is permeable only to them - to the cathode side. The electrons travel from the anode through an electrical conductor to the cathode. The resulting current flow, in turn, can be exploited. On the cathode side, two electrons reduce oxygen which then combines with two H^+-ions to form water (H_2O) [5], as shown in Eq. 1. This electrochemical process is schematically shown in Fig. 2.

$$2\,H^+ + O^{2-} \longrightarrow H_2O \tag{1}$$

3.2   Mixed Reality

The HoloLens is a device to realize mixed-reality applications. Mixed Reality is known to have the highest intersection of reality and virtual environment of all augmented reality approaches [20] due to a concept named *spatial mapping*. This procedure creates a model of the environment in the augmented reality device. Therefore, interactions of holograms and real-world objects become possible. Mixed Reality is basically used to display 3D models (1) for which a real-world model would be too large or small (e.g., in the domains automotive or architecture), (2) in medical use cases (assistance during a surgery), or (3) in industrial maintenance support setting. The HoloLens is equipped with various cameras and sensors, such as a depth sensor, a RGB camera, and an ambient light sensor. The holograms can be anchored to real-life objects, but infinite projections are not possible, neither to the distance nor to the proximity. With a weight of 579 g,

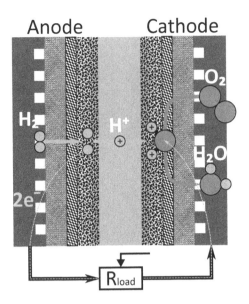

Fig. 2. Schematic of the electrochemical reaction inside a fuel cell

the HoloLens should not be used for a long period of time due to an unnatural head positioning. A HoloLens case study [25] found out that the heavy weight of the device degrades the user's comfort level. Finally, in an intensive use case, the battery lasts for about 2.5 h, which also inhibits a long-time usage.

3.3 Message Queuing Telemetry Transport Protocol

The Message Queuing Telemetry Transport protocol (MQTT) is a light-weight machine to machine communication protocol. It uses a publish-subscribe [6] pattern, including the use of *topics*. According to [16], publish/subscribe systems are wide-spread in distributed computing. Hereby, a topic can be considered as a black board for messages. Subscribers are informed about changes to these topics and new messages (e.g., sensor values) can be pushed to these topics. A distribution server, denoted as *broker*, is responsible to forward messages to subscribed clients. MQTT offers a *Quality of Service* (QoS) level [19], for which the delivery of a specific message is guaranteed at most once, at least once, or exactly once. MQTT has, in contrast to OPC UA, no semantic structure and can therefore transport any kind of message. All these mechanisms make MQTT a suitable communication protocol for IoT use cases.

4 Prototype

The realized prototype shall allow for the monitoring of the current density distribution of a fuel cell in Mixed Reality. More specifically, the Microsoft HoloLens is used as MQTT client to display the current state of the fuel cell. The user, in turn, shall be enabled to interact with the holographic visualization through the MQTT interface. Following this, for example, the values of the fuel cell can be monitored and evaluated in real time. As a first step of the developed proto-type, the fuel cell was digitized using a 3D modeling tool (i.e., Blender, see also Fig. 3). Hereby, the arrows are animated to indicate input and output of the fuel cell gases. A cell grid represents all measuring points in the fuel cell. As a next step, the blender model can be attached with interaction logic. Therefore, we implemented a *tap to place method*, so that the model can be placed anywhere in the real world.

Fig. 3. 3D rendering of the fuel cell end plates with space for the current density distribution values for spatial visualization with the HoloLens device

Then, the values of the fuel cell are sent to the model for monitoring. We implemented the use cases *live monitoring* and *loading of a data set*. The main difference constitutes a possible replay and change of playing speed for the sec-ond use case. The control for the replay is provided by MQTT. As we integrated a MQTT client into the HoloLens application, MQTT can be used as a remote control to set the current frame or frame rate. Using the replay mode, it is possi-ble to get a quick overview of the temporal behavior by viewing the sensor values in a time-lapse mode. In contrast, when using the live mode, it is possible to be alerted via a sound or sending the alert to any IoT device that can implement the MQTT protocol (e.g., smartphones, machines, or computers). The realized architecture of the prototype is shown in Fig. 4.

5   Threats to Validity

The following limitations need to be considered for the work at hand. First, the weight of the HoloLens smart glass cannot be neglected. Intensive use might

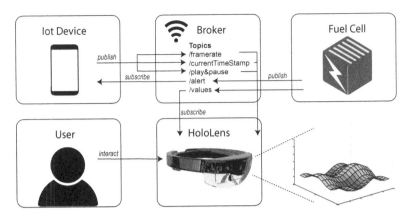

Fig. 4. Prototype architecture

cause headaches or dizziness. Second, the further connectivity of the HoloLens might be a problem. The options for data science analyses in Mixed Reality are not sufficiently evolved so far. However, the possibilities of data analytics in augmented reality, denoted as immersive analytics, are more and more investigated [12]. Third, a general problem of distributed systems is the network security [26] and network stability. In our approach, we solely rely on the connectivity and security features implemented by MQTT. Finally, new interaction methods, provided by the HoloLens, also result in new challenges. When using the HoloLens via voice commands, it is essential to be aware of a user bias. For the voice commands, studies have shown that speech recognition performs worse for women compared to men (cf. [21,24]). Despite these limitations, the strength of our prototype was that we combined a fuel cell, an IoT protocol, and a contemporary smart glass to an interactive visualization approach.

6   Summary and Outlook

We presented a prototype for the monitoring of the current density distribution of a fuel cell in Mixed Reality. The architecture incorporates an IoT message protocol (MQTT) for a light-weight communication to any device that supports this protocol. The Microsoft HoloLens, which represents a mixed-reality smart glass, is used as MQTT client to display the current state of the fuel cell. The user can interact with the holographic visualization using voice commands or by sending commands via the MQTT interface. The values of the fuel cell can be either inspected in real time or by using a preloaded data set. The latter offers a quick inspection, e.g., by using a time-lapse function. Currently, the field of view for the HoloLens is limited to a small window. However, upcoming types of mixed-reality glasses will fix this limitation and offer new user interaction patterns. Moreover, the analytic part of our approach is currently limited to alerts. By including stream analytic approaches [1], we could provide a more powerful online analytic tool. In the presented approach, solely the current density distribution was used. However, the spatially resolved visualization of the temperature and humidity would be also promising targets. Moreover, other important values may be presented in an interactive head-up display in order to keep an eye on the condition of the fuel cell in an even more efficient manner. On top of this, it is conceivable to adapt the operating parameters by means of gestures or voice controls. Altogether, this work has shown that Mixed Reality can play an important role in different domains that are less considered so far.

Acknowledgements. The authors want to thank the German Federal Ministry for Economic Affairs and Energy for funding part of the presented work within the project SoHMuSDaSS (FKZ: 03ET6057C).

References

1. Andrade, H.C., Gedik, B., Turaga, D.S.: Fundamentals of Stream Processing: Application Design, Systems, and Analytics. Cambridge University Press, Cambridge (2014)
2. de Beer, C., Barendse, P.S., Pillay, P., Bullecks, B., Rengaswamy, R.: Classification of high-temperature PEM fuel cell degradation mechanisms using equivalent circuits. IEEE Trans. Ind. Electron. **62**(8), 5265–5274 (2015)
3. Chandan, A., et al.: High temperature (HT) polymer electrolyte membrane fuel cells (PEMFC) - a review. J. Power Sources **231**, 264–278 (2013)
4. Datta, S.K., Bonnet, C., Nikaein, N.: An IoT gateway centric architecture to provide novel M2M services. In: 2014 IEEE World Forum on Internet of Things (WF-IoT), pp. 514–519. IEEE (2014)
5. Ehret, O.: Wasserstoff und Brennstoffzellen: Antworten auf wichtige Fragen. NOW GmbH - Nationale Organisation Wasserstoff- und Brennstoffzellentechnologie, Berlin (2018)
6. Eugster, P.T., Felber, P.A., Guerraoui, R., Kermarrec, A.M.: The many faces of publish/subscribe. ACM Comput. Surv. **35**(2), 114–131 (2003)
7. Fuel Cell Technologies Office: Multi-Year Research, Development, and Demonstration Plan. Technical report, U.S. Department of Energy (2012)
8. Geske, M., Heuer, M., Heideck, G., Styczynski, Z.A.: Current density distribution mapping in PEM fuel cells as an instrument for operational measurements. Energies **3**(4), 770–783 (2010)
9. Goldsmith, D., Liarokapis, F., Malone, G., Kemp, J.: Augmented reality environmental monitoring using wireless sensor networks. In: 12th International Conference Information Visualisation, pp. 539–544. IEEE (2008)
10. Golparvar-Fard, M., Peña-Mora, F., Savarese, S.: D4AR-a 4-dimensional augmented reality model for automating construction progress monitoring data collection, processing and communication. J. Inf. Technol. Constr. **14**(13), 129–153 (2009)
11. Gonzaga, J., Meleiro, L.A.C., Kiang, C., Maciel Filho, R.: ANN-based soft-sensor for real-time process monitoring and control of an industrial polymerization process. Comput. Chem. Eng. **33**(1), 43–49 (2009)
12. Gracia, A., González, S., Robles, V., Menasalvas, E., Von Landesberger, T.: New insights into the suitability of the third dimension for visualizing multivariate/multidimensional data: a study based on loss of quality quantification. Inf. Vis. **15**(1), 3–30 (2016)

13. Hannelius, T., Salmenpera, M., Kuikka, S.: Roadmap to adopting OPC UA. In: 6th IEEE International Conference on Industrial Informatics, pp. 756–761. IEEE (2008)
14. Hoppenstedt, B., et al.: Techniques and emerging trends for state of the art equipment maintenance systems–a bibliometric analysis. Appl. Sci. **8**, 1–29 (2018)
15. Hoppenstedt, B., et al.: Holoview: exploring patient data in mixed reality. In: TRI/TINNET Conference 2018 (2018)
16. Hunkeler, U., Truong, H.L., Stanford-Clark, A.: MQTT-S–a publish/subscribe protocol for wireless sensor networks. In: 3rd International Conference on Communication Systems Software and Middleware and Workshops, pp. 791–798. IEEE (2008)
17. Hurter, C., McDuff, D.: Cardiolens: remote physiological monitoring in a mixed reality environment. In: ACM SIGGRAPH 2017 Emerging Technologies, p. 6. ACM (2017)
18. Lasi, H., Fettke, P., Kemper, H.G., Feld, T., Hoffmann, M.: Industry 4.0. Bus. Inf. Syst. Eng. **6**(4), 239–242 (2014)
19. Lee, S., Kim, H., Hong, D.K., Ju, H.: Correlation analysis of MQTT loss and delay according to QoS level. In: International Conference on Information Networking, pp. 714–717. IEEE (2013)
20. Milgram, P., Takemura, H., Utsumi, A., Kishino, F.: Augmented reality: a class of displays on the reality-virtuality continuum. In: Telemanipulator and Telepresence Technologies, vol. 2351, pp. 282–293. International Society for Optics and Photonics (1995)
21. Nicol, A., Casey, C., MacFarlane, S.: Children are ready for speech technology-but is the technology ready for them. Interaction Design and Children, Eindhoven, The Netherlands (2002)
22. Orfanidi, A.: Electrocatalytic Investigation of High Temperature PEM Fuel Cells Alin Orfanidi, December 2014
23. Rangwala, S., Dornfeld, D.: Sensor integration using neural networks for intelligent tool condition monitoring. J. Eng. Ind. **112**(3), 219–228 (1990)
24. Rodger, J.A., Pendharkar, P.C.: A field study of the impact of gender and user's technical experience on the performance of voice-activated medical tracking application. Int. J. Hum Comput Stud. **60**(5–6), 529–544 (2004)
25. Zhang, L., Dong, H., El Saddik, A.: Towards a QoE model to evaluate holographic augmented reality devices: a hololens case study. IEEE MultiMedia (2018)
26. Zhang, Z.K., Cho, M.C.Y., Wang, C.W., Hsu, C.W., Chen, C.K., Shieh, S.: IoT security: ongoing challenges and research opportunities. In: 7th International Conference on Service-Oriented Computing and Applications, pp. 230–234. IEEE (2014)

Virtual Environment for Training Oil & Gas Industry Workers

Carlos A. Garcia[1], Jose E. Naranjo[1], Fabian Gallardo-Cardenas[2], and Marcelo V. Garcia[1,3(✉)]

[1] Universidad Tecnica de Ambato, UTA, 180103 Ambato, Ecuador
{ca.garcia,jnaranjo0463,mv.garcia}@uta.edu.ec
[2] Escuela Politecnica de Chimborazo, ESPOCH, 060155 Riobamba, Ecuador
felix.gallardo@espoch.edu.ec
[3] University of Basque Country, UPV/EHU, 48013 Bilbao, Spain
mgarcia294@ehu.eus

Abstract. Oil and Gas industries have grown year by year in the automation field, trying to improve their processes and productivity, and at the same time trying to ensure the safety of their workers, machinery and company. To achieve these goals, it is necessary to periodically train instrumentation technicians and field operators; however, it must be taken into account that most of industrial equipment is really expensive. Due to this it has been necessary to use tools such as Virtual Reality (VR) and Augmented Reality (AR), which optimize and reduce both the time and the cost of training. This research proposes the development of a virtual training system for the commissioning, calibration and mounting HART transmitters in dynamic and dangerous environments for the human being such us Oil & Gas process facilities.

Keywords: Oil & Gas industry · Virtual reality · Virtual training · Engineering education

1 Introduction

In the last ten years, several simulation systems developed in Virtual Reality have gained strength when it comes to training in hostile environments with specialized equipment. The industry has incorporated such tools or technological software to improve their processes as well as their productivity. Simulators have been created to work just like it would be in a physical or real environment, according to the needs inside this type of industries [1].

Virtual Reality is the creation of computer generated 3D interactive worlds, with the intention of giving the user the illusion of being immerse in a different reality. It started in 1956, when Morton Heilig created the first Virtual Reality machine [2]. Nowadays Virtual Reality is accessible to all people and includes several application fields that can range from the simulation of industrial processes to the training of medical personnel.

L. T. De Paolis and P. Bourdot (Eds.): AVR 2019, LNCS 11614, pp. 379–392, 2019.
https://doi.org/10.1007/978-3-030-25999-0_32

The use of 3D virtual environments in the simulation field allows for advances in the development of various systems, whose effectivity and costs are meant to improve, whatever the field of application. The goal is to collect previous analysis and study for a later implementation of a 3D virtual environment that will be useful for simulating industrial work environments, in which the operators can be trained and develop skills before the interaction with the real industrial environment [3].

A software used in recent researches is Unity Pro 3D^{TM}, which is one of the first simulators that reproduces an automaton's performance, it was developed by Schneider Electric, this tool will be used to virtually create the industrial instruments combined with a virtual environment, which will allow to simulate some industrial process [4].

The aim of the industrial scenes or environments model is to commit the operators to the enlargement of their knowledge in instrument handling and the development of the processes they will be facing in the industry [5,6].

As described in previous paragraphs, the present research work contributes to the adaptation of industrial virtual environments, which will allow for the training of instrumentation technicians and operators in the commissioning of equipment into industrial processes. This new proposal aims to innovate and offer optimal solutions, also it will contribute future research works looking to improve the virtual education of novice engineers, technicians and operators.

The main goal of this article is to develop a training software based in virtual reality for instrumentation technician in Oil & Gas industry who will learn in a virtual way the commissioning of HART instrumentation before the installation and operation. This virtual environment gives the necessary information to carry out a complete simulation of the industrial instruments calibration and configuration, all of this with the goal of reduce economic losses that could take place due to the wrong calibration and configuration of these instruments in a real process environment, allowing the feasibility of making calibration mistakes and knowing how to correct them.

This article is divided into 7 sections including the introduction. In Sect. 2 the related works are detailed. In Sect. 3 the state of technology is analyzed. The case study is presented in Sect. 4. The implementation proposal is explained in Sect. 5. The analysis of the results of this investigation states in Sect. 6. Finally, Sect. 7 develops the conclusions and future work.

2   Related Works

In this section, researches and works directly related to the areas that have worked with virtual reality and simulators are analyzed. Also, the focus and reach of the research proposed in this work is described from a technical point of view, which will allow having a reference concerning real industrial environments.

During the last years, research about the design of simulators that aid in teaching and learning in the educational area have been published, but still a simulator allowing development of skills involving industrial operators does not exist.

Potkonjak et al. in their investigative work [7], states that her objective is to propose a virtual course evaluation model that includes the greater possible amount of variables intervening, and its application to a virtual course. The proposed model includes the study materials quality analysis, the teacher's performance and the technological environment quality. In this work, it is only treated a virtual platform that helps in performing evaluations and quality analysis, but it does not cover a virtual industrial environment. The present investigation pretends to create these non-existing environments in many of the investigations performed earlier.

Pelargos et al. in research [8], talks about simulation and online games as a tool for educational immersion, which he says should be approached in scientific manner, that way giving answers to new and great questions in such unusual interpersonal, working and even emotional relations, that comes off the new communication tools and define them as a complex phenomenon. Immersive education [9] is a new learning platform that combines interactive 3D graphics, videogames, simulation, virtual reality, webcams, digital media and online classrooms. Once again, it can be noted that it is not about simulation of industrial operations, which is what the present work seeks to establish, because technology must be taken advantage of to apply these type of simulators to aid to the preparation and training of industrial operators.

Dengel et al. in his article [10], talks about simulators and their potential in university teaching. This article presents some of the results of a research work performed around the introduction of technologies to the university teaching, especially in the simulations case. In particular, the article pretends to announce diverse implications about the incorporation of simulators in relation with their potential to favor learning processes. Again, it is not about an industrial simulator, because the intention is to operate a simulator to develop potential in university learning, though not in the working area.

Marquez et al. in their research article [11], which main goal is about a simulation environment in Unity Pro 3D^{TM}, applied to a bottling system which has proposed a problem, programming an optimal solution and a docent module, bound to be test bank for the students; with all of that an operator screen that fully represents the installation has been added. Additionally, the development and graphical evolution have been tested, both in the simulation module and in the PLC, to observe the existing differences. This investigative work comes very close to what the present article looks for, the difference is that additional to simulating an industrial process, the performance of control instruments such as a temperature transmitter and a pressure transmitter will be simulated.

In summary, these works coincide in the application of simulators, mostly for development and learning, but not for their application in industrial processes. Whereby the present article seeks to highlight the part of industrial operators training, who will perform it through the application of a virtual environment and the virtual development of the instruments to be used in them.

3   State of Technology

The difficulty in training and learning of the manipulation of intrinsic instruments has awakened the interest of developing a wide range of devices for the virtual management of those instruments applied in industry [12].

In the industrial field, the training has been approached like an educational short-term process that develops in a systematic and organized way, through which is possible to acquire aptitudes, knowledge and skills relevant to the defined goals of the environment in which the operators work. Therefore, in this area, virtual reality has taken a very important role with realistic 3D models, simulations and visualizations with specific purposes that allow for the handling of new equipment, that can present problems if not handled the correct way, and cause damage to the equipment and generate economic losses to the industry.

3.1   Unity Pro 3D^{TM}

Is a software tool developed by Unity Technologies in 2005. It is a completely integrated development engine that provides a complete functionality create both three-dimensional and two-dimensional games as well as simulations for its many platforms. Developers, students, designers, corporations, researchers, etc, use this platform because it allows reduction in time, effort and costs, in many branches. Unity Pro 3D^{TM}is one the existing versions that eases software development for a large range of platforms, while being very attractive for its benefits for developers and researchers. User benefit from cross-platform native C++ performance with the Unity-developed backend IL2CPP (Intermediate Language To C++) scripting [13].

3.2   Virtual Reality

Group of images designed in third dimension with realistic appearance generated through computing technology, giving the user the sensation of being immerse in an environment through devices sending sensitive signals to the user. Virtual reality can be defined as a computing system which generates representations of reality in real time, which in fact are nothing more than illusions as it is a perceptive reality with no physical support that exists only inside of a computer [14].

The simulation that the virtual reality develops can refer to virtual scenes, creating a world that only exists inside the computer. Additionally, it allows working in a completely virtual world, disconnecting us from reality and introducing us into the virtual world that had been created. Applications found nowadays are activities form everyday life, reconstruction of cultural heritage, medicine, crowd simulation and presence sensation [15].

3.3   Virtual Simulator

System that reproduces a machine's behavior under certain conditions. It allows the person responsible of operating the system to practice without the need to use a real machine. Simulators can be found in the professional field or as a leisure and entertainment instrument. In the first case, these devices are essential for the training of people with great responsibility at their charge. A simulator allows training until the experience and skill necessary to perform professionally is acquired [16].

3.4 Virtual Environments

Space that contains restricted accesses, conceived and designed for the people who access it to develop skill and knowledge integration processes, through telematics systems. A virtual environment is created by a group of computing tools that make didactic interaction between the user and the created ambient possible. All of that in a simulated fashion without any physical interaction.

A virtual environment must be accessible and easily understandable for its use. Lastly, these are not limited to education or training, but they are a complementary tool to presence-requiring formation in many cases [17,18].

4 Case of Study

The aim of this study case is to development a virtual environment training software to be used by industrial technicians and operators based on augmented and virtual reality. This type of training contribute significantly to the worker performance and strengthens their knowledge. These virtual training centers are an efficient and cheap alternative for handling and learning about industrial equipment.

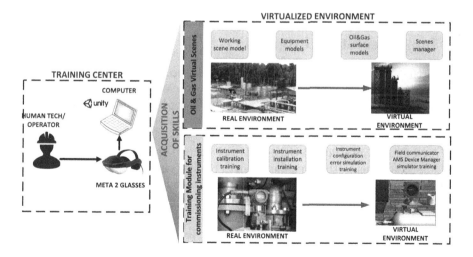

Fig. 1. Training system architecture.

In order to achieve maximum immersion in the simulation procedure a hardware component based on real-time vision is included. For this, the META 2^{TM} Developer kit is used, which is a head-mounted display as shown in Fig. 1.

The virtual Oil & Gas Central Process Facilities environment elements were built, using some models such as surface buildings models, working scene models and equipment models. The training system platform consists of these elements and has many basic functions, such as space management, collision detection, and scene build tools. Trainees can interact with the equipment in the scene created by the system. Comparing with the traditional training mode, the virtual Oil & Gas environment is more abundant and vivid, which is more natural and harmonious.

Scenes manager model defines the behavior of a scene of virtual environment. The programming language used is C language for the representation and the communication of the 3D scenes, through command coded sequences to respond to the inputs and make sure that the game events are run in the right moment, depending on the function of each scene.

Furthermore, the Oil&Gas training system for commissioning instruments is develop to teach techniques of configuration HART instruments, contains instrument calibration training, instrument installation training, instrument configuration error simulation training and Field communicator AMS Device Manager simulator training.

This research offers an alternative that gains acceptability as well as reduces possible equipment damage and training expenses. This paper focuses on the implementation of immersive learning environments that can allow the instrumentation technicians to interact with a virtual world. As can be seen in Fig. 1 the user will be able to interact among several scenes that allow him to become familiar with each industrial equipment, appreciate its electrical connection, know its functional operation and learn the manual configuration.

5   Implementation Proposal

The virtual environment presented in this article, pretends to replicate the original commissioning process and adhere to reality as closely as possible so that the operator doesn't notice the difference, then, the proposal becomes a very important element for developing the skills of the operator or for preparing a new worker in the use of the equipment. For this proposal, Unity Pro 3D ^{TM} software was used for its versatility in the creation of virtual environments. This software provides components that handle a fairly realistic industrial event simulation that, by simply adjusting some parameters with code instructions can simulate the process of installation and commisioning temperature as well as pressure HART instruments.

5.1   Virtual Environment Design Guidelines

The virtual environment allows the instrumentation technician and operator to interact constantly with the environment, therefore the following features are needed: (i) First, is to represent the establishment stage in which all the elements that will take part in the virtual environment are defined. That implies designing geometric models that represent the environment in 3D. (ii) Second, the interaction that provokes the system to respond immediately to the events generated by the user through the navigation of the virtual environment of the sub-menus. (iii) Third, the user must got the sensation of being inside the virtual environment, perceiving the 3D objects of the environment visually through different screens. (iv) Fourth, the system must replicate, as exactly as possible, the functionality of the HART transmitter. (v) Fifth and finally, the interaction between the device ad its different types of assemble that allow the operator to manipulate each of the objects in the virtual environment learning system to verify if its working and fulfilling the training goal.

5.2   Training System Development

Unity Pro 3D^{TM}is a graphic engine for PC that comes packed as a tool for creating interactive applications, visualizations and animations in 3D and real time. Its programming language uses Scripts. The images are created using FBX file extension, so that they can be imported to the Unity virtual platform. Scripting is used for system creation, this functionality allows to build the environment's behavior based on a C programming language. As displayed in Fig. 2 a programming sheet is created for the transmitter equipment automatic assembly when the operator is using the training mode.

```
void Update()
{
    if (Input.GetKeyDown (KeyCode.DownArrow))
    {
        KeyHit = true;
    }
    if (KeyHit==true){
            currentLerpTime += Time.deltaTime;
            if (currentLerpTime >= lerpTime)
            {
                currentLerpTime = lerpTime;

            }
            float Perc = currentLerpTime / lerpTime;
            wall.transform.position = Vector3.Lerp(startPos, endPos, Perc);
    }
```

Fig. 2. Automatic assembly scene sheet.

5.3   Scene Sequence Design

The training system is made up of two options for the operator to choose; the first option is to become familiar to the instruments and the seconds is to start the training in which, the operator is allowed to give a series of instructions for the industrial equipment assembly. Figure 3(a) shows what was previously explained.

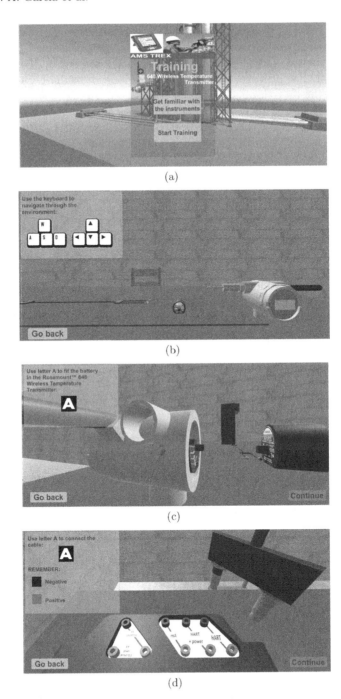

Fig. 3. VR interface. (a) Interface to start virtual training. (b) Basic steps of connection for HART instruments. (c) Interface transmitter equipment assembly. (d) Instrument polarity connection interface

This scene shows each part of the device for familiarization purposes, the scene allows to navigate through the keyboard keys and incorporates a zoom feature with the (W, A, S, D) keys, while also showing a dialog box with the instructions that the operator must follow. Figure 3(b) shows what was previously mentioned.

The instrument commissioning training is showed in the second option of the virtual environment. User follow an instructive document showing them the correct way of install the HART instrument Fig. 3(c).

Once completed the HART module install connection, the system allows to continue with the training, enabling the cable connection to AMS TREX^{TM}virtual device manager terminals is teach. This device has terminals to (i) measure current on a 4–20 mA current loop. (ii) Power and connect to a HART device terminals that can measure the current output of a connected transmitter or control the current input to a connected positioner. Finally (iii) Externally-powered HART device terminals that have an optional loop resistor for enabling HART communications on 4–20 current loop and optional current control for moving a positioner. As shown in Fig. 3(d).

Once instrument installation and AMS TREX^{TM}device manager connection is completed the configuration and calibration training start. The Home screen appears on the AMS TREX^{TM}virtual unit. The Home screen displays the installed applications and the status bar at the top of the screen, as detailed in Fig. 4. Home screen has: (i) the Field Communicator module that simulate configuration of HART instruments. (ii) Loop Diagnostics module that simulate measure loop current and voltage. (iii) Settings module that simulate the adjust settings for the Trex unit and (iv) Help module that shows help topics that describe the hardware and applications on the Trex unit.

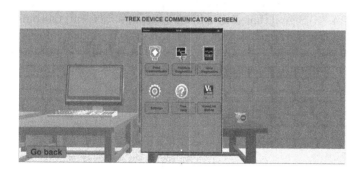

Fig. 4. Interface field communicator

Field Communicator module enables a screen that shows a device dashboard menu that display the following options: Device Setup, Primary variable (PV), Analog output (AO), PV lower range value (LRV) and PV upper range value (URV). This screen are organized by tabs for example: (i) Overview displays a view a graphical representation of the process variable, (ii) Service Tools shows

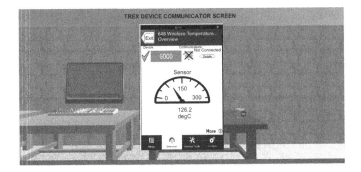

Fig. 5. Dashboard menu of virtual training module

a view and modify options for alerts, variables, and maintenance and (iii) Configure permits configure the device parameters. See Fig. 5.

Next, user starts configuration training, this is on Service Tool option that accesses every configurable parameter for the connected virtual device. Functions in this menu can include tasks such as characterization, configuration, as well as sensor and output trims. The architecture of this tool is shown in Fig. 6. The architecture of the field communicator used in this research (AMS TREX^{TM}) is composed by a first scenes called Device Setup where the user can see the parameters previously configured. Device Setup has two sub-menus; (i) Basic Setup goes through the necessary steps in order to see the basic setup of a HART transmitter. For example here the user can change the instrument tag, change engineering units of measurements, write range values among others. (ii) Detailed Setup has functions such as (i) Sensor trim which is a two-point sensor calibration where two end-point pressures are applied, and all output is linearized between them. (ii) Signal that can be linear or square root. (iii) Output Condition command changes the response time of the transmitter; higher values can smooth variations in output readings caused by rapid input changes. Technicians should determine the appropriate damping setting based on the necessary response time,

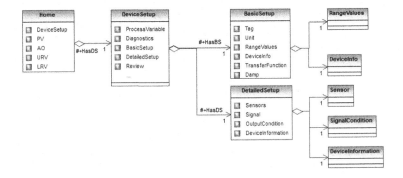

Fig. 6. UML class diagram of virtual training environment.

signal stability, and other requirements of the loop dynamics within system. (iv) Device information shows all configuration of transmitters did by user. An example of this architecture in design of scenes in Service Tool model can be seen in Fig. 7.

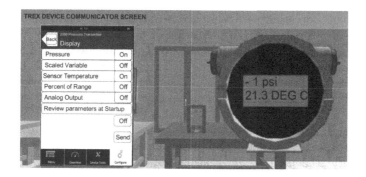

Fig. 7. Basic setup menu interface

6 Result and Discussion

The case study focuses on the training of instruments technicians and industrial operators in Oil & Gas field through a virtual environment which provides the opportunity to develop the skills of the users of this research (see Fig. 8(a) and (b)). This virtual 3D tool developed for the training, will help the operator to be trained in an industrial environment that allows him the commissioning of instruments, thus giving the operator the opportunity to make mistakes during commissioning process. It allow him to acquire the skills to perform an effective job when facing the real working environment in the Oil & Gas industry.

(a)   (b)

Fig. 8. User of virtual tool for instruments commissioning. (a) User interaction with VR training tool. (b) META Developer user glasses view of VR training tool.

The virtual training tool has been validated by users such as: senior and junior instrumentation technicians, senior and junior field operators and university students. Training times depend on the user's experience, with 20 h being the training time required for an advanced level and focusing on closing practical gaps to improve productivity. For an intermediate level the average number of hours will be 40 h where training will focus on delivering knowledge of normal equipment operation and basic maintenance issues. The results obtained through the development of the simulation are totally satisfactory as it can be seen on Fig. 8, where after testing the research population were asked about their experience during the training.

Due to the software allows full immersion of a simulated and controlled environment, different kind of users can manipulate the communicator device AMS TREX^{TM}in an easy and safe way. In total, 70% of senior technicians, 90% of junior technicians, 50% of senior operators and 90% of junior operators answered that the software is among the useful and very useful category. Finally, 90% of the university students consider that the software has been useful to improve their skills in the commissioning of industrial instrumentation. See Fig. 9.

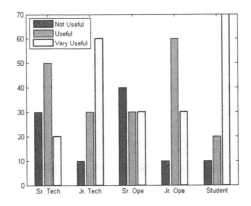

Fig. 9. Users survey of virtual reality software.

The key result of this article is to show the design, construction and fine-tuning stages of an immersive didactic environment to instruct workers in Oil & Gas processes in which the commissioning of instrumentation within these processes intervenes. This immersion environment shows some scenes very similar to those that an worker can face in real life. Such an environment serves to generate an accumulation of experiences of the operator, to build a knowledge and immersive learning. That is to say, it solves the problem of the damages that can be caused when workers commission instrumentation, without any experience reducing this way the costs that would be generated by the repair or in the worst of the cases the acquisition of a new instruments which represents loss of time and money for the industry.

7   Conclusion and Future Work

Since it is a training software of a simulation focused on the training of Oil & Gas workers its implementation is much cheaper and safe, because of the high prices of acquiring a second device to be used only for training when it can be possible to damage a thousands of dollars instrument due to non-proper manipulation and lack of experience.

Due to the advance of technology virtual environments will be having more actualization's in time which will allow to be more efficient and real. So they can guarantee a more immersive and full-quality environment for the user. Since the experience can become more realistic overtime it may be possible to put the user on specific scenarios that will allow to develop its skills for any situation.

The implementation of this technology will help not only to the training of workers for the use of any kind of devices, but at the same time it will guarantee the security of the user and the simulated instrument and the process. A non-proper use the certain devices combined with the inherent dangers of the procedure can cause human and economic losses or possibly accidents that can affect the environment.

The next step will be the implementation of virtual environments to a way of test to measure the degree of satisfaction that users will have, after having manipulated the virtual environment created, which would allow to measure the level of learning acquired by those who manipulate this system, thus allowing to set up statistical data of the performance presented by it.

Acknowledgment. This work was financed in part by Universidad Tecnica de Ambato (UTA) and their Research and Development Department under project CONIN-P-0167-2017.

References

1. Naranjo, J.E., Ayala, P.X., Altamirano, S., Brito, G., Garcia, M.V.: Intelligent oil field approach using virtual reality and mobile anthropomorphic robots. In: De Paolis, L.T., Bourdot, P. (eds.) AVR 2018. LNCS, vol. 10851, pp. 467–478. Springer, Cham (2018). https://doi.org/10.1007/978-3-319-95282-6_34
2. Towey, D., Walker, J., Austin, C., Kwong, C.-F., Wei, S.: Developing virtual reality open educational resources in a sino-foreign higher education institution: challenges and strategies. In: 2018 IEEE International Conference on Teaching, Assessment, and Learning for Engineering (TALE), Wollongong, Australia, pp. 416–422. IEEE (2018)
3. Naranjo, J.E., Lozada, E.C., Espín, H.I., Beltran, C., García, C.A., García, M.V.: Flexible architecture for transparency of a bilateral tele-operation system implemented in mobile anthropomorphic robots for the oil and gas industry. IFAC-PapersOnLine **51**, 239–244 (2018)
4. Cardoso, A., et al.: VRCEMIG: a virtual reality system for real time control of electric substations. In: 2013 IEEE Virtual Reality (VR), Lake Buena Vista, FL, pp. 165–166. IEEE (2013)

5. Demers, M., Mbiya, N., Levin, M.F.: Industry and academia collaboration in the design of virtual reality applications for rehabilitation. In: 2017 International Conference on Virtual Rehabilitation (ICVR), Montreal, QC, Canada, pp. 1–2. IEEE (2017)

6. Jing, X.: Design and implementation of 3D virtual digital campus - based on Unity3D. In: 2016 Eighth International Conference on Measuring Technology and Mechatronics Automation (ICMTMA), Macau, China, pp. 187–190. IEEE (2016)

7. Potkonjak, V., et al.: Virtual laboratories for education in science, technology, and engineering: a review. Comput. Educ. **95**, 309–327 (2016)

8. Pelargos, P.E., et al.: Utilizing virtual and augmented reality for educational and clinical enhancements in neurosurgery. J. Clin. Neurosci. **35**, 1–4 (2017)

9. Cochrane, T., Cook, S., Aiello, S., Aguayo, C., Danobeitia, C., Boncompte, G.: Designing immersive mobile mixed reality for paramedic education. In: 2018 IEEE International Conference on Teaching, Assessment, and Learning for Engineering (TALE), Wollongong, Australia, pp. 645–650. IEEE (2018)

10. Dengel, A., Mazdefrau, J.: Immersive learning explored: subjective and objective factors inuencing learning outcomes in immersive educational virtual environments. In: 2018 IEEE International Conference on Teaching, Assessment, and Learning for Engineering (TALE), Wollongong, Australia, pp. 608–615. IEEE (2018)

11. Marquez, M., Mejias, A., Herrera, R., Andujar, J.M.: Programming and testing a PLC to control a scalable industrial plant in remote way. In: 2017 4th Experiment@International Conference (exp.at 2017), Faro, Portugal, pp. 105–106. IEEE (2017)

12. Back, M., et al.: The virtual factory: exploring 3D worlds as industrial collaboration and control environments. In: 2010 IEEE Virtual Reality Conference (VR), Boston, MA, USA, pp. 257–258. IEEE (2010)

13. Unity. https://unity3d.com/es/unity

14. Lorenzo, G., Pomares, J., Lledó, A.: Inclusion of immersive virtual learning environments and visual control systems to support the learning of students with Asperger syndrome. Comput. Educ. **62**, 88–101 (2013)

15. De Paolis, L.T., Bourdot, P. (eds.): AVR 2018. LNCS, vol. 10850. Springer, Cham (2018). https://doi.org/10.1007/978-3-319-95270-3

16. Lee, E.A.-L., Wong, K.W.: A review of using virtual reality for learning. In: Pan, Z., Cheok, A.D., Müller, W., El Rhalibi, A. (eds.) Transactions on Edutainment I. LNCS, vol. 5080, pp. 231–241. Springer, Heidelberg (2008). https://doi.org/10.1007/978-3-540-69744-2_18

17. Donalek, C., et al.: Immersive and collaborative data visualization using virtual reality platforms. In: 2014 IEEE International Conference on Big Data (Big Data), Washington, DC, USA, pp. 609–614. IEEE (2014)

18. Bo, T., Hongqing, Z., Ning, W., Yuangang, J., Guowei, J.: Application of virtual reality in fire teaching of mining. In: 2012 7th International Conference on Computer Science & Education (ICCSE), Melbourne, Australia, pp. 1079–1081. IEEE (2012)

Virtual Training for Industrial Process: Pumping System

Edison P. Yugcha[(⊠)], Jonathan I. Ubilluz[(⊠)],
and Víctor H. Andaluz[(⊠)]

Universidad de las Fuerzas Armadas ESPE, Sangolquí, Ecuador
{epyugcha, jiubilluz, vhandaluz1}@espe.edu.ec

Abstract. The article presents a virtual environment of a pumping system oriented to training of users that interact with industrial processes. The application of it was performed in a graphic engine Unity 3D, where shows two training environments: *(i) Electro pumps laboratory,* simulates control operations of control for manipulating many configurations from centrifugal pumps in individual, serie or parallel in order to visualize by an HMI the physical parameters such as: pressure, flow and temperature; *(ii) Industrial environment* the user prepares in a complementary way how to know industrial processes in a practical and realistic way. In order for making the virtual application immersive and interactive, the modeling of the electrical characteristics of the pumping system was carried out.

Keywords: Virtual environment · Pumped system · Unity 3D · Immersive and interactive

1 Introduction

Nowadays, the virtual technology has been implemented in several environments such as: education, medicine and especially at industries, this one allows to develop the catchment level and with it the understanding of the work environment that all staff in the industry must have [1]. The industrial processes always require to get support, calibration and adjustments of each system's instrument, where a lot of them are dangerous to the operator who doesn't have enough training [2], in order to avoid that the company has required to different alternatives where the process does not represent any danger for the operator and machine, *e.g.,* industrial digitalization where the operator can visualize the behavior of the machine. The use of virtual entertainment results a better experience for the operator, duo to the fact that it can interact with elements of the system and at the same time to watch all different changes [3]. By opting virtual technology for information and management of industrial equipment, it presents better results and reduces costs compared to the use of usual technologies [4].

At present induction videos and virtual tours are used by the industry for training all the staff, which seems to be very theoretical because It does not do any interaction with the other industrial processes, to satisfy the demands, RV and AR technologies are being developed, those ones allow to improve the training scenarios [5]. Some applications stand out in the field of design optimization, maintenance, process control and operator

© Springer Nature Switzerland AG 2019
L. T. De Paolis and P. Bourdot (Eds.): AVR 2019, LNCS 11614, pp. 393–409, 2019.
https://doi.org/10.1007/978-3-030-25999-0_33

training *e.g.*, there is a virtual application with focuses on the operator training, to simulate experiences in the decision-making that takes place in the field work [6], only focuses on operator training and not on the interaction with the process. The RV allows the user to interact with elements of the virtual world with a realistic approach, in this sense it can perform training scenarios for the operator's simulation under specific conditions [7]. On the other hand, the AR shows in the real world fictitious elements generated in a virtual way, which one can be used through software [8].

Petroleum processes tend to work continuously, with periodic maintenance planning and no work stoppages, duo to that it has led to staff training through virtual scenarios that sever to evaluate possible cases that can happen with the real equipment, by applying the RV it is not necessary the presence of the operator in the industry [9]. One if the methods of virtual in the pumping system is the extraction of crude oil by means of centrifugal pumps from the production wells and displacing them by pipes with the objective of the discharge is at ground level [10]. Besides there's an area of industrial maintenance where the operator can manipulate the pitching and receiving traps of the pipe scrapers (PIG) in order to strengthen knowledge [11]. The detection of virtual modules in existing plant topologies by means of extended search algorithms and the definition of functional patterns, it is now possible to identify all virtual modules within a plant topology [12]. By the necessity of modernizing the industry's facilities digital models are implemented for simulation in a virtualized environment, which allows to automatically convert engineering documents obtained specifically from pipe diagrams and instrumentation. The resulting models can serve as a basis to support engineering tasks that require simulations [13]. The applications that were mentioned take focus on the training of professionals but not the educational field.

This work presents a virtual application aimed at training users in the pumping system, in order to solve this trouble, it has been developed two scenes that allow immersion and interaction through the graphic engine Unity 3D and through mathematical models to do the system analysis. *(i) Electro pump laboratory* has disposed of several pumping system modules where the user can simulate control operations, supervision and data acquisition; how opening or closing of valves to configure VFD parameters, perform configurations of centrifugal pumps in individual, serie or parallel, and visualize temperature data in a graphical interface (HMI) that are obtained by modeling the electrical characteristics, while in the *(ii) Industrial environment*, allows to know industrial processes in a practical and realistic way, so that the user is trained to perform maintenance practices through didactic instructions, in this method it is able to see the risks that can occur when an incorrect maneuver is performed. Finally, to obtain a more realistic application is implemented a safety signage, surround sounds and the use of external devices to interact with the environment.

The article is structured in 6 sections including the Introduction. Section 2 describes the development of the virtual application using blocks; In Sect. 3 it is represented by stages for the development of the virtual environment. Section 4 shows the behavior of the system. Section 5 shows the results obtained in the investigation; to finally, establish the conclusions in Sect. 6.

2   Problem Formulation

For the development of the virtual environment of a pumping system, *scripts* are used, those ones are the principal part of the application because they allow to link different stages such as the *input and output devices*, through the use of virtual reality glasses and optical controls, the *mathematical model* of the system is performed in the Matlab Software and by using shared memories it's linked to Unity 3D software for sending and receiving data, with it the user will be able to interact with different work scenes that contain the virtual environment to establish a greater realism in the interaction and immersion between the user and the 3D virtual pumping system.

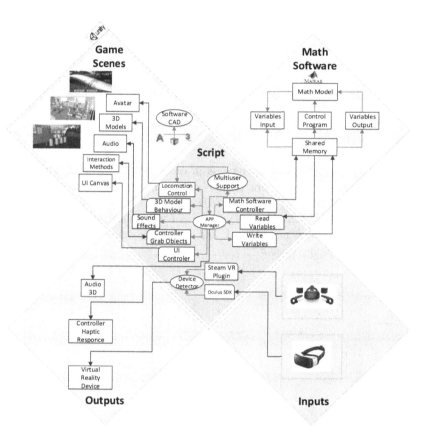

Fig. 1. Block diagram of interrelation between components

Figure 1 shows different blocks for application's development. The first section called SCENES, represents the content of GAME OBJECTS for the edition of pumping system variables. In this phase the interaction between the user and the application is configured by the Controller Grab Objects which allows to take over and supervise the virtual environment behavior.

The **Scripts** block indicates the communication between the Unity 3D AppManager, mathematical software and the input and output devices. Through a set of GUI extensions allow the game developer to insert windows, panels, buttons among others to visualize the behavior of electrical variables, when interacting with the user.

The **mathematical software** covers the mathematical modeling of the electrical characteristics of the pumping system, for which input variables of the frequency inverter (speed) and the type of configuration of the hydraulic circuit are used by the centrifugal pumps that are controlled through a HMI, where the user can monitor the process, to obtain the physical variables of output such as flow, pressure and temperature. Finally, the connection between Unity 3D and Matlab is made through shared memories that allow sending and receiving data.

In the **entry stage**, two devices are considered: *(i) Oculus SDK,* allows the user to immerse for the control of movements and the angle of vision. *(ii) Controller Grab Object*, is responsible for the control of user functions, while in the **output stage** the sounds that are generated in the virtual environment are emulated when the pumping system process starts, it also shows the output variables through an HMI implemented in the virtual laboratory, so that the user has a more realistic sensation.

3   Development

In Fig. 2, it's shown the pumping process in order to be virtualized, for which the instrumentation and pipe diagrams (P&ID) are used, which allow maintenance and modification of the process [14]. The P&ID diagram is developed in the AutoCAD P&ID software that contains 2D CAD designs of mechanical and electronic components such as: centrifugal pumps, tank and measuring equipment that emulate the operation of the pumping system.

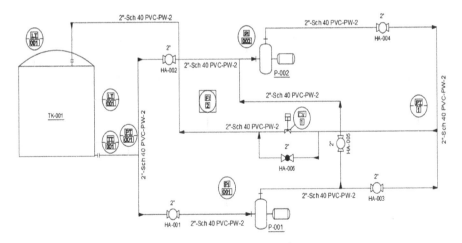

Fig. 2. P&ID diagram of pumping system

In Fig. 3, the different stages for the digitization of tools and instrumentation equipment are described, those one will be visualized in the virtual environment.

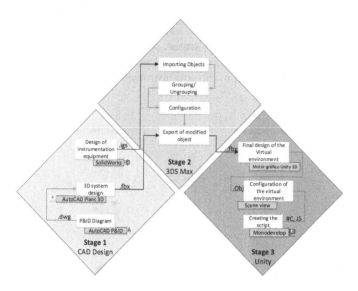

Fig. 3. Virtualization stages

Stage 1. In the CAD design phase, AutoCad Plant 3D software is used to design and model piping systems for industrial processes, the advantage is that components to be implemented are governed by catalog standards. The design of all components of the pumping system must be exported with the extension *. FBX, *e.g.,* centrifugal pumps, valves, pipes, among others, as shown in Fig. 4.

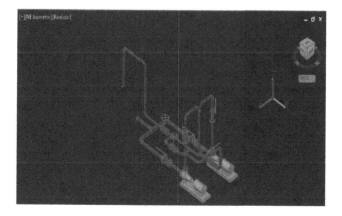

Fig. 4. 3D pumping system design

Stage 2. In this stage we used SolidWorks software that allows the creation and assembly of 3D elements, these designs must be exported with the extension. IGS because it's compatible with the 3DS MAX software, this one allows to group or uncouple the parts of the model in groups without incurring in the relationships of position and movement created, finally to export the created file to UNITY 3D with extension FBX, *e.g.* Flow meter, flow sensor, pressure, instrumentation components, laboratory accessories, and others. Figure 5, shows the comparison between the virtualized instruments with respect to the real instruments.

Fig. 5. Shows the comparison between virtualized instruments with respect to real instruments.

Stage 3. In the third stage, the virtual environment is developed using the Unity 3D Graphic Engine, where the virtual simulation is carried out through different scenes, which show the training modes. The configuration of the HTC VIVE and Gear VR devices allow a better interaction between the user to explore and interact with the pumping system, by a programming in scripts such as *e.g., (i) **Physical Events*** On Collision (Enter, Stay, Exit) and On Trigger (Enter, Stay, Exit); events that occur when colliding and physically interacting with the object. *(ii) **The event controller*** for the user to interact with the environment is imported Package of Steam VR Plugin provided by the Asset Store. The user can move around the virtual environment using the HTC teleportation system. In addition, in the main window of Unity is located the Asset in which folders are created for animations, scripts, textures and materials to organize each event of the pumping system, see Fig. 6.

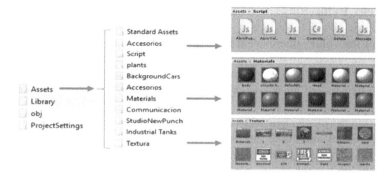

Fig. 6. Distribution of parameters for the development of the virtual environment

By giving more realism to the virtual environment textures and physical properties are added, also, scripts are used for the animation of the process to be carried out *e.g.,* the opening and closing of the control valve with linear actuator, as indicated in Fig. 7.

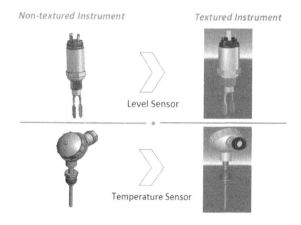

a) Characteristics of textures in virtual instruments

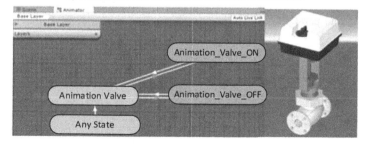

b) Animation of instruments

Fig. 7. Instruments in the Unity 3D editor

4   System Behavior

Because of the virtual environment has similarity with reality, the data of the virtual laboratory are linked with the data generated in the mathematical processing, where the variables and operating conditions that involve the pumping system are specified, at this it will be obtained an answer in real time to the variable modifications that made some user.

The interchange of input and output data is done in a distributed way through different partitions in a shared memory, duo to the fact that they allow to process more information in an orderly and dynamic between the software Matlab and Unity 3D. In Fig. 8 the bidirectional interaction between the *pressure (P), flow (Q), temperature (T)* and *valve position (ON/OFF)* data is indicated, in order to emulate the behavior of the plant through mathematical processing which varies according to the actions performed by the user in the virtual environment.

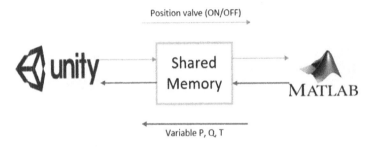

Fig. 8. Data interaction between Unity 3D Software and Matlab

Mathematical modeling is important because it allows to simulate the interaction between the pumping system and the user. This is done using MathWorks Simulink software that offers a wide variety of mathematical engineering elements, in this case the modeling of a centrifugal pump and valves was obtained, as shown in Fig. 9. In the Matlab - Simulink software [15], the frequency inverter (VFD) is programmed to modify the electrical parameters of the centrifugal pump such as frequency, nominal speed, power, current, and in the same way the configuration of the valves in individual, serie and parallel.

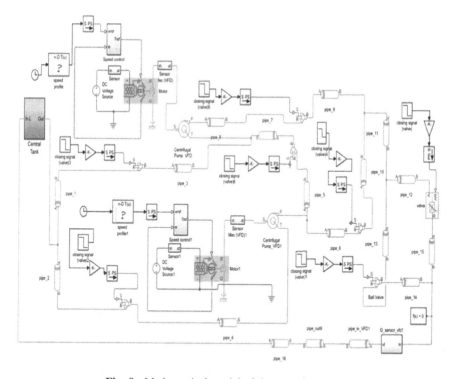

Fig. 9. Mathematical model of the pumping system

5   Results and Discussion

In this section, the results reached in the virtual application developed are shown, as a novel alternative for the training of users in industrial processes, because real laboratories do not emulate large processes due to lack of equipment, instruments, budget or infrastructure. The virtual training module in the area of Pumping System allows the user to know in a practical way the industrial processes and the consequences when performing incorrect maneuvers.

Fig. 10. Scene selection

The virtual environment shows two training modes that allow user interaction and immersion and can be selected according to the need for learning, as shown in Fig. 10.

Virtual Laboratory Mode, in order to change the conventional teaching methodology, additional operating instructions are used in the form of dialog boxes that the user can observe during the development of the laboratory practice in the area of pumping system. In addition, the laboratory is equipped with electrical and mechanical tools such as, e.g., Screwdrivers, wrenches, clamps, centrifugal pumps, frequency inverter, valves and measuring equipment that allows to emulate maintenance tasks. In this training mode the user can assemble pumping systems according to the configuration of the centrifugal pumps, *i.e.,* individual, serie or parallel, see Fig. 11.

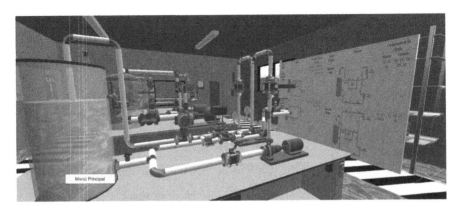

Fig. 11. Virtual lab

The virtual application data is obtained from the real pumping system as shown in Fig. 12, through the LocalHost base that receives the data from the PLC, with the use of Xampp software real process information is stored in LocalHost and through MySQL manage the data in Unity 3D, with a unidirectional communication.

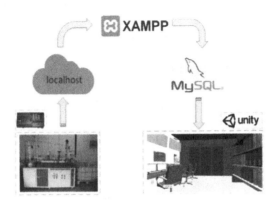

Fig. 12. Communication between the physical system and the virtual environment

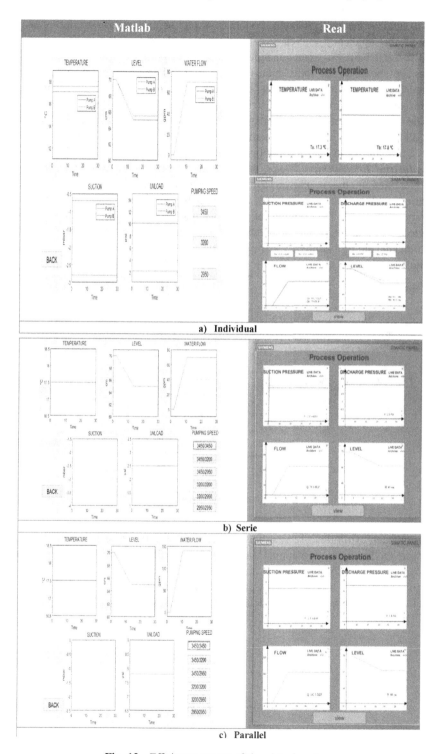

a) Individual

b) Serie

c) Parallel

Fig. 13. Efficiency curves of the virtual system

To validate the mathematical model, an experimental comparison test was used between the efficiency curves generated in the Matlab graphical editor and the LocalHost database with data generated by the actual pumping system. In Fig. 13, the curves are shown in (a) Individual, (b) Serie or (c) Parallel, the serie configuration maintains the flow while the parallel configuration increases due to the change in the suction and discharge pressure of each pump.

The curves generated in the real system and in the Matlab software, have similarity what allows to verify that the mathematical model applied in the virtual application is correct, and allows to interpret the operation of the real system.

Industrial Environment Mode, the route in the environment does not replace the physical training, but it allows to know maintenance tasks that it doesn't have a planning of the industry, it is difficult to have access. The virtual application presents a great advantage for the training of students and new workers who do not have knowledge in real industrial processes, with the implantation of the virtual industrial environment, the probability of learning in the area of the pumping system increases. As shown in Fig. 14.

Fig. 14. Industrial environment

In order to avoid that users, suffer any accident both the virtual laboratory and the industrial environment, operational, safety and prevention information is implemented according to the INEN-ISO 3864-1 standard [16], this one allows to avoid situations of risk in any training mode, they are strategically located in different places with the objective that any user can easily detect them [17], as shown in Figs. 15 and 16.

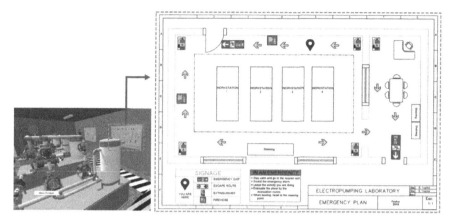

Fig. 15. Evacuation plan in the laboratory

Fig. 16. Safety signs in the industrial environment

In the training modes situations of risks are development that are activated when the user does not comply with the operating protocols, in this way, more realism is given to the virtual environment. An emergency situation occurs when the user causes accidental changes to the work plan or to perform different instructions in the process *e.g., (i) Serial configuration:* valve 3 is closed and causes a pressure rise in the suction pipe, *(ii) Parallel configuration:* raises the excess pressure in the suction and discharge pipes when valves 3 and 4 are closed. In both situations, if the problem is not controlled, it can cause the pipes to explode, mechanical and electrical faults in the centrifugal pump. These emergency situations are visualized when the pipe changes to a red color, see Figs. 17 and 18.

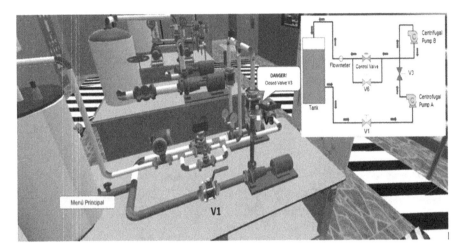

Fig. 17. Emergency situation in serial configuration (Color figure online)

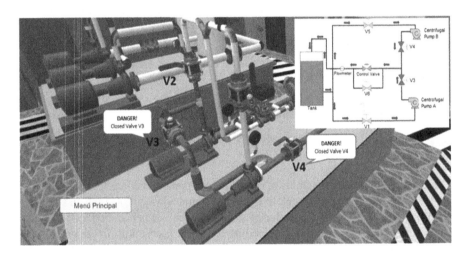

Fig. 18. Emergency situation in parallel configuration (Color figure online)

The application of this one presents the training modes with the inclusion of an avatar that simulates an assistant who is determined to make instruction to the users in specific places into the virtual mode. With this a better immersion is obtained, in order that the user can take guided tours for leaning or training in the area of pumping system, as indicated in Fig. 19.

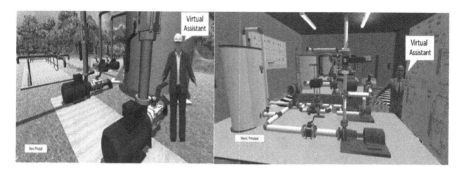

Fig. 19. Virtual assistant

To determine the efficiency of the virtual application, usability test was used to different users in chairs *e.g.,* Fluid Mechanics, Plant Design, Industrial Instrumentation and Process Control to determine the experience in the virtual training environment, as indicated in Table 1.

Table 1. The usability of the virtual application

Questions
1. ¿Have you ever used devices HTC VIVE y GEAR VR that allow to immersion and interact in virtual environments?
2. ¿The use of the training module is intuitive?
3. ¿Does it result easy the management of devices in the virtual environment?
4. ¿Are you able to perform the operation given by the virtual assistant?
5. ¿Could you identify risk conditions and emergency situations in a real industrial environment?
6. ¿Are the operating modes in the Pump System clear?
7. Would you recommend the training module as an additional tool to the theoretical training that is given in the educational and industrial field?
8. ¿Did the virtual application help you understand and improve skills in industrial processes?
9. ¿Do the safety signals placed in the virtual environment comply with the established standards?
10. ¿Does the incorporation of new technologies (RV) allow the development of practices with problems of real industrial processes?

Figure 20, indicates the standard deviation of the users surveyed for each question, these results show that the virtual application is accepted by users as a training module in the Pumping System area, where the achievement of the application is evaluated between the range of regular (1) and excellent (5).

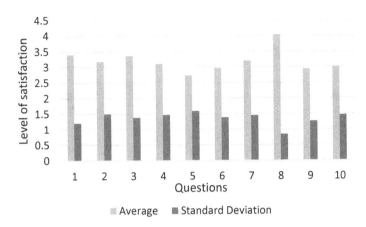

Fig. 20. User learning level within the virtual environment

6 Conclusions

The article presents the development of a virtual application containing pumping plants and 3D measurement and control instruments designed in the software AutoCad Plant 3D and SolidWorks respectively, the files are exported to the Unity 3D graphics engine to give realism to the working environment and through manipulation devices such as HTC VIVE and Gear VR allow the user to immerse and interact realistic with the pumping process.

Acknowledgements. The authors would like to thanks to the Corporación Ecuatoriana para el Desarrollo de la Investigación y Academia–CEDIA for the financing given to research, development, and innovation, through the CEPRA projects, especially the project CEPRA-XI-2017-06; Control Coordinado Multi-operador aplicado a un robot Manipulador Aéreo; also to Universidad de las Fuerzas Armadas ESPE, Universidad Técnica de Ambato, Escuela Superior Politécnica de Chimborazo, Universidad Nacional de Chimborazo, and Grupo de Investigación ARSI, for the support to develop this work.

References

1. Chiluisa, M.G., Mullo, R.D., Andaluz, V.H.: Training in virtual environments for hybrid power plant. In: Bebis, G., et al. (eds.) ISVC 2018. LNCS, vol. 11241, pp. 193–204. Springer, Cham (2018). https://doi.org/10.1007/978-3-030-03801-4_18
2. Andaluz, V.H., Castillo-Carrión, D., Miranda, R.J., Alulema, J.C.: Virtual reality applied to industrial processes. In: De Paolis, L.T., Bourdot, P., Mongelli, A. (eds.) AVR 2017. LNCS, vol. 10324, pp. 59–74. Springer, Cham (2017). https://doi.org/10.1007/978-3-319-60922-5_5
3. García-Peñalvo, F.J., Cruz-Benito, J., Griffiths, D., Achilleos, A.: Tecnología al Servicio de un Proceso de Gestión de Prácticas Virtuales en Empresas: Propuesta y Primeros Resultados del Semester of Code. In: IEEE-ES, pp. 52–59 (2015)

4. Chicaiza, E.A., De la Cruz, E.I., Andaluz, V.H.: Augmented reality system for training and assistance in the management of industrial equipment and instruments. In: Bebis, G., et al. (eds.) ISVC 2018. LNCS, vol. 11241, pp. 675–686. Springer, Cham (2018). https://doi.org/10.1007/978-3-030-03801-4_59

5. Zovko, M.E., Dillon, J.: Humanism vs. competency: traditional and contemporary models of education. Educ. Philos. Theor. **50**, 554–564 (2017)

6. Yu, Y., Duan, M., Sun, C.H., Zhong, Z., Liu, H.: A virtual reality simulation for coordination and interaction based on dynamics calculation. Ships Offshore Struct. **12**, 873–884 (2017)

7. Gonzaga, L., et al.: Immersive virtual fieldwork: advances for the petroleum industry. In: IEEE Computer Society, pp. 561–562 (2018)

8. Hutton, C., Suma, E.: A realistic walking model for enhancing redirection in virtual reality. In: IEEE Virtual Reality (VR), pp. 183–184 (2016)

9. Xu, W., Wei, D., Lei, C.: Control for the centrifugal pump in the simulation platform of power plants, pp. 264–267. Springer Nature Switzerland (2016)

10. Koltun, G., Kolter, M., Vogel-Heuse, B.: Automated generation of modular PLC control software from P&ID diagrams in process industry. In: IEEE Institute of Automation and Information Systems, Technical University of Munich, Munich, Germany, pp. 978-985 (2018)

11. Andaluz, V.H., et al.: Oil processes VR training. In: Bebis, G., et al. (eds.) ISVC 2018. LNCS, vol. 11241, pp. 712–724. Springer, Cham (2018). https://doi.org/10.1007/978-3-030-03801-4_62

12. Arroyo, E., Hoernicke, M., Rodríguez, P., Fay, A.: Automatic derivation of qualitative plant simulation models from legacy piping and instrumentation diagrams. Comput. Chem. Eng. **92**, 112–132 (2016)

13. Hahn, A., Hensel, S., Hoernicke, M., Urbas, L.: Concept for the detection of virtual functional modules in existing plant topologies. In: IEEE 14th International Conference on Industrial Informatics, pp. 820–825 (2017)

14. Toghraei, M.: Principles of P&ID development: the tips provided here will streamline efforts to develop piping & instrumentation diagrams. Chem. Eng. 62 (2019)

15. Saguarduy, J.: MathWorks (2016). https://la.mathworks.com/matlabcentral/fileexchange/56568-pump-speed-control-water-hammer-pressure-waves

16. INEN. Simbolos gráficos. Colores de seguridad y señales de seguridad (2013). https://www.aguaquito.gob.ec/wp-content/uploads/2018/01/IN-3-NORMA-TECNICA-NTNINEN-ISO-3864-12013-S%C3%8DMBOLOS-GR%C3%81FICOS-COLORES-DESEGURIDAD-Y-SE%C3%91ALES-DE-SEGURIDAD.pdf

17. NORMALIZACION, I.E.: INEN. Recuperado el 12 de 02 de 2019 (05 de 2013). https://www.ecp.ec/wp-content/uploads/2017/10/INEN_ISO_3864

Virtual Training on Pumping Stations for Drinking Water Supply Systems

Juan E. Romo$^{(\boxtimes)}$, Gissela R. Tipantasi$^{(\boxtimes)}$, Víctor H. Andaluz,
and Jorge S. Sanchez

Universidad de las Fuerzas Armadas ESPE, Sangolquí, Ecuador
{jeromo, gdtipantasi, vhandaluzl, jssanchez}@espe.edu.ec

Abstract. This article offers the development of a realistic and intuitive virtual environment for training on pumping stations for drinking water supply systems. The environment has structures and equipment found in pumping stations that simplify the user interaction with the industrial process, instrumentation, monitoring and control of the station. The environment has been created using photogrammetry techniques, CAD tools and the UNITY 3D graphic engine that combined with MATLAB exchanges information in real time to execute the simulation of the process, achieving a high degree of realistic immersion. Experimental tests allow the operator to interact with the virtual environment and achieve maximum experience in the integrated station. Through experimental tests, the immersion of the training environment system is verified.

Keywords: Virtual environment · Pumping systems · Photogrammetry · Unity 3D

1 Introduction

The training of professionals in the industry is a determining factor to maximize the efficiency of production processes and services offered by companies, estimating that the skills gained in the training sessions will be put into practice, not only in the operation of equipment but also in decision-making when facing adverse events [1]. The operational staff of an industry represents the starting point of its productivity. This is the reason why it is necessary that the company provides its employees with the appropriate knowledge and materials to ensure quality in the execution of procedures [1, 8–10]. For these reasons, the training in the operation of industrial plants must meet the needs of operators in terms of knowledge about the instrumentation and structure of the plant, as well as mitigation strategies for faults, thus minimizing losses due to human error [2, 3, 11].

Traditionally, one of the most used training methods is usually limited to user guides and verbal instructions, which do not significantly contribute for the future performance of operators. In other cases, personnel and logistical support are added to recreate scenarios in a specific space, which do not represent the workplace as it should be [2]. In addition, low occurrence situations can be rather difficult to simulate, as in the case of equipment failures or catastrophic events [2, 4, 10]. Then, when the training is simple and has a low impact on operators, the risk of human error increases and consequently the level of danger in the workplace.

© Springer Nature Switzerland AG 2019
L. T. De Paolis and P. Bourdot (Eds.): AVR 2019, LNCS 11614, pp. 410–429, 2019.
https://doi.org/10.1007/978-3-030-25999-0_34

In this context, Information and Communication Technology <ICT> has revolutionized several fields of engineering and education [3, 12]. Virtual reality is one of the most used alternatives at present because of the potential it has demonstrated in these areas and the application fields for it grows according to the development and implementation of new technologies [5, 12, 15]. The realism and immersive capacity that virtual reality provides and makes it an ideal means for knowledge acquisition to be effective in the learning process or industrial training. It is also able to stimulate the user's senses to reproduce sensations that only could be experienced in real situations, which makes possible to evaluate these reactions and decision making to assess their abilities with a reliable degree of effectiveness [6, 7, 15]. Several virtual reality applications have been developed for training in fields such as medicine, manufacturing, construction, energy systems and a long et cetera [3, 13, 14]. In these cases, the user must achieve predetermined goals and follow working sequences to comply with certain tasks related to the training topic.

This work presents the development of a virtual reality system for training on pumping stations for drinking water supply systems of EP-EMAPA. The components of the pumping station are designed based on its P & ID diagram, using of photogrammetry techniques and Computer-Aided Design software. The characterization and animation of the objects is implemented in UNITY3D, in addition safety signs are added as well as sounds and lightness to achieve the desired realism and upgrade user experience. The dynamics and control of the plant are simulated using its transfer function in MATLAB, which works as a data manager for the emulation of process variables in the virtualized plant. Finally, methodical fault simulations are implemented in some zones so that the user can use their skills in emergency management. This makes possible to evaluate their speed of response when unfavorable situations occur. The system provides an adaptive learning process to be used according to the user's profile and the level of skills required. It also provides flexibility to be used at any time and the possibility of evaluating optimally the skills gained.

The present work describes the following stages: Sect. 2 describes the structure of the system; the virtualization of the environment using CAD software. The photogrammetry technique is detailed in Sect. 3; Sect. 4 describes the simulation and emulation of the process in UNITY3D and methods developed with MATLAB; Sect. 5 contains the results and Sect. 6 presents the conclusions and future applications of the system.

2 System Structure

This article describes the development of an interactive virtual environment for training in pumping systems for the distribution of drinking water through the virtualization of the plant and the emulation of the industrial process giving as a result a tool for training new operators in a safe environment and free of occupational hazard. In addition, it has the ability to detect potential danger caused by the inadequacy of the operator when controlling or manipulating the variable of interest of the industrial process. It even allows safety instruction in industrial plants through the interpretation of preventive and informative signals located in strategic locations of the pumping station.

The industrial processes control, hazard detection and safety education are developed through events that occur when the operator interacts with the pumping process, this allows the simulation of the behavior of the process and critical situations, which provide an interactive training environment for the user.

The events in the pumping process occur when the variable of interest exceeds its operating limits or when a control action is performed on the process. The training system uses both software and hardware tools which allow the virtualization of the training system the closer to reality. This virtualization of the environment is developed by means of photogrammetry techniques that generate the digitalization of the terrain, whereas the P&ID diagrams provide the necessary information that allows the virtualization of the pipes and instruments of the pumping station. The simulation of the process behavior is generated by the mathematical model developed through the analysis of components and variables acting in the process of the plant. The exchange of data with MATLAB together with the mathematical modeling produces a reliable behavior of the variables of the virtual environment. The realism and immersion of the system is granted by the virtual reality devices, facilitating the user the interaction with the process, instruments, monitoring, and control of the station and detection of hazards. The architecture of the virtual training system offered in this paper is shown in Fig. 1.

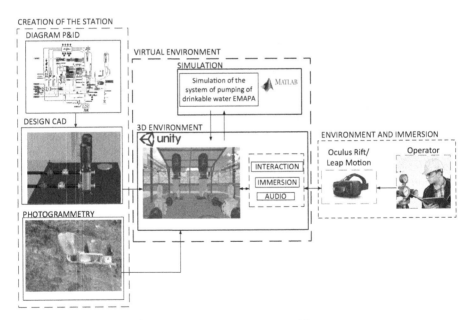

Fig. 1. Virtual training system architecture

This training system presents three main stages in its architecture: (i) *Creation of the station,* the virtual station is designed, from photos, P&ID diagrams and sounds found in the real station. In order to develop realistic 3D models of bombs, pipes, transmitters, valves and tanks in CAD software the three-dimensional survey of the area is achieved by using a UAV to capture several photos of the area, then they are classified and edited for reconstructing the area using photogrammetry techniques; (ii) *Virtual Environment,* the environment of the pumping station is implemented thanks to the UNITY3D graphics engine using different animations that allow inter-action with the environment. MATLAB generates the simulation of the industrial process by means of the mathematical model obtained from the real behavior of the pumping station. Finally, (iii) *Interaction and Immersion,* it consists of virtual reality devices that allow the operator to interact with the virtual environment achieving the maximum possible experience of immersion and interaction in the integrated station.

3 Virtualization of the 3D Environment

This section describes the process used for the virtualization of the pumping plant through techniques of photogrammetry, CAD and UNITY 3D design, as well as avatar design, animations and sound effects that provide a high degree of immersion in the training environment.

3.1 Photogrammetry

This technique creates the 3D model of the structure of the pumping station from photographs taken with the UAV Mavic Air, which are processed in the Agisoft PhotoScan software to obtain a 3D reconstruction of the environment. The pho-togrammetry technique for the reconstruction of the 3D environment of the pumping station is described in Fig. 2.

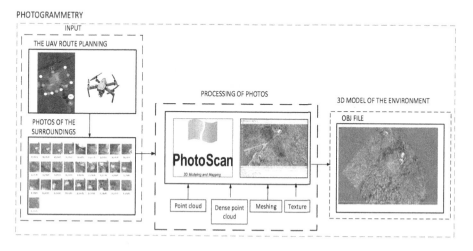

Fig. 2. Description of photogrammetry

In the input stage, the acquisition of photos is provided by the UAV Mavic Air, which first requires the route planning of the area of the pumping station to be rebuilt. Afterwards it captures the photos. This job is done by the Dronedeploy application, which obtains approximately 200 photos classified according to their capture distance in 3 groups that provides the appropriate information for the image processing stage.

The Agisoft PhotoScan software is used in the image processing stage; these processes requires the uploading of the photos in the software to perceive their quality and eliminate duplicate data, in order to generate a scattered points cloud that creates the base of the 3D model of the pumping station. The scattered points cloud allows the generation of a dense points cloud, which carries the largest amount of information from the pumping plant to generate a mesh and texture of the environment giving a more realistic visualization aspect. The output stage provides a 3D model in an .OBJ file with its respective textures that is used in the virtualization of the pumping station as the external part of the environment.

3.2   CAD Design

The CAD design allows virtualizing the internal structure of the pumping station, which consists of a system of pipes and industrial equipment.

The pumping process consists of 5 pumps, 2 of them are set up vertically and 3 horizontally and they are connected to the tanks, thanks to a pipelines system. The creation and three-dimensional interconnection of the piping systems and instruments in the pumping system is done by the AutoCAD Plant 3D software aimed to the design of industrial plants, see Fig. 3.

Fig. 3. Desing using AutoCAD plant 3D

The virtual environment of the pumping plant to perform training the closer to the reality requires a high level of detail in order to have high impact training. For this reason it is necessary to use other types of design software such as SolidWorks which allows the design of customized instruments, which gives even more realism to the pumping station and consequently to the training process. This is the reason why, additional components have been designed in SolidWorks, such as a control valve, see Fig. 4.

Fig. 4. Control valve designed in SolidWorks

3.3 Design in UNITY 3D

The design obtained in AutoCAD Plant 3D and the photogrammetry developed in Agisoft PhotoScan, are imported in FBX and OBJ format, respectively, Unity 3D, which are scaled and located inside the plant in their corresponding place. The design in format (* .fbx) must be characterized by the creation of materials, which will be assigned to each of the objects, unlike the photogrammetry that requires only the allocation of its texture. Photogrammetry, the design in AutoCAD Plant 3D, the creation of scripts and animations, the allocation of surround sound to existing equipment and the design of levels of ranks and privileges, make up the design in UNITY 3D, see Fig. 5. This provides a Training environment with greater realism for the user of the pumping plant.

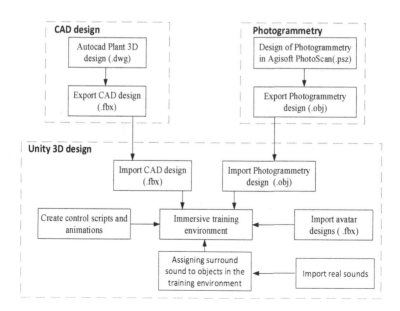

Fig. 5. Design process in unity 3D

3.4   Design of User Levels and Privileges

The training environment provides a menu that has the ability to select between 4 user levels and privileges determined in order to limit the activities it has and thus provide an specific learning method for each level of the pumping plant for distribution of drinking water, for this reason, avatars are created, which interact with the different processes of the pumping plant, these are created in Adobe Fuse (see Fig. 6), which is a program that allows to easily perform the Auto-Rigger when exporting them to the MIXAMO web page for the creation of the animations. The avatars have different types of clothing and accessories according to their different user levels and privileges within the plant. For this reason, a start scene was created in the training application (see Fig. 7) that allows selecting the different user levels and privileges with their respective avatars. The user levels and privileges that the training system offers are as follows: (i) *Supervisor, Engineer – High Level*, can manipulate manual valves, pump selectors and the HMI for the monitoring and control of the plant. In (ii) *Operator – Medium Level*, can manipulate manual valves and pump selectors. Finally, (iii) *Visitor - None*, sees the whole process, but cannot manipulate anything. This information is presented through an audio message and also in text boxes.

Fig. 6. Avatar created in Adobe Fuse

SCENE OF CHOICE OF AVATARS

SELECTION MENU

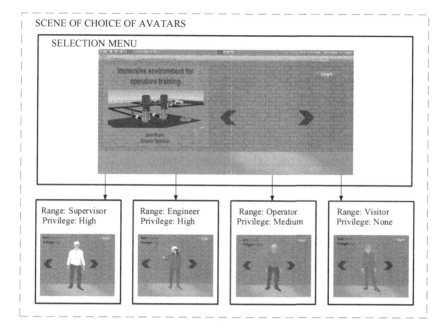

| Range: Supervisor
Privilege: High | Range: Engineer
Privilege: High | Range: Operator
Privilege: Medium | Range: Visitor
Privilege: None |

Fig. 7. Selection menu for avatar

3.5   Industrial Safety Signals and Immersive Ambient Sounds

Industrial Safety Signals. The interpretation and recognition of preventive and informative industrial safety signals provide the user with an environment conducive to training and assistance in accident management. For this reason, industrial safety signs are implemented in hazardous locations, to provide the user with the necessary information during the experience of an accident event that is generated in the training environment. In the Fig. 8, shows an industrial safety signal (informative), which indicates the existence of a fire extinguisher in a flammable environment, which allows the mitigation of a fire in a timely manner.

Fig. 8. Safety signals in a flammable environment

Immersive Ambient Sounds. Realism in the training environment is increased by providing characteristic sounds to industrial pumps, which are aggregated in the limited areas by the pumps and surround sound (see Fig. 9). With the purpose of obtaining a progressive sound attenuation effect as the user moves away or closer to the industrial equipment. The sounds assigned to the different elements of the immersive environment are obtained from real industrial processes with the purpose of improving the interaction with the equipment that the pumping plant possesses.

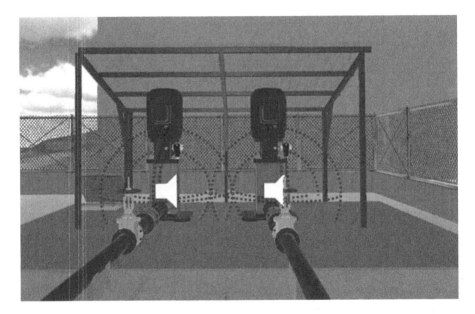

Fig. 9. Surround sound of pumping pumps

4   Process Simulation

This section presents the development of a mathematical modeling of the pumping station and the design of a controller that performs the simulation of the process through MATLAB.

4.1   Mathematical Model

Mathematical modeling is acquired through the schematic diagram of the process of pumping illustrated in Fig. 10, conformed by five pumps in parallel; which are divided into two groups: (i) S*krocket of chlorine*, formed by twin pumps 1 and 2 which are in charge of supplying enough chlorine to the water purification process; and finally (ii) *skyrocket of water*, formed by 3 and 4 identical bombs, as well as the auxiliary pump 5, which supplied water to the process. The inputs of the process are the voltages applied to the inverters of the pumps, while its output is the flow of chlorine and water.

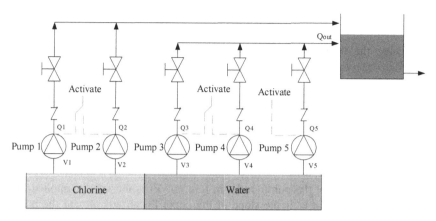

Fig. 10. Diagram of the pumping process

To know the relationship that exists between the variator input voltage measured in volt (V) that connects to the pump and the flow that this provides to the system measured in litres/second (l/s), it is necessary to do a trial with the data acquired from the pumping plant.

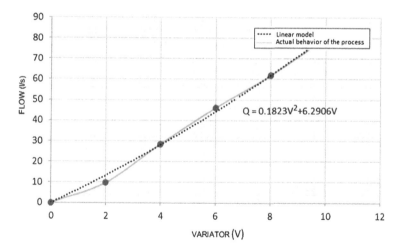

Fig. 11. Pump 1 voltage-flow ratio

Figure 11 shows the relationship between the tension applied to the frequency variator and the output flow, where the information is provided both of the actual behavior of the plant, as well as its linear model through the method of linear regression. So the approximate relationship that exists between tension (V) applied to the frequency converter and the flow rate (Q) is able to provide is:

$$Q = 0.1823V^2 + 6.2906V$$

Knowing the linearity between the tension and the flow is set two points of operation in order to obtain their respective earrings; obtaining a 70 flow equal to 52.96 (l/s) with a slope of 7,565 (l/s); while that with 80 flow is equal to 61.99 (l/s) with a slope of 7.748 (l/s), by which the average value of these earrings is therefore 7.656 (l/s).

Experimentally, the operation is described by the following equation [16]:

$$q_i = k_i V_i$$

where: q_i is the flow of the pump outlet in (cm^3/s); V_i is the voltage applied to the frequency converter (V); k_i is the slope of the curve linealizada around the point of operation having:

$$k_i = 7.656(l/s) = 7656\ cm^3/s$$

For dynamic behavior is a model test that a tension is introduced into the speed controller, is subjected to a voltage step pump and shows the behavior of the signal from the transmitter.

The dynamic of this system is typical of a first-order as the equation system:

$$G(s) = K/Ls + 1$$

So the model of each of the pumps pumping from the collected data system; and considering that pump 1 and 2 have the same characteristics as well as 3 and 4 twin pumps, you have the following equations:

$$G_{pump1}(s) = G_{pump2}(s) = 1.0614/1.544s + 1$$
$$G_{pump3}(s) = G_{pump4}(s) = 1.1723/1.746s + 1$$
$$G_{pump5}(s) = 0.964/1.04s + 1$$

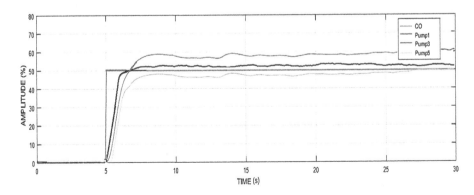

Fig. 12. Dynamic behavior on each flow response.

Figure 12 indicates the behavior of each of the flows, that by submitting to the pump at a given step of the response of flow input voltage follows a certain curve until it reaches its final value, from which the previous equations is valid.

Knowing the behavior dynamic of each one of the pumps of different groups, both the supply of chlorine and the water supply is obtained the following transfer functions:

$$G_{chlorine}(s) = 1.0614/1.544s + 1$$

$$G_{water}(s) = 2.902s + 2.136/1.816s^2 + 2.786s + 1$$

4.2 Ratio Control

This control algorithm allows to control the relationship between the chlorine flow variable and the water flow variable, in Fig. 13 it can be seen that there is only one control variable and two measurement variables, for this reason, this is a system of two inputs and one output, where a flow relation (FC) controller is placed.

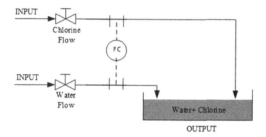

Fig. 13. System for the dosificatio of water

For controlling the dose of the flow of water and chlorine, the ratio control scheme is implemented see Fig. 14, where each variable of the process is regulated by the controllers C1 and C2, the result obtained from the first process (water flow) is multiplied by a ratio factor (a) and adopted as the set point to control the second process variable (chlorine flow), that is, $r2(t) = ay1(t)$ [17].

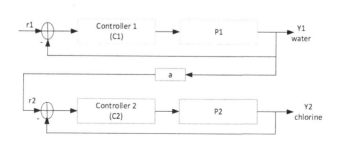

Fig. 14. Ratio control scheme

For the simulation and control of the process of drinking water purification, it is required the exchanging of significant process data between UNITY3D and MATLAB. For this task to be done it is necessary the design of an HMI and the respective programming (see Fig. 15), to subsequently implement an interconnection method through the TCP/IP communication protocol.

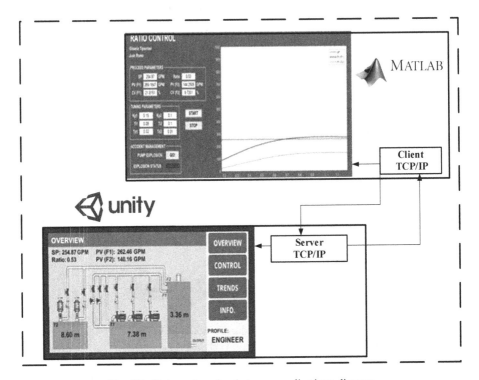

Fig. 15. Data exchanging between applications diagram

5   Analyses and Results

The training environment developed is oriented towards the correct behavior of the control algorithm that governs the water purification process and in the generation of emergency events that may occur in the drinking water distribution system. In this way, the user will have an integrated and effective training to control the different equipment and emergencies that occur in the pumping plant.

The Fig. 16 presents a menu that allows the selection of the different privileges and ranges that the application possesses, in order to choose the type of preparation that is required to receive.

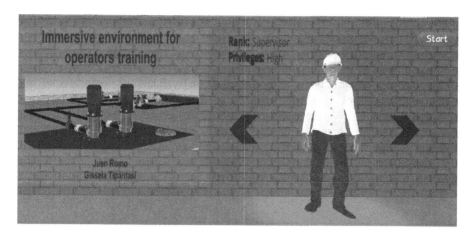

Fig. 16. Privilege selection menu

The virtual environment offers a visual and auditory explanation of the activities that are allowed to be performed within the plant, depending on the privilege and range selected, see Fig. 17.

Fig. 17. Activities allowed for the avatar

The training system for personnel consists of events, both in the control algorithm and in emergencies present in water purification.

Control Algorithm for Water Purification. In order to demonstrate the correct functioning of the control algorithm applied to water purification, the following experiment is presented, which consists in establishing a ratio of 0, 4, to dose chlorine to 40% and water to 60%, in addition to placing a set point 504.07 gallons per minute (GPM), in order to corroborate the process data that validate the correct tuning and operation of the control algorithm. In Fig. 18 the data of the control variable (CV) and the process variable (PV) are counteracted, both in the water flow (F1) and in the chlorine flow (F2) presented in the Matlab software and in the virtual environment HMI. (a) Validates the operation of the control algorithm to counteract Matlab software data and virtual environment HMI. (b) Validates the correct tuning of the control algorithm to compare the Matlab software trends and the virtual environment HMI.

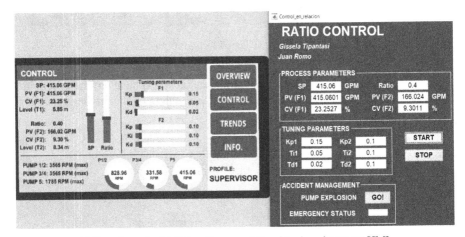

(a) Data from the Matlab software and the virtual environment HMI.

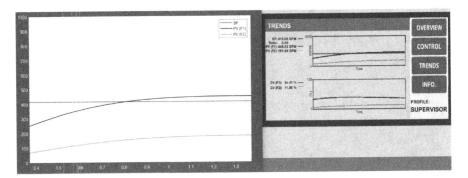

(b) Trends in Matlab software and virtual environment HMI.

Fig. 18. Control algorithm data contrast

The F1 and F2 provide changes of levels in the water tank (T1), chlorine tank (T2) and in the tank containing the mixture, all this can be seen both in the virtual environment HMI and in the indicator transmitters level (LIT) that each tank has, see Fig. 19.

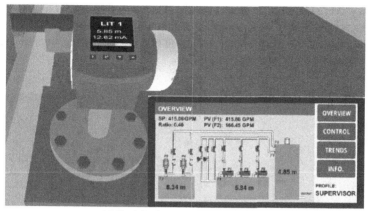

(a) Check level T1 in both LIT1 and Overview.

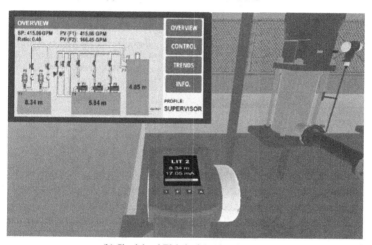

(b) Check level T2 in both LIT2 and Overview.

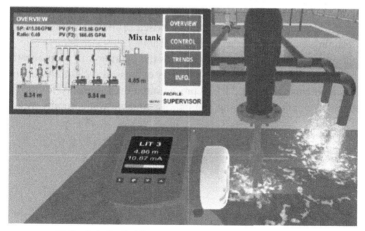

(c) Check the level of the mix tank in both the LIT3 and the Overview.

Fig. 19. Level contrast between LIT and overview

Emergency Management in the Process. This event focuses on generating an industrial emergency and the measures that the user must take to mitigate the emergency. The accident that occurs is the explosion of a pump, causing a training situation for the user against emergencies, see Fig. 20.

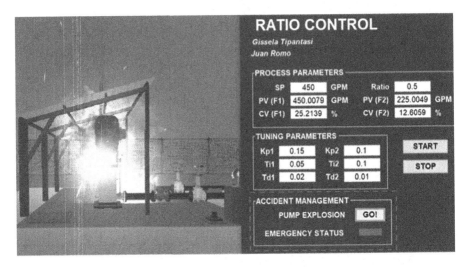

Fig. 20. Explosion of the bomb 3

To end the accident that is evidenced in the emergency status located in the accident manager, immediate corrective actions are performed described through informative tables that must be executed by the user, see Fig. 21.

The results presented below indicate the validity of the usability of virtual training environments, to carry out an adequate recognition and management of pumping stations for drinking water supply systems. For this purpose, the SUS summarized evaluation method is used [19]. Generally providing the style scale that generates a single number, represented by an average composed of the usability of the application for virtual training environments as shown in Table 1.

The total number, obtained from the sum of the operation in each question, results in 36.53; based on this result, the SUS score is expressed by a multiplication of 2.5, which means a percentage of 91.33%, representing a high usability for the recognition and proper management of pumping stations for potable water supply systems.

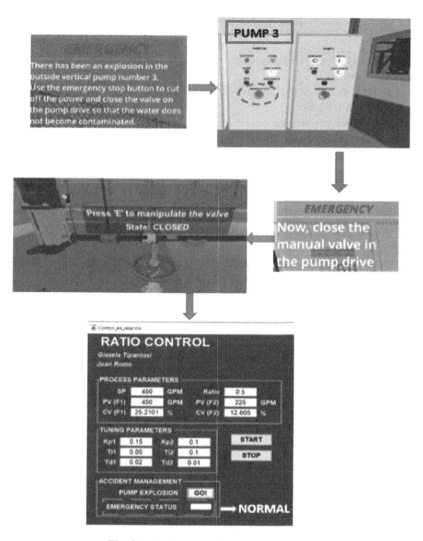

Fig. 21. Actions to mitigate the explosion

Table 1. Results of the questionnaire.

Questions	Punctuation					Operation
Would you like to use this virtual training environment frequently?				3	12	3.80
Did you find that the virtual training environment is unnecessary and complex?	13	2				3.87
Was it easy to use the virtual training environment?				3	12	3.80
Did you need a technician to be able to use this application?	10	3	2			3.53
Is the controller in relation used for the dosing of water and chlorine efficient for drinking water purification?		1	2		12	3.53
Is the virtual training environment adequate to become familiar with the equipment that the pumping station has?	12	1	2			3.67
Did the training environment provide some perception into the purification process of drinking water carried out in the pumping station?			1	2	12	3.73
Found the application difficult to use?	13		1	1		3.67
Was it easy to manipulate the avatars within the application?			2	2	11	3.60
Did you need to learn many things before using the application?	8	4	3			3.33
Total						36.53

6   Conclusion

The use of the virtual environment provides adequate training to staff in case of emergencies and in the implementation of control techniques that allow simulating the response of the process, which together with the virtual keyboard created generates a large amount of information on the most relevant variables of the process.

The virtual environment provided by the application provides realism in the structure of the 3D reconstruction of the pumping plant; as well as in the design of the 3D objects implemented for the realistic and continuous training of the pumping station for the distribution of drinking water. Finally, the virtual environment application provides useful knowledge to field operators in both industrial safety and process control, which in turn reduces risks.

References

1. Stork, A., et al.: Enabling virtual assembly training in and beyond the automotive industry. In: 2012 18th International Conference on Virtual Systems and Multimedia (2012)
2. Carruth, D.: Virtual reality for education and workforce training. In: 2017 15th International Conference on Emerging eLearning Technologies and Applications (ICETA) (2017)

3. Galvan-Bobadilla, I., Ayala-Garcia, A., Rodriguez-Gallegos, E., Arroyo-Figueroa, G.: Virtual reality training system for the maintenance of underground lines in power distribution system. In: Third International Conference on Innovative Computing Technology (INTECH 2013) (2013)
4. Prasolova-Forland, E., Molka-Danielsen, J., Fominykh, M., Lamb, K.: Active learning modules for multi-professional emergency management training in virtual reality. In: 2017 IEEE 6th International Conference on Teaching, Assessment, and Learning for Engineering (TALE) (2017)
5. Wong, S., Yang, Z., Cao, N., Ho, W.: Applied RFID and virtual reality technology in professional training system for manufacturing. In: 2010 IEEE International Conference on Industrial Engineering and Engineering Management (2010)
6. Li, M., Li, L., Jiao, R., Xiao, H.: Virtrul reality and artificial intelligence support future training development. In: 2017 Chinese Automation Congress (CAC) (2017)
7. Rosero, M., Pogo, R., Pruna, E., Andaluz, V.H., Escobar, I.: Immersive environment for training on industrial emergencies. In: De Paolis, L.T., Bourdot, P. (eds.) AVR 2018. LNCS, vol. 10851, pp. 451–466. Springer, Cham (2018). https://doi.org/10.1007/978-3-319-95282-6_33
8. Bluemel, E., Hintze, A., Schulz, T., Schumann, M., Stuering, S.: Virtual environments for the training of maintenance and service tasks. In: Proceedings of the 2003 International Conference on Machine Learning and Cybernetics (2003)
9. Gomes de Sá, A., Zachmann, G.: Virtual reality as a tool for verification of assembly and maintenance processes. Comput. Graph. 23(3), 389–403 (1999)
10. Cai, L., Cen, M., Luo, Z., Li, H.: Modeling risk behaviors in virtual environment based on multi-agent. In: 2010 The 2nd International Conference on Computer and Automation Engineering (ICCAE) (2010)
11. Turner, C., Hutabarat, W., Oyekan, J., Tiwari, A.: Discrete event simulation and virtual reality use in industry: new opportunities and future trends. IEEE Trans. Hum.-Mach. Syst. 46(6), 882–894 (2016)
12. de Jong, T., Linn, M., Zacharia, Z.: Physical and virtual laboratories in science and engineering education. Science 340(6130), 305–308 (2013)
13. Fuyu, W., Guoming, C.: The virtual experiment system of sucker-rod pumping development and its application. In: 2010 International Conference on Computing, Control and Industrial Engineering (2010)
14. Gavish, N., et al.: Evaluating virtual reality and augmented reality training for industrial maintenance and assembly tasks. Interact. Learn. Environ. 23(6), 778–798 (2013)
15. Webel, S., Bockholt, U., Engelke, T., Gavish, N., Olbrich, M., Preusche, C.: An augmented reality training platform for assembly and maintenance skills. Robot. Auton. Syst. 61(4), 398–403 (2013)
16. Perez-Cisneros, M.A., Readman, M.: Control systems principles white papers library. Hamilton Institute at the National University of Ireland & TQ Equipment Ltd., Nottingham, U.K., pp. 18–26, Revistas Arbitradas (2005)
17. Izquieta, S.P.: Control de Tanques Acoplados (Tesis de pregado). Universidad Pública de Navarra, Pamplona (2011)
18. Shinskey, F.G.: Process Control Systems-Application. Design and Tuning. McGraw-Hill, New York (1996)
19. Sauro, J., Lewis, J.R.: When designing usability questionnaires, does it hurt to be positive In: Proceedings of the SIGCHI Conference on Human Factors in Computing Systems, pp. 2215–2224. ACM, May 2011

Virtual Training System for an Industrial Pasteurization Process

Alex P. Porras$^{(\boxtimes)}$, Carlos R. Solis$^{(\boxtimes)}$, Víctor H. Andaluz$^{(\boxtimes)}$,
Jorge S. Sánchez$^{(\boxtimes)}$, and Cesar A. Naranjo$^{(\boxtimes)}$

Universidad de las Fuerzas Armadas ESPE, Sangolquí, Ecuador
{apporras, crsolis, vhandaluz1, jssanchez,
canaranjo}@espe.edu.ec

Abstract. This article presents the development of a training system in a virtual environment, a pasteurization plant, which allows interaction with the industrial process. The environment has structures, equipment and other instrumentation that is presented in a real pasteurizer plant. Thus, the proposed system manages textures and movements in a Unity 3D graphic environment, specially designed to develop monitoring and manipulation skills of this industrial process. The interface and the animation of the process are developed in Unity 3D software, together with the modeling of all the elements; in CAD design software. The experimental tests allow the operator to interact with the virtual environment and have knowledge of the different stages of pasteurization in the dairy industry.

Keywords: Virtual reality · Unity 3D · Pasteurization

1 Introduction

Virtual Reality has opened up field with supporting applications in design, industrial training and commercial activities. With virtual reality a very detailed perspective of structures and installations can be shown and it facilitates the design and appreciation of an environment [1]. Virtual reality systems are proposed as an economic advantage, as the user is given intuitive knowledge of the structure and functioning of a real system [2]. Thus, virtual applications have participated in the creation of various support systems, both in education, medicine, military training, tourism and museography, leisure and entertainment, and industrial applications and engineering.

In Virtual Reality in Industrial Applications and Engineering can create a specialized virtual environment, running in real time, which can provide critical and high-priority information of a process [3]. In terms of industrial training, a virtual system can serve as an effective and very useful instructional module. Virtual industrial training seeks to implement collaborative environments that facilitate the participation of several users in the development of processes where they acquire skills and become familiar with the activities that will be carried out in real life. Through virtual reality a series of very realistic scenarios are presented, the same ones that show industrial infrastructure and real processes. These virtual processes integrate real parameters and units, which are very effective in contributing knowledge to new users of the facilities [4]. With a complete virtual environment, reliable design criteria and key points in

© Springer Nature Switzerland AG 2019
L. T. De Paolis and P. Bourdot (Eds.): AVR 2019, LNCS 11614, pp. 430–441, 2019.
https://doi.org/10.1007/978-3-030-25999-0_35

quality maintenance can be given, in this way the virtuality of the equipment becomes a very useful and effective tool for maintenance processes [5].

In recent years the success of the industries has involved the ability of the working staff both to perform and improve their daily activities and to solve technical problems that arise in a process. The dairy industry does not make the exception since here it handles an industrial process called pasteurization, which involves parameters and variables such as temperature and fluid level. The pasteurization process consists of the thermal variation of liquids (generally foodstuffs such as milk), with the intention of eliminating or reducing the germs that exist in these, that is to say, it consists of taking the liquid to a high temperature and in a matter of seconds changing it to a low temperature, obtaining the elimination of bacteria present in the milk.

Also with the advance of technology, in the virtual industrial training, we seek to implement collaborative environments that facilitate the participation of several users in the development of real processes of the industry. This type of systems can be exposed through web portals, which facilitates their use, allowing participation in virtual tasks through a web browser commonly installed on any computer [6].

In order to improve industrial training methods, this work proposes the creation of a virtual didactic tool, effective for the practice and manipulation of a milk pasteurization process. The implementation of a virtual system of a pasteurization process is presented as a solution to the difficulties of staff training, where they can familiarize themselves with the process and learn from it, in a safe way and without having the need to access the physical plant, which frees us from occupational risks. This means that thanks to the tools and characteristics that Virtual Reality has, it is possible to build an environment similar to the real one and capture it in a computer to the point that the users do not notice the difference.

The structure and instrumentation developed in the system are developed based on P&ID diagrams of a dairy pasteurizer industry, which includes the animation of valves, switches, pumps and motors, all necessary for the operation of the process.

The present work is divided into the following sections, including; 1 introduction, 2 methodology used for the development of the system, 3 Virtualization of the environment through CAD software, animation of the process in Unity 3D, 4 analysis of results, and finally, 5 presents conclusions and future work.

2   Methodology

This work is designed for the training of operators in an industrial pasteurization process, through virtualization and emulation of the process of a pasteurizing plant. The system focuses on the interaction of the operator with the virtual plant. The objectives of this work are to design and test a virtual reality environment for the manipulation and learning of the operation and processes of a pasteurizing plant, which will allow the users or operators of the plant to acquire knowledge based on the experimentation that takes place in the virtual plant and this under safe conditions for both people and industry and is also immune to accidents and damages that generate losses to the company.

The system is adapted to the daily environments of the company, where you can evaluate the efficiency of the plant in high and low working conditions, and thus avoid possible failures; it is also designed to run from a personal computer to specific devices for virtual use (Oculus, Joystick, RV Helmets, etc.).

The work is carried out in stages, which includes the creation, animation and interaction of the pasteurizer plant process. Figure 1 shows the general scheme of the implementation methodology.

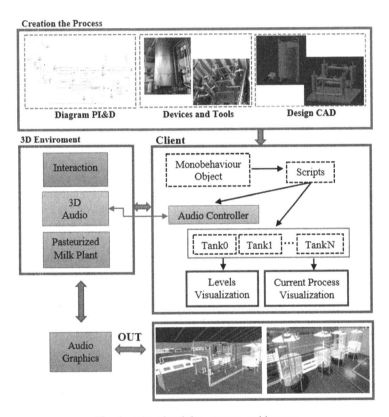

Fig. 1. Virtual training system architecture

3   Virtualization of the 3D Environment

3.1   Investigation

For the implementation of a VR system it is necessary to integrate a certain number of components and parameters, which work independently, but have the same purpose, which is to achieve the best immersion of the user in a virtual environment. For this reason it is very important to take into account all the details that are presented in the real environment.

For all these reasons, it is necessary to carry out an investigation about the dairy industries, where you will become familiar with the stages of the pasteurization process as well as with the devices and instrumentation that are handled in the plant. For this purpose, it is taken as support, research technical sheets of the industries that carry out this process, P&ID and HMI diagrams of pasteurization processes (see Fig. 2) and the safety standards that are executed in these processes, with the sole purpose of having a representation of the real process in a virtual environment [7]. The technique chosen for the virtualization of the industrial pasteurization plant is based on the interaction with programs that allow the creation of 3D equipment and instrumentation (Blender, Unity 3D, Adobe Fuse), and the animation and functionality of the process.

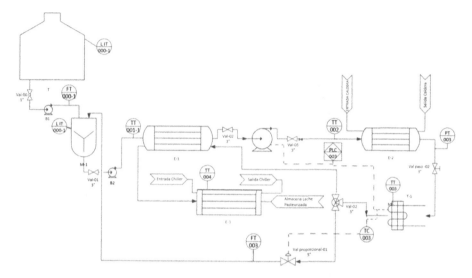

Fig. 2. P&ID diagram [13]

3.2 CAD Design

The virtualization process begins with the references previously obtained. To design a milk pasteurization plant, it is necessary to have the PI&D plans of the process, to be able to carry out the conversion of the objects from 2D to 3D, (see Fig. 3) and of the technical manuals of the operation, to be able to detail the movements of the objects and to be able to give them the different animations of their utility, that is to say to give them a high level of detail to the objects, this by means of computerized drawing techniques carried out in Blender.

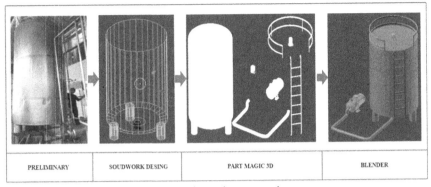

(a) Reception and storage tank.

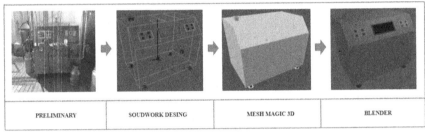

(b) Milk Homogenizer

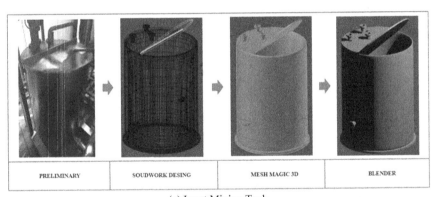

(c) Input Mixing Tank

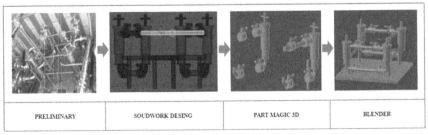

(d) Milk filtering instrument

Fig. 3. CAD design of process instrumentation.

The details that are given to the objects are of great import, they may not represent a change in their functionality, but if they improve the experiences in the interaction with the users, as a result the appropriate components and instrumentation of the process are generated [14].

In addition, the avatars were designed using Adobe Fuse software, characteristic for their ease of development (see Fig. 4), which are differentiated by the type of access they are granted, *i.e.*, a Scada system was applied for user access, where each has different privileges.

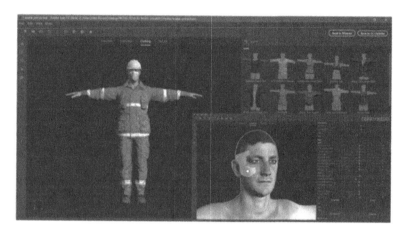

Fig. 4. Creating Avatar

3.3 Environment Creation

The plans of the civil infrastructure are necessary to make a virtual survey the most similar of the industry, where walls, doors, windows, etc. are detailed. The Pasteurization industry is located in the city of Salcedo in the province of Cotopaxi, Ecuador. In addition, it is important to mention that textures are imported to improve the design of the virtual environment, (see Fig. 5) ensuring a friendly environment for users to feel more how-two and familiar. All these objects were designed in Blender and imported into Unity 3D in a fbx file.

Fig. 5. Virtual environment of the pasteurizing plant (Color figure online)

Thinking about improving the immersion experience in the virtual environment, different colors and activation spaces (celestial circle) were assigned to the devices that act in the plant, which goes according to its operation [8, 9]. Red color for a pump in "off state" and green color for a pump in "on state". Another important element are the control valves. Movements were also assigned according to their functionality, a right turn for "closed state" and a left turn for "open state". For a switch, the "state on" button is displayed pressed and a switch in "state off" button is displayed protruding.

3.4   Unity 3D Platform Integration

The Unity 3D graphic engine is a software of great features, compatible with several elements that are not included in the program, the same ones that allow you to improve and expand the applications that are realized in this one [10]. This is why the pasteurization process in the virtual environment is fully suited to the real plant environment, including animations, movements, and sounds. In order to include the animations of the process it is necessary to generate commands of handling and programs, by means of code VBScript, to, this it is necessary to generate the executable file and to tie it to the platform Unity 3D [11, 12].

These animations include the visualization of the level of the tanks of reception, mixture and storage of the milk. It also details the activation of an alarm in the event that the liquid level has exceeded the level of the tank producing a spill (see Fig. 6). Another aspect and the most important in the pasteurization process is the change of temperature to which the milk is submitted, this animation consists of the change of color of the pasteurization instrument, in high temperatures Red color and in low temperatures blue color, clearly appreciating this process (see Fig. 6b).

(a) Animation of milk reception tanks.

(b) Animation of the pasteurizing plant

Fig. 6. Unity 3D Plataform. (Color figure online)

4 Analysis of Results

This section presents the virtualization of the training system of the pasteurization process, which focuses on the interaction of the user with the process, allowing to obtain a great experience of the actions carried out in the different stages that pasteurized milk entails. These actions focus on the operation of the control valves, for the passage or not of the milk, between the different tanks (reception, storage and mixing). As well as the activation of switches for the on and off of the pumps connected to the tanks and to the homogenization instruments. The different stages of the virtualized pasteurization process are detailed below.

i. *Milk reception*
At this stage, the milk from the supplier trucks is stored in the reception tanks at room temperature by means of the connected pipe (see Fig. 7), which is sent to an impurity filter and then the reception value in the tank is measured (see Fig. 8).

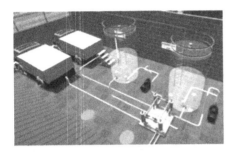

Fig. 7. Trucks and reception tanks   **Fig. 8.** Reception filtering

ii. *Storage of cold milk*

At this stage, the milk temperature is changed by a heat exchanger at a temperature of 8 °C (see Fig. 9), for subsequent storage in other tanks (see Fig. 10).

Fig. 9. Heat exchanger and cold water tank   **Fig. 10.** Reception filtering

iii. *Mixture of preservatives*

This stage is in charge of sending a percentage of milk from the cold milk tanks to a small tank where inputs for pasteurized milk are added (see Fig. 11).

Fig. 11. Input tank

iv. *Pasteurization*

This stage is carried out by means of heat exchangers, the thermal process of raising the temperature to approximately 78 °C and lowering it to 8 °C see Fig. 12, with the intention of reducing the presence of pathogenic agents (such as certain bacteria, protozoa, moulds, yeasts, etc.) that may be contained in the milk. Due to high temperatures the vast majority of bacterial agents die. All this change of temperature is shown in the virtual environment animation.

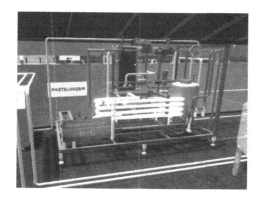

Fig. 12. Pasteurizer

v. *Homogenisation*

After the previous stage, a homogenizer (see Fig. 13.) is sent by pipeline to spray the whole milk through small nozzles under pressure; the size of the fat globules is reduced to a size in which the cream is no longer separated.

Fig. 13. Homogenizer

vi. *Storage of pasteurized milk*

In this final stage, the pasteurized milk is stored through a pipe to a tank with its respective filling measurement value, and then sent to another packaging process (see Fig. 14).

Fig. 13. Pasteurized milk storage tanks

5 Conclusions

The virtualization of the pasteurization plant through 3D modeling of the elements and instrumentation present in the industrial process, allows to obtain a great audiovisual experience in the immersion of the users with the virtual environment. The advantage of this is the qualification and training that the users obtain about the pasteurization process without exposing themselves to manipulation and operation risks, present in the real process equipment. At the same time, it becomes a tool to become familiar with the stages handled by the pasteurizing industry; in addition, making the user develop his skills in an efficient way and improve the performance with the virtual immersion.

In future works the 3D virtualization of an industrial process linked to the mathematical modeling of the dynamics of the elements that intervene in it, allows to emulate the real functioning of this industrial process, in addition it allows us to perform control algorithms in which we can optimize the operational resources; to manipulate control data at different levels of privileges, with a multi-user web server system.

Acknowledgements. The authors would like to thanks to the Corporación Ecuatoriana para el Desarrollo de la Investigación y Academia–CEDIA for the financing given to research, development, and innovation, through the CEPRA projects, especially the project CEPRA-XI-2017-06; Control Coordinado Multi-operador aplicado a un robot Manipulador Aéreo; also to Universidad de las Fuerzas Armadas ESPE, Universidad Técnica de Ambato, Escuela Superior Politécnica de Chimborazo, Universidad Nacional de Chimborazo, and Grupo de Investigación ARSI, for the support to develop this work.

References

1. de Paiva Guimarães, M., Dias, D.R.C., Mota, J.H., Gnecco, B.B., Durelli, V.H.S., Trevelin, L.C.: Immersive and interactive virtual reality applications based on 3D web browsers. Multimedia Tools Appl. **77**(1), 347–361 (2018)
2. Andaluz, V.H., et al.: Multi-user industrial training and education environment. In: De Paolis, L.T., Bourdot, P. (eds.) AVR 2018. LNCS, vol. 10851, pp. 533–546. Springer, Cham (2018). https://doi.org/10.1007/978-3-319-95282-6_38

3. Castillo-Carrión, D., Miranda, R.J., Andaluz, V.H., Alvarez, V.M.: Control and monitoring of industrial processes through virtual reality. Am. Sci. Publishers **24**(11), 8894–8899 (2018)
4. Cobo, L., Castillo, F., Castillo, C., Velosa, J.: Methodological proposal for use of virtual reality VR and augmented reality AR in the formation of professional skills in industrial maintenance and industrial safety. In: Online Engineering & Internet of Things, **22**, 987–1000 (2017). ISBN 978-3-319-64352-6
5. Choi, S.S., Jung, K., Noh, S.D.: Virtual reality applications in manufacturing industries: past research, present findings, and future directions. Concurr. Eng. **23**, 40–63 (2015)
6. Burle, M., Michael, M.: Virtual learning: videogames and virtual reality in education. In: Virtual and Augmented Reality; Concepts, Methodologies, Tools and Applications. IGI Global, pp. 1067–1087 (2018)
7. Carter,C., El-Rhalibi, A., Merabti, M., Price, M.: The creation and web-based deployment of cross-platform 3D games. In: UMSA, pp. 12–16 (2014)
8. Porcelli, I., Rapaccini, M., Espíndola, D., Pereira, C.: Technical and organizational issues about the introduction of augmented reality in maintenance and technical assistance services. IN: IFAC Proceedings, pp. 257–262 (2013)
9. Orduña, P., Almeida, A., López-de-Ipiña, D., Garcia-Zubia, J.: Learning analytics on federated remote laboratories. In: IEEE Global Engineering Education Conference EDUCON, pp. 299–305 (2014)
10. Dini, G., Mura, M.: Application of augmented reality techniques in through-life engineering services. In: Procedia CIRP 38, pp. 14–23 (2015)
11. Harshfield,N., Chang, D., Rammohan: A unity 3D framework for algorithm animation. In: CGAMES, Louisville, KY, pp. 50–56 (2015)
12. Kim, S., Suk, H., Kang, J., Jung, J., Laine, T., Westlin, J.: Using unity 3D to facilitate mobile augmented reality game development. In: IEEE World Forum on Internet of Things (WF-IoT), Seoul, pp. 21–26 (2014)
13. Murillo, A.: Diseño e implementacion de un sistema de realidad virtual para una planta pasteurizadora de leche. Universidad la Salle, Bogota (2016)
14. Andaluz, V.H., Pazmiño, A.M., Pérez, J.A., Carvajal, C.P., Lozada, F., Lascano, J., Carvajal, J.: Training of tannery processes through virtual reality. In: De Paolis, L.T., Bourdot, P., Mongelli, A. (eds.) AVR 2017. LNCS, vol. 10324, pp. 75–93. Springer, Cham (2017). https://doi.org/10.1007/978-3-319-60922-5_6

Virtual Environment for Teaching and Learning Robotics Applied to Industrial Processes

Víctor H. Andaluz$^{(\boxtimes)}$, José A. Pérez$^{(\boxtimes)}$, Christian P. Carvajal$^{(\boxtimes)}$, and Jessica S. Ortiz

Universidad de las Fuerzas Armadas ESPE, Sangolquí, Ecuador
{vhandaluz1, jsortiz4}@espe.edu.ec,
joansll@hotmail.com, chriss2592@hotmail.com

Abstract. This paper presents a structured application with virtual environments that emulate industrial processes that interact with the virtualized ScorBot ER-4U robot with the help of CAD software and texturing using other software. The manipulator is controlled from Matlab in response to the established modeling, communication is generated in two ways since both the real and the virtual manipulator act at the same time. Manipulator libraries allow data to be obtained in the encoders located in each motor and set the manipulator in a desired position. The real movement is replicated in the virtual environment for which a shared memory is used that allows access to a record inside the RAM of the computer that can be accessed from any program once in dynamic-link library (dll) is loaded.

Keywords: Shared memory · Manipulator kinematic model ·
Trajectory tracking · Virtualization

1 Introduction

The training stations have been of great benefit within the study centers to complement the theoretical within control processes that include industrial elements such as flow control, speed [1, 2] With the advance of applicable technologies for teaching, they have consolidated a great gap that existed between the theoretical and the practical thanks to virtual reality which offers a number of tools. Simulators are developed for different areas such as automotive for the assembly of automotive components [3, 4], in electrical for installation of high voltage lines from distribution stations [5–7], in nuclear to recreate manuals of procedures [8], in production processes such as tanneries to train new personnel [9]. The main objective of the simulators is to develop the skills of the person entering it in a virtual world that interacts in the same way as the real one but without being in any danger. In the field of medicine, it is used for rehabilitation through gadgets compatible with platforms [10] or to treat diseases such as autism or phobia [11] by the level of impression generated by the user in virtual reality [12, 13]. In robotics they test the performance of the proposed controllers in a virtual environment [14, 15].

Products produced in the industry represent a complex process due to the learning curve, for which new training options are taken for industrial training in different teams

© Springer Nature Switzerland AG 2019
L. T. De Paolis and P. Bourdot (Eds.): AVR 2019, LNCS 11614, pp. 442–455, 2019.
https://doi.org/10.1007/978-3-030-25999-0_36

that are virtualized and programmed to interact with the environment with the data that is generated by virtual sensors and generate response to the change of state in the actuators or final control elements.

This work is organized in 6 sections including the introduction. Section 2 describes the formulation of the problem in the learning process. The CAD design, the development of the virtual environment with its animation and the exchange of data with the manipulator is presented in Sect. 3. Modeling and control of the manipulator are details in Sect. 4 and finally the experimental results are presented in Sect. 5 while the conclusions are shown in Sect. 6.

2   Problem Formulation

Currently automation continues to gain more ground in the industrial sector for many factors including time, money. Future professionals within the learning classrooms are trained with the same industrial teams putting the theory into practice, controls or sequences of fictitious industrial processes are recreated, losing the capacity for logical development through observation as the student performs a predefined process, it suppresses creativity and familiarization by not being in an industrial environment where you can engineer to automate a process.

To consolidate the learning, the application of the following steps: *(i) virtualization of the manipulator*, in this step it is necessary to use a CAD software to reconstruct the manipulator with the same dimensions and with all its mechanical elements fixed and mobile in order to that can perform the movements that the real manipulator; *(ii) virtual environment*, it is essential to create a virtual environment in which you can observe the operation of the manipulator and can interact this in different processes. The Unity3D game engine allows us to interact with devices that enrich the person's immersion in a virtual environment. *(iii) controllers*, the advanced control algorithms allow us to know the kinematics and dynamics of a robot, when entering the model within the mathematical software tool MATLAB which resolves matrices and processes data without appreciable delays among some of its benefits becomes of great utility to generate the control data to perform a desired task. *(iv)* Implementation, we can observe the response in a 3D virtual environment of our control algorithm based on a virtualized

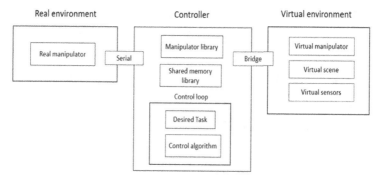

Fig. 1. Data interchange between controller and environments.

manipulator with equal dimensions and whose response to the control will be compared with the real one when performing the same task (Fig. 1).

In context, the following sections detail the advanced control algorithm of the application developed in MATLAB and shown in real time in a virtual environment developed in the Unity3D gaming engine. The application allows the manipulator to interact with the virtual environment.

3 Structure System

In every system, active and passive elements must interact within an environment governed by certain changes according to inputs and outputs that in this case emulate an industrial process, the scheme of the system is described in Fig. 2.

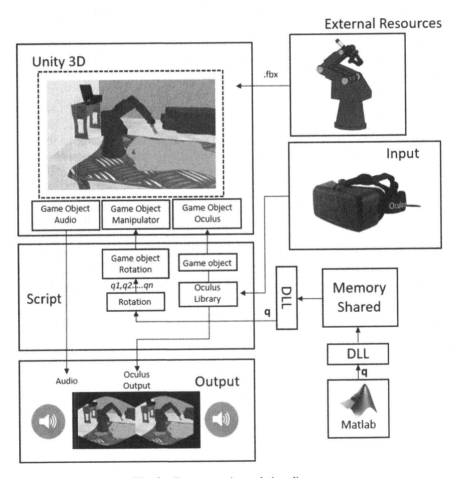

Fig. 2. Component interrelation diagram.

3.1 Manipulator Virtualization

To replicate a process, the elements must have the same characteristics as the real one. The manipulator must have the same movement characteristics in degrees of resolution for each axis, since there are positions in which the real manipulator cannot be located Fig. 3. The CAD design software SOLIDWORK facilitates the design of the model with great accuracy since it allows prototypes to be tested thanks to the tools it offers for the analysis of models in terms of resistance, forces, among others.

Fig. 3. CAD model ScorBot.

The prototype has characteristics similar to the real one but it is still an inanimate object without movement, to give properties of movement with respect to an axis that is called pivot 3DMAX is used, in each joint of the manipulator parts the pivots that represent each axis are positioned of movement but individually since the complete hierarchy of the object is disassociated.

The hierarchy is handled with the logic that an object can be the son of another object that is called a parent. This creates a chain effect, since automatically when this object moves the other objects inherit the movement. The appearance of a real object gives us the texturing for which BLENDER is very good ally to have a catalog of textures and nuances. Unity3D is compatible with certain type of format of objects among them *.fbx that Blender gives us the facility to export it (Fig. 4).

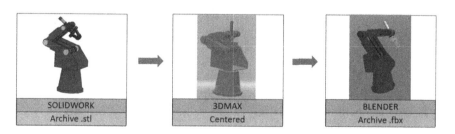

Fig. 4. 3D design diagram

3.2   Development Environment

The scenes were designed taking into consideration the type of logic that the process must have, among which stand out fixed sequences and tasks with continuous movement represented in a function.

This scene represents a process of palletizing boxes typically found in any industry, the student must be able to visualize sequences to locate the boxes parallel to some distance from each other, the end of the manipulator must hold the box that comes in the conveyor belt and locate it on the pallet that will later be moved by a forklift that is at the end of the conveyor belt which ends this process (Fig. 5).

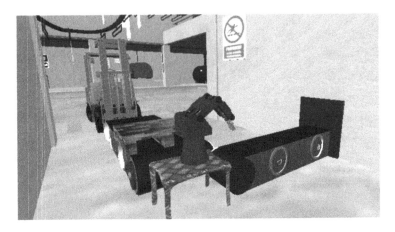

Fig. 5. Palletization scene

The following scene simulates a cutting process in which the operative end must interact with the object to be cut, it is a task of continuous movement in which cuts must be made in a rectangular shape simulating the pieces of leather that adorn the jeans with their logo. By touching the end of the manipulator to the leather it is generating a black stroke to visualize the trajectory that it realized (Fig. 6).

Fig. 6. Scene of cut

3.3   Animation

The objects in the environment are animated by the scripts that are executed when they are attached to the object, each object has its Transform properties within which we can vary position, rotation and scale with respect to the three dimensions (x, y, z). The movement characteristics of the manipulator are already established, it is only necessary to vary the angles of each axis to simulate the movement of the manipulator. The direction of movement is given by adding or subtracting the angle from its pivot point (Fig. 7).

Fig. 7. Line render.

The interaction with the objects of the environment is given by functions that facilitate us Unity among them the line renderer that graphs a line between an array of points becoming the last point the first of the next pair being able to graph figures of different shapes (Fig. 8).

Fig. 8. Rigid body property.

Objects in the environment can be affected by gravity or an external force when attaching the Rigid body property, the end of the manipulator generates a clamping effect by colliding the colliders of both objects and triggering an event.

3.4 Communication

The application requires two communication channels to control and replicate the movements of the manipulator, within the same script libraries are loaded so that they can exchange information by traversing a control loop in a defined time Fig. 9.

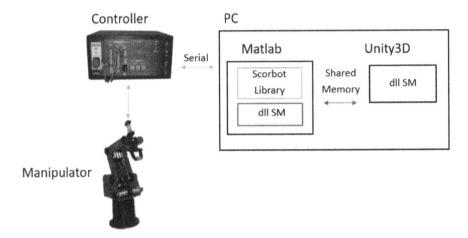

Fig. 9. Communication diagram.

The Scorbot manipulator has its controller that communicates serially with the computer. To communicate with this, we use a library that allows us to obtain all the data for reading and writing the encoders located on each axis of the robot. A frame is established for the exchange of information in a defined time for which functions that return flags are used to verify the correct execution.

Unity3D to perform the animations requires the data generated by in control for this the shared memory is used that creates memory spaces in the RAM so that different programs have access to the data stored in an assigned record, once the record is created we can read and write for this we must call the dll in the software that wants to use the records. To load the dll the location path is located, the memory is opened to access and the reading or writing functions are called depending on the type of data. Once the functions are established, it is only necessary to call them by passing the registration name parameters and position.

4 Modeling and Control

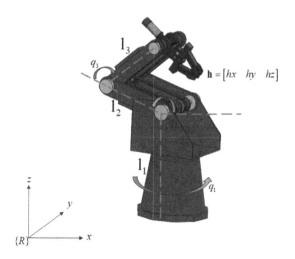

Fig. 10. Scheme of the manipulator

The manipulator configuration is defined by a vector $\mathbf{q}$ of n independent coordinates, called generalized coordinates of the manipulator, where $\mathbf{q} = [\,q_1 \quad q_2 \quad \cdots \quad q_n\,]^T = [\mathbf{q_a}]$ and $\mathbf{q_a}$ represents the generalized coordinates of the robotic arm. The configuration $\mathbf{q}$ is an element of the manipulator configuration space; denoted by $\mathcal{N}$. The location of the end-effector of the manipulator is given by the m dimensional vector $\mathbf{h} = [\,h_1 \quad h_2 \quad \cdots \quad h_m\,]^T$, where $\mathbf{h}$ define the position and the orientation, respectively, of the end-effector of the manipulator in $\mathcal{R}$. The set of all locations constitutes the aerial mobile manipulator operational space, denoted by $\mathcal{M}$ (Fig. 10).

4.1 Manipulator Kinematic Model

The kinematic model of the manipulator gives the location of the end-effector $\mathbf{h}$ as a function of the robotic arm configuration. The instantaneous kinematic model of the manipulator gives the derivative of its end-effector location as a function of the derivatives of the robotic arm configuration.

$$\dot{\mathbf{h}}(t) = \frac{\partial f}{\partial \mathbf{q}}(\mathbf{q_a})\mathbf{v} \qquad (1)$$

where, $\dot{\mathbf{h}} = [\,\dot{h}_1 \quad \dot{h}_2 \quad \cdots \quad \dot{h}_m\,]^T$ is the vector of the end-effector velocity, $\mathbf{v} = [\,v_1 \quad v_2 \quad \cdots \quad v_{\delta_n}\,]^T = [\mathbf{v_a^T}]^T$ is the control vector of mobility of the manipulator. Its dimension δ_n is the dimensions of the control vector of mobility associated to the robotic arm. Now, after replacing $\mathbf{J(q)} = \frac{\partial f}{\partial \mathbf{q}}(\mathbf{q_a})$ in the above equation, we obtain

$$\dot{\mathbf{h}}(t) = \mathbf{J(q)v}(t) \tag{2}$$

where, $\mathbf{J(q)}$ is the Jacobian matrix that defines a linear mapping between the vector of the manipulator velocities $\mathbf{v}(t)$ and the vector of the end-effector velocity $\dot{\mathbf{h}}(t)$. The Jacobian matrix is, in general, a function of the configuration $\mathbf{q}$. Those configurations where $\mathbf{J(q)}$ is rank-deficient are termed singular kinematic configurations.

4.2 Kinematic Control

The controller proposed to solve the trajectory tracking problem of the manipulator for to follow the cutting profile is to calculate at every time $\tilde{h}_x(t), \tilde{h}_y(t)$ and $\tilde{h}_z(t)$; use these measures to drive the robotic arm in a direction which decreases the control errors. The proposed kinematic controller is based on the kinematic model of the manipulator. Hence following control law is proposed,

$$\mathbf{v} = \mathbf{J}^{\#}\left(\dot{h}_d + \mathbf{L}\tanh\left(\mathbf{k\tilde{h}}\right)\right) \tag{3}$$

where $\dot{h}_d$ is the velocity of the manipulator for the controller; $\tilde{\mathbf{h}}$ is the control errors represented by $\tilde{\mathbf{h}} = \begin{bmatrix} \tilde{h}_x & \tilde{h}_y & \tilde{h}_z \end{bmatrix}$; $\mathbf{J}^{\#}$ is the matrix of pseudoinverse kinematics for the a manipulator; while that $\mathbf{L}$ and $\mathbf{k}$ are a matrices of control gains where the elements in this matrices are, $l_x > 0$, $l_y > 0$ and $l_z > 0$ for $\mathbf{L}$ and for $\mathbf{k}$ is $k_x > 0$, $k_y > 0$ and $k_z > 0$ are a gain constants of the controller that weigh the control error respect to the inertial frame $< \mathcal{R} >$; and the $\tanh(.)$ represents the function saturation of maniobrability velocities in the manipulator.

5 Experimental Results

The application simulates an industrial environment in which the manipulator must perform certain tasks within the established processes. The proposed application runs on a Core I7-7500U desktop computer running at 2.7 GHz, 8 GB of RAM and a RADEON (TM) R7 M445 1 GB Dedicated Graphics. The manipulator is a ScorBot ER-4U robotic arm with 6 articulations (axes) for its movement, made by cd motors. It has a USB-Controller that serves to operate the arm and accessories. It is connected to the computer by USB connector (Fig. 11).

Fig. 11. ScorBot ER-4U.

Experiment 1: Cutting Process

This process represents the cutting of a leather, the final product in a tannery. The manipulator has a blade in the gripper, when it comes in contact with the leather, now begins to cut a line that represents the path followed by the manipulator.

The figure shows the stroboscopic movement in the X-Y-Z space of Unity3D. This reflects the correct functioning of the control when following a desired trajectory. The real and simulated movement of the robot is compared as shown in the following images Fig. 12, 13, 14.

Fig. 12. Stroboscopic manipulator movement in Unity.

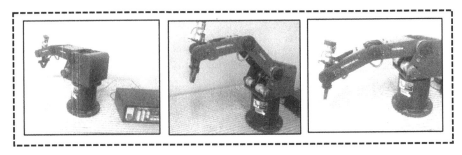

Fig. 13. Trajectory tracking captures.

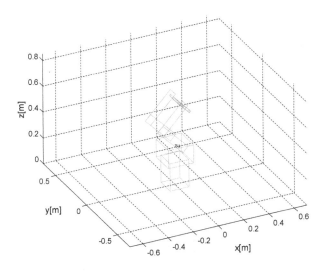

Fig. 14. Manipulator ScorBot in Matlab.

Experiment 2: Palletization Process.

The palletization process requires a fixed sequence to place the boxes on the pallet. The Fig. 15 shows the movement of the manipulator in the space X-Y-Z of Unity3D to follow an established sequence that must mainly hold the box that comes in the conveyor belt and place it on the pallet at a certain distance from each other. The correct functioning of the real movement replicated in the simulation can be observed in the snapshots Fig. 16.

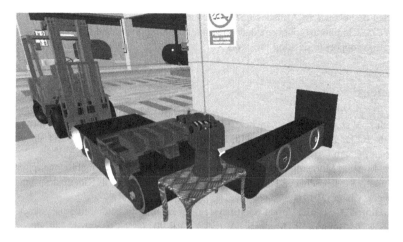

Fig. 15. Sequence of movement boxes

Fig. 16. Capture of the sequence of movement boxes.

6   Conclusions

This paper proposes an application that presents industrial processes in which the ScorBot ER-4U robot that was virtualized to interact in the virtual environment is required. Two internal communication channels were used with the shared between Unity-Matlab and an external one by serial port with the robot controller. The movements are replicated in real time, comparing the good performance of the proposed application.

Acknowledgements. The authors would like to thanks to the Corporación Ecuatoriana para el Desarrollo de la Investigación y Academia–CEDIA for the financing given to research, development, and innovation, through the CEPRA projects, especially the project CEPRA-XI-2017-06; Control Coordinado Multi-operador aplicado a un robot Manipulador Aéreo; also to

Universidad de las Fuerzas Armadas ESPE, Universidad Técnica de Ambato, Escuela Superior Politécnica de Chimborazo, Universidad Nacional de Chimborazo, and Grupo de Investigación ARSI, for the support to develop this work.

References

1. Andaluz, V.H., Ortiz, J.S., Sanchéz, J.S.: Bilateral control of a robotic arm through brain signals. In: De Paolis, L.T., Mongelli, A. (eds.) AVR 2015. LNCS, vol. 9254, pp. 355–368. Springer, Cham (2015). https://doi.org/10.1007/978-3-319-22888-4_26
2. Ortiz, J., et al.: Modeling and kinematic nonlinear control of aerial mobile manipulators. In: Zeghloul, S., Romdhane, L., Laribi, M. (eds.) Computational Kinematics, pp. 87–95. Springer, Cham (2017)
3. Quevedo, W.X., et al.: Virtual reality system for training in automotive mechanics. In: De Paolis, L.T., Bourdot, P., Mongelli, A. (eds.) AVR 2017. LNCS, vol. 10324, pp. 185–198. Springer, Cham (2017). https://doi.org/10.1007/978-3-319-60922-5_14
4. Ortiz, J, et al.: Teaching-learning process through VR applied to automotive engineering. In: Proceedings of the 2017 9th International Conference on Education Technology and Computers - ICETC 2017 (2017). https://doi.org/10.1145/3175536.3175580
5. Hernandez, Y., Ramirez, M.: Virtual reality systems for training improvement in electrical distribution substations. In: 2016 IEEE 16th International Conference on Advanced Learning Technologies (ICALT) (2016). https://doi.org/10.1109/icalt.2016.141
6. da Silva Netto, A., de Fatima Queiroz Vieira, M.: Virtual reality training environment a proposed architecture. In: 2010 IEEE International Conference on Virtual Environments, Human-Computer Interfaces and Measurement Systems (2010). https://doi.org/10.1109/vecims.2010.5609364
7. Galvan-Bobadilla, I., Ayala-Garcia, A., Rodriguez-Gallegos, E., Arroyo-Figueroa, G.: Virtual reality training system for the maintenance of underground lines in power distribution system. In: Third International Conference on Innovative Computing Technology (INTECH 2013) (2013). https://doi.org/10.1109/intech.2013.6653713
8. Ciger, J., Sbaouni, M., Segot, C.: Virtual reality training of manual procedures in the nuclear sector. In: 2015 IEEE Virtual Reality (VR) (2015). https://doi.org/10.1109/vr.2015.7223455
9. Andaluz, V.H., et al.: Training of tannery processes through virtual reality. In: De Paolis, L. T., Bourdot, P., Mongelli, A. (eds.) AVR 2017. LNCS, vol. 10324, pp. 75–93. Springer, Cham (2017). https://doi.org/10.1007/978-3-319-60922-5_6
10. Andaluz, V.H., Patricio, C., José, N., José, A., Shirley, L.: Virtual environments for motor fine skills rehabilitation with force feedback. In: De Paolis, L.T., Bourdot, P., Mongelli, A. (eds.) AVR 2017. LNCS, vol. 10324, pp. 94–105. Springer, Cham (2017). https://doi.org/10.1007/978-3-319-60922-5_7
11. Ortiz, J.S., et al.: Realism in audiovisual stimuli for phobias treatments through virtual environments. In: De Paolis, L.T., Bourdot, P., Mongelli, A. (eds.) AVR 2017. LNCS, vol. 10325, pp. 188–201. Springer, Cham (2017). https://doi.org/10.1007/978-3-319-60928-7_16
12. Mourning, R, Tang, Y.: Virtual reality social training for adolescents with high-functioning autism. In: 2016 IEEE International Conference on Systems, Man, and Cybernetics (SMC) (2016). https://doi.org/10.1109/smc.2016.7844996
13. Carvajal, C.P., Proaño, L., Pérez, J.A., Pérez, S., Ortiz, J.S., Andaluz, V.H.: Robotic applications in virtual environments for children with autism. In: De Paolis, L.T., Bourdot, P., Mongelli, A. (eds.) AVR 2017. LNCS, vol. 10325, pp. 175–187. Springer, Cham (2017). https://doi.org/10.1007/978-3-319-60928-7_15

14. Castellanos, E.X., García-Sánchez, C., Llanganate, W.Bl., Andaluz, V.H., Quevedo, W.X.: Robots coordinated control for service tasks in virtual reality environments. In: De Paolis, L. T., Bourdot, P., Mongelli, A. (eds.) AVR 2017. LNCS, vol. 10324, pp. 164–175. Springer, Cham (2017). https://doi.org/10.1007/978-3-319-60922-5_12

15. Andaluz, V.H., Carvajal, C.P., Pérez, J.A., Proaño, L.E.: Kinematic nonlinear control of aerial mobile manipulators. In: Huang, Y., Wu, H., Liu, H., Yin, Z. (eds.) ICIRA 2017. LNCS (LNAI), vol. 10464, pp. 740–749. Springer, Cham (2017). https://doi.org/10.1007/978-3-319-65298-6_66

StreamFlowVR: A Tool for Learning Methodologies and Measurement Instruments for River Flow Through Virtual Reality

Nicola Capece[1], Ugo Erra[1(✉)], and Domenica Mirauda[2]

[1] Dipartimento di Matematica, Informatica ed Economia,
Università degli Studi della Basilicata, Potenza, Italy
{nicola.capece,ugo.erra}@unibas.it
[2] Scuola di Ingegneria, Università degli Studi della Basilicata, Potenza, Italy
domenica.mirauda@unibas.it

Abstract. Virtual Reality potentialities can be exploited for several types of contexts beyond the mainstream video game industry. One of the most interesting application fields concerns its use in education and teaching especially in the contexts where security and safety are essential elements. The measurement of river flow is required for river management purposes including water resources planning, pollution prevention, and, flood control. The teaching of this complex task requires to transfer well-known methodologies as well as careful attention while performing the training in situ. In this paper, we propose a virtual reality tool called StreamFlowVR to improve the learning process for the river flow instruments and measuring methodologies. The basic idea is to create a rich user experience in a secure environment where students can understand the correct methodologies. We believe that the use of virtual reality benefits the students in learning hydraulic phenomenon faster and more confident respect to the traditional approach on a real river.

Keywords: Virtual reality · Hydraulic education ·
River flow measurements · Learning hydraulic instrumentation

1 Introduction

Virtual reality (VR) mimics the real environment through an advanced human-computer interface, which enables users of interactive software and hardware to be translated into an imaginary place simulating their physical presence. Researchers used the term VR first in the 1960s, discussing the employment of aiding tools, such as Head Mounted Display (HMD) and special glasses, improving the interactive VR experience [13].

Although there are more and more applications of VR in several contexts, its use in education and training as well as in other sectors different from the

© Springer Nature Switzerland AG 2019
L. T. De Paolis and P. Bourdot (Eds.): AVR 2019, LNCS 11614, pp. 456–471, 2019.
https://doi.org/10.1007/978-3-030-25999-0_37

video game industry represent a challenge [10, 29]. Among such contexts there are medical education field [15, 25], in which a virtual simulation of surgical operations can improve the skills of medical students without risk to patients; for mesh painting combined with human computer interaction [2]; in the computer science field, for the software system and complex data structure comprehension [3, 4, 8]; for the learning of the combined arms and to train soldiers to shoot [17]; in the flight and vehicular simulation application [7] *etc.* VR has been used also in primary education [27] in which the immersion in 3D environment, the interaction with the 3D objects and the involvement in the virtual world exploration can play an important rule for the education process.

In the educational sector, the VR applications can be divided into two main categories, *e.g.*, non-immersive and immersive [30]. In particular, in the non-immersive category, the user's visualization is through a "window", represented by the flat screen of a computer, and through either conventional computer peripheral devices (*e.g.*, mouse or keyboard) or advanced devices similar to the ones used in the real control (*e.g.*, machine operation consoles or vehicle control cockpits). In the immersive category, instead, the user's interaction occurs through different types of devices, such as the HMD and the CAVE (Cave Automatic Virtual Environment). Both non-immersive and immersive VR can provide learners with a safe and monitored virtual environment, in which to develop their skills without the real-world, sometimes dangerous consequences. These environments present a scientific concept to students in situations that cannot be entirely and perfectly recreated in reality, making studying even more interesting and sometimes producing faster learning results.

In hydraulic engineering education, practical experience in the field is an essential part of the learning where the students can better understand some physical phenomena and improve research methods and analysis techniques explained during the frontal lesson. However, such field activities are not always possible because some university courses have limited duration and are mainly taught in the winter season when the weather conditions impair the regular performance of practical classes. In addition, such activities can involve a large number of students who, when monitored by only one teacher, can get easily distracted and acquire incorrect or inaccurate concepts. Finally, the open-air activities can sometimes be dangerous and thus limit the possibility of integrating traditional classes with field practice. Additional difficulties are due to the use of sophisticated equipment, complex measurement techniques, and lack of territorial knowledge. In order to overcome these issues, and in support of the physical laboratory of the Basilicata University School of Engineering, a virtual reconstruction of a river is proposed for introducing hydraulic engineering students to the measuring methodologies of the fluid discharge in an open channel cross-section.

2   Related Work

In engineering education field, the most common VR application is based on 3D Virtual Laboratories (3D-VLs), which substitute some practical classes and reduce the difficulties related to traditional educational methods, such as the danger of using products or machines, overcrowding, and timetable schedule availability of laboratories. Therefore, 3D-VLs allow students to obtain an experience very close to the real one and, in some cases, results similar to those gained in the real environment, with the opportunity of continuously evaluating the learning process. In fact, students increase their knowledge and skills through the technical practice and improve their comprehension of abstract concepts and complex three-dimensional graphics, as well as production, manufacturing, operative, and assembly processes [9]. A good example of the above-mentioned experiences is a study by Messner et al. [18], in which the undergraduate students of Architectural Engineering were even able to plan the construction of a nuclear power plant without much knowledge and within a very short time frame (1 h), thanks to a CAVE-like projection system that allowed increasing visualisation capacities, usually restricted in the traditional planning techniques.

In 2008, a virtual learning platform, complementary to a physical one, was developed at the Department of Chemical Engineering of Oregon State University, including an advanced 3D graphical user interface, an instructor Web interface with integrated assessment tools, and a database server [14], aimed at explaining and analysing the chemical vapour deposition.

Villar-Zafra et al. [32] developed a multiplatform virtual laboratory to easily describe the dynamical behaviour of the magnetic levitation, an interesting process in control engineering which, due to its instability and nonlinearity, could be explained only through analytical modelling. Different experiments were performed and, in all of them, the students could see the levitating ball movements in real time. Abdulwahed and Nagy [1], at Loughborough University, UK, implemented a Process Control Virtual Laboratory (PCVL) based on an Armfield PCT40 tank filling process, which combines the three access modes (Hands-On, Virtual, and Remote) in one unifying software package (the TriLab). In this way, the students were introduced to the principles of control engineering, such as the main components and instruments of a feedback loop, the concept of open-loop control, feedback control, PID (Proportional-Integral- Derivative) control, and PID tuning. Considering the opinion of both students and teachers, Vergara et al. [31] presented a 3D interactive virtual laboratory including a universal testing machine for enhancing the learning of mechanical characterisation of materials. Dinis et al. [6] developed an immersive VR prototype interface that allowed students of the Integrated Master's in Civil Engineering to automatically edit Building Information Modelling (BIM) elements, linking game engine input and BIM authoring software. Despite the increase of VR laboratories in the engineering field, there is still little mention of its employment in the hydraulic engineering educational branch. In fact, most studies have been performed on the process of developing schematic diagram-based 2D virtual hydraulic circuits and 3D virtual hydraulic equipment, mainly aimed at supporting courseware to facilitate the physical experiment teaching practice [11,33], or on the flood events management through AR-VR applications [21,22] in support to mobile workforces. None of the mentioned studies focussed on the virtual recreation of a real hydraulic laboratory. Therefore, the present paper proposes a new VR application based on the virtual reconstruction of a natural environment, such as the river, allowing the improvement of engineering students' performance during hydraulic field practice aimed at the monitoring and control of open channel flows. This new technology application, in support of traditional practical classes, managed to increase the students' motivation and improve their skills, and simultaneously reduce the costs, time, and potential dangers that continuous physical experiments would involve.

3   Background

The proposed tool was developed through the well known Unity 3D game engine and the head-mounted display (HMD) Oculus Rift. The scene Graph is based on a *GameObjects* hierarchy. In particular, each *GameObject* has a spatial position and orientation and can be connected with one or more scripts, sound, textures, animations, 3D models, *etc.* The available programming languages are C# and JavaScript. In our application we use C#. Unity 3D allows modifying the script's variable at run-time through its editor user interface allowing to visualize quickly and easily the effect of these changes. In this way, it is possible to modify such values without having to specify them in the code. In order to keep the compatibility with other HMD devices such as HTC Vive, the virtual reality visualization and interaction was developed using the SteamVR SDK version 2.0. In particular, in such SteamVR version was introduced an abstraction layer between Unity 3D and the hardware. The one-to-one hardware mapping of the input system was replaced with an action based system. Such a system is oriented to user actions rather than the buttons to press. SteamVR is able to detect what are the inputs to use for a specific action. The SteamVR actions configuration is browser-based and all changes are stored in several JSON files. SteamVR is used as Unity 3D plugin which contains a set of scripts useful to manage the sensors of the devices.

The VR scene visualization is possible through the Oculus Rift which consists of an OLED panel for each eye with 1080×1200 as resolution. Each panel has a refresh rate of 90 Hz and $110°$ as a field of view which is well adapted to the user's field of view. Oculus Rift has rotational and positional tracking on 6 degrees of freedom, in order to well manage the tracking of the user through the Oculus's Constellation tracking system [28]. Constellation is the positional tracking system of the HMD based on infrared tracking sensors. It is possible to configure the real world user space called room scale [26] by using three or more constellation sensors. For our work, we used only two constellation sensors because we obtain a good tracking precision in our context keeping the scene always in front of the user. To interact with the VR scene we use the Oculus Touch motion controller. Each controller is based on 6 degrees of freedom and is associated with a specific user hand. The controllers are tracked through the Constellation system and are shown in the VR scene. Each controller can detect the fingers user gestures during the gaming time. Moreover through the controllers the user can move across the scene for along distances using the analog thumbstick.

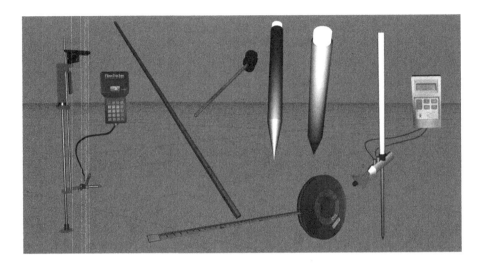

Fig. 1. River Flow Instruments. From left to right: the first sensor is called SonTek Flow Tracker which is able to measure the river water speed and depth; graduated rod for depth measurements; meter tape, useful to measure the cross-section; the hammer and pickets useful to mark the river banks and support the cross-section measurement; finally the last instrument is the Current Meter which is able to measure the river water speed.

To simulate the river flow measurements we created the 3D models of the instruments by using Autodesk 3D Studio Max 2019 and Adobe Photoshop CC 2018. In particular, we created a 3D model of SonTek Flow Tracker ultrasound sensor, Current Meter, Meter Tape, Hammer, Pickets and Graduated Rod as shown in Fig. 1. The models were created inspired by the original instruments.

4 Learning River Flow

In general, in the water sector, the quantitative monitoring of rivers is an essential step for environmental purposes both for addressing mid and long-term surveillance and control activities and for civil protection aims in terms of flood forecasting and risk mitigation [19]. In particular, for hydraulic engineering students, the use of such measuring techniques allows them to understand the importance of accurate monitoring activities and the survey modalities of input data for numerical and analytical modelling of fluvial phenomena [12,20,24]. VLs improve the students' operational abilities and the mastering of certain skills, which would normally require numerous field experiments and repetitive training. Therefore, the virtual experiment can be the advance test of the physical experiment, ensuring a higher success rate of future practical operations. Another advantage of VR interaction is that the one-to-one teaching-learning relationship, in which students are the active part, lets them take initiative and manage the practice more autonomously. The evaluation of water discharge is introduced during the course through the use of standard techniques, according to the rules ISO 748/1997[1] and ISO 1100-2/1998[2] and through the use of simple (current meters) towards complex (acoustic and laser sensors) equipment. Operatively, computations require dividing the section areas into several verticals and a further subdivision of each vertical into discrete points, in order to evaluate the mean velocity of the flow along each vertical. The number of verticals and the distribution inside the cross section is chosen according to the section width, riverbed geometry, and flow regimes and characteristics, while the measurement points are fixed according to the measurement methodology used, that is, at the ford or from the bridge. The main objective of this measuring technique is to obtain a correct evaluation of the mean velocity for each vertical, which is related to a reliable reconstruction of flow field obtained through velocity point measurements in several marks of open channel cross sections, generally distributed from the bottom up to the free surface flow. Depending on the velocity measure points, the mean velocity and, subsequently, the water discharge are calculated [23].

The tool here developed allows the student to perform the above-mentioned operations in detail, by acting in the safe and monitored recreated virtual environment. In fact, thanks to this VR experience carried out indoors, the students can learn to move around a natural channel, working more safely and reducing risks of accidents. Besides, such virtual laboratory requires relatively small space and, thus, fewer expenses since only a computer, a software, and a VR device are needed, compared to the traditional field experience which includes even transportation costs. Other reduced difficulties are those related to the equipment wear and tear due to the continuous use during experiments. Last but not least, this tool could increase the learners' interest, by combining education with pleasure and making knowledge fun.

5   The StreamFlowVR Tool

Our proposed VR tool try to replicate the liquid river flow measurements procedures [16] in the open canals. In particular the measurement procedure is based on the international standards ISO 748/1997 (See footnote 1) and ISO 1100-2/1998 (See footnote 2). These standards involves several steps: *(i)* choice of the measurement zone; *(ii)* measurement the cross-section; *(iii)* the choice of measuring verticals; *(iv)* depth measurement of each individual canal; *(v)* speed measurement on each vertical.

---

[1] ISO 748: 1997, Measurement of liquid flow in open channel – Velocity-area methods.
[2] ISO 1100-2: 1998, Measurement of liquid flow in open channel - Part 2: Determination of the stage-discharge relation.

5.1 Measurement Zone

Although the virtual scene represents a simulated reality, we have tried to respect as much as possible the instructions in the international standards ISO 748/1997 (See footnote 1) and ISO 1100-2/1998 (See footnote 2). In particular, the choice of the measurement zone depends on several factors: one of the most important is the straightness of the canal and the cross-section must be uniform with a slope that is able to minimize irregularities present in water speed distribution. To obtain a realistic measurement developed the landscape through the Unity 3D terrain editor. The editor allows defining the banks of the canal in a stable and well-defined manner so as to ensure a constant condition during the measurements. Figure 2 shows the virtual scene, in particular as explained in the standards, the water movement direction must be orthogonal to the cross-section and the water flow must be regular.

Fig. 2. The virtual scene. The bigger picture represents the section and the arrows show the water movement direction. The bottom left image represents the cross-section and the bottom right image represent a full overview of the scene in isometric visual.

The water was developed through a particular Unity 3D asset called AQUAS which allows creating a very realistic water simulation of the rivers and lakes. AQUAS allows to manage the water movement and its direction and also allows to manage the physical features of water such as small, medium and large waves tiling, speed and refraction. It is possible to change the deep water colour, the shore fade, the shore and depth transparency, the flow speed and direction, *etc.* The vegetation, trees and grass are developed by using a Unity 3D asset called Vegetation Studio, in particular, the trees are developed by using the speed tree plugin and consist of a set of conifers and broad-leaved trees and the grasses are mainly allium, fern, dune and rustic grass. Vegetation allows to increase the realism of the scene but we have been careful not to add vegetation to the river as explained in the standards. In fact, the cross-section must be entirely visible in width and should not be obstructed by trees, aquatic plants and other obstacles. AQUAS allows to manage the depth of water and ensure that it is sufficient for the immersion of the measuring instruments.

5.2   Measurement the Cross-Section

As suggest from standards, whenever the canal is not very wide, the width can be measured using a metric rope or a graduated wire. Instead, when the canal is very wide, the cross-section has to measured by using optical or electronic distance meters.

Fig. 3. An example of cross-section measurement. On the river banks are placed the pickets and the meter tape define the length of the cross-section.

Our tool allows measuring the river cross-section by focusing on a canal not too wide. As shown in Fig. 3, the user uses two pickets and a hammer to define the measurement area. To accomplish this task, the user grabs a picket with a hand and through the other hand grab the hammer and hit the top of the picket.

Such part of the picket is represented through a white cylinder and when it collides with the hammer a trigger is activated. Through this trigger, the picket is translated on its local vertical axis penetrating into the terrain. When the user touches the objects, they are highlighted in yellow colour and when the objects are used in the wrong way, their highlights turn red until they are positioned correctly. As shown in Fig. 4 the user has to place the pickets with the tip toward the terrain. The pickets must be in a vertical position to be correct and the highlight colour will turn green. After the user has stuck the pickets in the terrain on the banks of the river, he can use a graduated rod composed of two interactive elements: ring and the main body. Such elements are connected through a textured cable. On the texture, the centimetres and meters are represented in textual form and in real-world scale. To simulate a real functioning of the meter tape we use a simple cable Unity 3D asset and define a new material and texture for the cable that represents a simple meter. We use a dynamic texture mapping in order to visualize the pieces of texture corresponding to the centimeters displayed on the meter tape, as the cable extends. When the visualized texture is greater then one meter the texture will be repeated one time for each visualized meter.

Fig. 4. The red border image represents the wrong way to place the instrument by the user while the green border image represents the correct way. The yellow border represents the hand over detection on the instrument. (Color figure online)

In this way, the user can read the width of the river cross-section such as if he was really on a river.

5.3   Verticals Measurements

Choice of verticals is made in order to perform depth and speed measurement of the canal. Standards suggests considering the verticals close together, especially for the depth measurement, to define accurately the vertical section profile of the canal. Each vertical was measured considering its horizontal distance respect a fixed reference point placed on a bank of the canal in relation with the cross-section. Typically the distance among the verticals must not be greater than one-twentieth of the width of the section. For canals not too wide, with a regular profile, the number of verticals can be reduced. The virtual reality system allows the user to mark the vertical points by using white spheres. The spheres can be placed along the cross-section following the metric rope using the button of the Oculus Touch controller.

Depth Measurement of Individual Canals. River depth can be measured using several types of instruments, such as a rod, rope probes or other instruments. Generally, the rods were used when the canals are not too deep. Since our approach is based on the teaching of measurement methodology, we considered an ideal scenario in VR as a shallow canal. Our system allows measuring the depth on the verticals by using a graduated rod. In particular, such rod can be immersed in the water by the user along the verticals and the measurements are detected through a well-defined pipeline in the standards.

Fig. 5. Depth of water measurement. The image shows the correct position of the graduated rod for depth measurements. (Color figure online)

When the user grabs the rod, the highlight that wraps it turns red. When the user place in a correct vertical position the rod, its highlight colour turn green as shown in the Fig. 5. By using a Unity 3D script we detect when the rod collides with the terrain by blocking penetration and the depth value can be identified precisely.

Speed Measurement of Verticals. Water speed measurement is typically taken with depth measurement, especially in the unstable river bed. When the two measurements were taken at two different times, the water speed measurements have to be taken by using a sufficient number of verticals, with well defined horizontal distance based on the standards. To take the water speed we developed a 3D model of the Current Meter (see Fig. 1), that is composed of a height adjustable rod and a propeller. The real instrument sends an impulse signal to the console, which is generated from each revolution of the propeller through a permanent magnet. Through the console is possible to set a measurement timer. The console of the current meter shows the number of revolutions per second of the propeller and convert them in the speed measurement through a standard formula: $v = nk + \Delta$ where n is the number of revolutions per second and k, Δ are experimental constants based on the diameter and the step of the propeller and the typology of the measurement. This 3D model consists of a hierarchy of meshes and was designed from an original one. In particular, two cylinders are placed one inside the other and the second one can slide inside the first one allowing to stretch the rod to which the sensor is attached. In the 3D model hierarchy, the propeller is a child of the sensor and the propeller can rotate when it touches the water. The measurement starts when the user sets the timer and presses the start button of the console. The console was developed by using a parallelepiped and using a photo of the real console as texture.

The developed procedure can simulate a real measurement methodology as shown in Fig. 6. The standards defines how to place the instrument during the speed measurement. In particular, the instrument must be placed in a vertical position with the axes parallel to the water movement direction. The instrument must be placed before the measurement and maintained in a vertical position through the rod. To give the user the feeling of accuracy we used a red highlight when the user wrongs the position and green highlight when the user places the instrument in a correct position. Also, in this case, the instrument was brought in a virtual world through a 3D model developed based on the real instrument.

5.4   Ultrasound Sensor

The last developed instrument is an Ultrasound Sensor. Such a sensor is based on a SonTek FlowTracker instrument to detect the speed and the depth of the river water. Through the sensor is possible to detect the water speed while with the graduated rod that represents the principal support of the sensor, is possible to detect the water depth. The real sensor consists of a transmitter and two or three probe receivers in relation to the measurement of 2D or 3D speed components.

Fig. 6. An example of water speed measurement. Placed in the water there is a Current Meter and the user is holding the console with the current timer.

The sensor measures the water speed through a physical principle called Doppler effect. The transmitter generates a sound pulse with a known frequency, which is propagated through the water perpendicular to the axis of the instrument. The sound pulse is propagated through the measurement volume and is reflected in any direction from the particles contained in the volume itself. Only the sound pulse reflected towards the reception beams is captured by the receivers. The sensor is characterized by a system which performs the signal modulation and converts sound signals into electrical signals. These signals are sent through a high-frequency cable to the console that converts them into digital signals.

Also, in this case, the 3D model is a hierarchical model and consists of a base, two rods, one of which is extendable by the user and a sensor. The sensor is attached to the extendable rod in order to allow the user to change the vertical position of the sensor. From the console, the user chooses the speed measurement and set the timer as can be seen in the Fig. 7. In order to simplify the usability in VR we developed only some function of the real flow tracker. As for the other instruments to detect the correct position we used the green highlight and to detect the wrong position we used the red highlight.

Fig. 7. Ultrasound Sensor is placed in the water in the correct position. The user sets the timer and takes the river water speed through the console button. The depth can be seen through the rod of the 3D object. (Color figure online)

6 Conclusions

We have presented a VR tool useful teaching and learning of the liquid flow measurement based on international standards ISO 748/1997 (See footnote 1) and ISO 1100-2/1998 (See footnote 2). The system was developed using the Unity 3D game engine. This system was thought for hydraulic engineer students, and to validate our proposal the next step of our research will consist of experimentation. Such experimentation involves the students of the hydraulic engineer course of the University of Basilicata, in particular, the students with a theoretical background due to the attendance of the hydraulic classroom lessons. The participants will be divided into two groups. A first group will go on the real river to learn as take a fluvial measurement with a practical approach under the guidance of Professor and using the real measurement instruments. The second group will try on the lab the StreamFlowVR tool to take a fluvial measurement in a simulated river in VR scenario. After the simulated practical lesson also the second group will go on the real river to learn and use the real instrument always under the guidance of Professor. Both groups will report their feelings through a survey and there will be an evaluation of the student's behaviour on the real river to understand whether an effective improvement in learning the river flow measurements was achieved, thanks to the support of the VR. The final goal of our research is to demonstrate that the use of VR is a valid approach to teaching in the hydraulic context in order to learn and practice the river flow measurements in a safe and risk-free environment. An interesting idea could be to create a serious game that emphasizes the added pedagogical value of fun and competition [5]. In future, this approach can also be used by institutions specializing in river monitoring as a support for the training of their employees.

References

1. Abdulwahed, M., Nagy, Z.K.: Developing the TriLab, a triple access mode (hands-on, virtual, remote) laboratory, of a process control rig using LabVIEW and Joomla. Comput. Appl. Eng. Educ. **21**(4), 614–626 (2013)
2. Capece, N., Erra, U., Grippa, J.: GraphVR: a virtual reality tool for the exploration of graphs with HTC Vive system. In: 2018 22nd International Conference Information Visualisation (IV), pp. 448–453, July 2018
3. Capece, N., Erra, U., Romaniello, G.: A low-cost full body tracking system in virtual reality based on Microsoft Kinect. In: De Paolis, L.T., Bourdot, P. (eds.) AVR 2018. LNCS, vol. 10851, pp. 623–635. Springer, Cham (2018). https://doi.org/10.1007/978-3-319-95282-6_44
4. Capece, N., Erra, U., Romano, S., Scanniello, G.: Visualising a software system as a city through virtual reality. In: De Paolis, L.T., Bourdot, P., Mongelli, A. (eds.) AVR 2017. LNCS, vol. 10325, pp. 319–327. Springer, Cham (2017). https://doi.org/10.1007/978-3-319-60928-7_28
5. Chiara, R.D., Santo, V.D., Erra, U., Scarano, V.: Real positioning in virtual environments using game engines. In: Amicis, R.D., Conti, G. (eds.) Eurographics Italian Chapter Conference. The Eurographics Association, Aire-la-Ville (2007)
6. Dinis, F.M., Guimarães, A.S., Carvalho, B.R., Martins, J.P.P.: Virtual and augmented reality game-based applications to civil engineering education. In: 2017 IEEE Global Engineering Education Conference (EDUCON), pp. 1683–1688. IEEE (2017)
7. Dourado, A.O., Martin, C.: New concept of dynamic flight simulator, part I. Aerosp. Sci. Technol. **30**(1), 79–82 (2013)
8. Erra, U., Scanniello, G.: Towards the visualization of software systems as 3D forests: The CodeTrees environment. In: Proceedings of the 27th Annual ACM Symposium on Applied Computing, SAC 2012, pp. 981–988. ACM, New York (2012)
9. Flanders, M., Kavanagh, R.C.: Build-A-Robot: using virtual reality to visualize the Denavit-hartenberg parameters. Comput. Appl. Eng. Educ. **23**(6), 846–853 (2015)
10. Freina, L., Ott, M.: A literature review on immersive virtual reality in education: state of the art and perspectives. In: eLearning & Software for Education, no. 1 (2015)
11. Gao, Z., Liu, S., Ji, M., Liang, L.: Virtual hydraulic experiments in courseware: 2D virtual circuits and 3D virtual equipments. Comput. Appl. Eng. Educ. **19**(2), 315–326 (2011)
12. Greco, M., Mirauda, D.: An entropy based velocity profile for steady flows with large-scale roughness. In: Engineering Geology for Society and Territory. River Basins, Reservoir Sedimentation and Water Resources, vol. 3, pp. 641–645. Springer, Switzerland (2015)
13. Hall, R.J., Miller, J.: Head mounted electrocular display: a new display concept for specialized environments. Aerosp. Med. **34**, 316 (1963)

14. Koretsky, M.D., Amatore, D., Barnes, C., Kimura, S.: Enhancement of student learning in experimental design using a virtual laboratory. IEEE Trans. Educ. **51**(1), 76–85 (2008)
15. Kuehn, B.M.: Virtual and augmented reality put a twist on medical education. JAMA **319**(8), 756–758 (2018)
16. Liptak, B.G.: Flow Measurement. CRC Press, Boca Raton (1993)
17. Maxwell, D.B.: Application of virtual environments for infantry soldier skills training: we are doing it wrong. In: Lackey, S., Shumaker, R. (eds.) VAMR 2016. LNCS, vol. 9740, pp. 424–432. Springer, Cham (2016). https://doi.org/10.1007/978-3-319-39907-2_41
18. Messner, J.I., Yerrapathruni, S.C., Baratta, A.J., Whisker, V.E.: Using virtual reality to improve construction engineering education. In: American Society for Engineering Education Annual Conference & Exposition (2003)
19. Mirauda, D., Greco, M., Moscarelli, P.: Practical methods for water discharge measurements in fluvial sections. In: Brebbia, C,A. (ed.) River Basin Management VI, WIT Transactions on Ecology and the Environment, vol. 146, pp. 355–367. Wessex Institute of Technology, Ashurst, UK (2011)
20. Mirauda, D., De Vincenzo, A., Pannone, M.: Statistical characterization of flow field structure in evolving braided gravel beds. Spat. Stat. (2017). https://doi.org/10.1016/j.spasta.2017.10.004
21. Mirauda, D., Erra, U., Agatiello, R., Cerverizzo, M.: Applications of mobile augmented reality to water resources management. Water **9**(9), 699 (2017). https://doi.org/10.3390/w9090699
22. Mirauda, D., Erra, U., Agatiello, R., Cerverizzo, M.: Mobile augmented reality for flood events management. Int. J. Sustain. Dev. Plann. **13**, 418–424 (2018). https://doi.org/10.2495/SDP-V13-N3-418-424
23. Mirauda, D., Pannone, M., De Vincenzo, A.: An entropic model for the assessment of streamwise velocity dip in wide open channels. Entropy **20**, 69 (2018). https://doi.org/10.3390/e20010069
24. Moramarco, T., Ammari, A., Burnelli, A., Mirauda, D., Pascale, V.: Entropy theory application for flow monitoring in natural channels. In: Proceedings of iEMSs 4th Biennial Meeting, pp. 430–437. International Congress on Environmental Modelling and Software (iEMSs 2008), Barcelona, Catalonia, Spain (2008)
25. Moro, C., Štromberga, Z., Raikos, A., Stirling, A.: The effectiveness of virtual and augmented reality in health sciences and medical anatomy. Anat. Sci. Educ. **10**(6), 549–559 (2017)
26. Peer, A., Ponto, K.: Evaluating perceived distance measures in room-scale spaces using consumer-grade head mounted displays. In: 2017 IEEE Symposium on 3D User Interfaces (3DUI), pp. 83–86, March 2017
27. Piovesan, S.D., Passerino, L.M., Pereira, A.S.: Virtual reality as a tool in the education. International Association for Development of the Information Society (2012)
28. Suznjevic, M., Mandurov, M., Matijasevic, M.: Performance and QoE assessment of HTC Vive and Oculus Rift for pick-and-place tasks in VR. In: 2017 Ninth International Conference on Quality of Multimedia Experience (QoMEX), pp. 1–3, May 2017
29. Velev, D., Zlateva, P.: Virtual reality challenges in education and training. Int. J. Learn. Teach. **3**, 33–37 (2017)
30. Vergara, D., Rubio, M.P., Lorenzo, M.: On the design of virtual reality learning environments in engineering (2017)

31. Vergara, D., Rubio, M., Prieto, F., Lorenzo, M.: Enhancing the teaching/learning of materials mechanical characterization by using virtual reality. J. Mater. Educ. **38**(3–4), 63–74 (2016)
32. Villar-Zafra, A., Zarza-Sánchez, S., Lázaro-Villa, J.A., Fernández-Cantí, R.M.: Multiplatform virtual laboratory for engineering education. In: 2012 9th International Conference on Remote Engineering and Virtual Instrumentation (REV), pp. 1–6. IEEE (2012)
33. Wong, T., Bigras, P., Cervera, D.: A software application for visualizing and understanding hydraulic and pneumatic networks. Comput. Appl. Eng. Educ. **13**(3), 169–180 (2005)

Author Index

Abdi Oskouie, Mina   I-292
Abel, Mihkel   I-237
Acosta, Aldrin G.   II-239
Acosta, Aldrin   II-335
Adamo, Nicoletta   I-10
Alboul, Lyuba   II-256
Albuquerque, Georgia   II-71
Aleksandrov, Jan   I-36
Alhakamy, A'aeshah   II-179
Alvarez V., Marcelo   II-310
Alvarez-M., Edison   I-91
Andaluz, Gabrilea M.   I-138
Andaluz, Víctor H.   I-138, I-351, I-362,
    II-146, II-239, II-335, II-393, II-410,
    II-430, II-442
Angeloni, Renato   II-199
Arra, Mumen Abu   II-295
Averbukh, Natalya   I-60
Averbukh, Vladimir   I-60

Baert, Patrick   II-111
Balseca, M.   II-310
Barba, Maria Cristina   I-394, II-264
Barbieri, Loris   II-99
Beer, Martin   II-256
Behlen, Manuel   II-71
Bergamasco, Massimo   I-274
Bodensiek, Oliver   II-71
Bohak, Ciril   I-36
Bonilla, Edison L.   I-138
Bonino, Brigida   I-43
Borecký, Jiří   I-105
Bouck-Standen, David   II-208
Boulanger, Pierre   I-292
Brice, Daniel   I-120
Bundt, Lennart   II-208
Bustillo, Andres   I-385

Caiza, Gustavo   I-330
Calapaqui, Cinthya   I-330
Capece, Nicola   II-456
Carrozzino, Marcello   I-254, I-274
Carvajal, Christian P.   II-442

Cattari, Nadia   II-170
Celi, Simona   I-376
Chasi, Christian P.   I-362
Checa, David   I-385
Chin, MeiTing   II-277
Cirulis, Arnis   II-126
Cisternino, Doriana   II-264
Clini, Paolo   II-199
Condino, Sara   I-344, I-376
Corallo, Angelo   II-221
Covino, Attilio   I-394
Covre, Nicola   II-23
Cruz-Neira, Carolina   I-71
Cutolo, Fabrizio   II-170
Cuzco, Giovanny   II-239

D'Errico, Giovanni   I-394, II-264
Daineko, Yevgeniya   I-150
De Cecco, Mariolino   II-23
De Luca, Valerio   I-394, II-264, II-348
De Paolis, Lucio Tommaso   I-394, II-264,
    II-348
De Troyer, Olga   II-295
Di Bitonto, Pierpaolo   I-394
Di Gestore, Simona   I-394
Dib, Hazar N.   I-10
Dragoni, Aldo Franco   II-199
Dzardanova, Elena   II-230

Erazo, Yaritza P.   I-362
Erra, Ugo   II-456
Esposito, Marco   II-221
Ewais, Ahmed   II-295

Ferrari, Mauro   I-376
Ferrari, Vincenzo   I-344, I-376, II-170
Flores, Leonardo A.   I-351
Flotyński, Jakub   I-220
Fontana, Umberto   I-376, II-170
Fornaser, Alberto   II-23
Fortunier, Rolnd   II-111
Frontoni, Emanuele   I-203, II-319
Fuentes, Esteban M.   I-175

Gallardo-Cardenas, Fabian   II-379
Garcés, E.   II-310
Garcia, Carlos A.   I-91, I-330, II-379
Garcia, Marcelo V.   I-91, I-330, II-379
García-Magariño, Iván   I-175
Gatto, Carola   II-264
Gavalas, Damianos   II-230
Gesi, Marco   I-344
Giannini, Franca   I-43
Granizo, R.   II-310
Greci, Luca   I-313, II-137
Guanoluisa, Gissela M.   I-351
Guanoluisa, J.   II-310
Gulfraz, Ali Raja   II-3
Gvozdarev, Ilya   I-60

Herczeg, Michael   II-84, II-208
Hicks, Ben   II-51
Hoppenstedt, Burkhard   II-43, II-63, II-371
Howard, John E.   I-376
Hudák, Marián   II-161

Ipalakova, Madina   I-150
Ivan, Ioan-Alexandru   II-111

Ježek, Bruno   I-105

Kammerer, Klaus   II-43, II-63, II-371
Kasapakis, Vlasios   II-230
Kechiche, Marwene   II-111
Kent, Lee   II-51
Kong, Li   II-3
Korečko, Štefan   II-161
Kose, Ahmet   I-237
Kraus, Martin   I-26

Latta, María A.   I-362
Levchuk, Georgy   I-60
Li, Ying   II-277
Longobardi, Alessandro   I-274
Lu, Yaping   II-13
Lupinetti, Katia   I-43

Ma, Pengfei   II-3
Magliaro, Serena   I-394
Magnor, Marcus   II-71
Mahroo, Atieh   II-137
Malinverni, Eva Savina   I-203
Mameli, Marco   I-203

Marino, Emanuele   II-99
Marolt, Matija   I-36
Marra, Manuela   II-221
Mayorga, Oscar A.   I-138
Mazhar Çelikoyar, M.   I-303
McRoberts, Thomas   I-120
Melkozerov, Leonid   I-60
Mikhaylov, Igor   I-60
Mirauda, Domenica   II-456
Mondellini, Marta   I-313
Monti, Marina   I-43
Mora-Aguilar, Jorge   II-335
Moreno, Hugo Oswaldo   II-239
Munoz-Arango, Juan Sebastian   I-71

Naranjo, Cesar A.   II-335, II-430
Naranjo, Jose E.   I-91, II-379
Nisiotis, Louis   II-256
Nunnari, Fabrizio   I-394, II-23

Ohlei, Alexander   II-84, II-208
Ortiz, Jessica S.   I-138, II-442

Palacios-Navarro, Guillermo   I-175
Paladini, Giovanna Ilenia   I-394, II-264, II-348
Palmitesta, Giovanni   I-254
Paolanti, Marina   I-203, II-319
Pascarelli, Claudio   II-221
Perez, J.   II-310
Pérez, José A.   II-442
Petlenkov, Eduard   I-237
Piazza, Roberta   I-376
Pierdicca, Roberto   I-203, II-319
Pilatasig, Jimmy A.   I-351
Pizzagalli, Simone   I-313
Porras, Alex P.   II-430
Positano, Vincenzo   I-376
Potenza, Ada   I-394
Pryss, Rüdiger   II-43, II-63, II-371
Puggioni, Mariapaola   II-319
Pusda, Fernando R.   II-146

Quattrini, Ramona   II-199

Rafferty, Karen   I-120
Ramon, Lydia   I-385
Regalado, Fabricio   I-330
Reichert, Manfred   II-43, II-63, II-371

Reiners, Dirk   I-71
Reunanen, Markku   I-26
Rivas, D.   II-310
Romi, Mohammed   II-295
Romo, Juan E.   II-410
Ruof, Jona   II-43

Sacco, Marco   I-313, II-137
Saltos, Lenin F.   I-330
Sánchez, Carlos R.   I-138
Sánchez, Jorge S.   I-138, II-410, II-430
Santana, Jaime A.   II-239, II-335
Sarzosa, Darwin S.   II-335
Sasso, Michele   II-319
Scarlato, Francesco   I-254
Schena, Annamaria   I-394
Schmid, Michael   II-371
Scholta, Joachim   II-371
Sernani, Paolo   II-199
Shen, YiRun   I-3
Slabý, Antonín   I-105
Snider, Chris   II-51
Sobota, Branislav   II-161
Solis, Carlos R.   II-430
Sonntag, Dörte   II-71
Spiliopoulou, Myra   II-63
Strugała, Dominik   I-158, I-220
Sun, Xuelian   II-3
Sylaiou, Stella   II-230

Tamayo, Mauricio   II-239
Tanderup, Stefan H.   I-26
Tecchia, Franco   I-254, I-274

Tepljakov, Aleksei   I-237
Tichy, Matthias   II-43
Tipantasi, Gissela R.   II-410
Topsakal, Elif   II-286
Topsakal, Oguzhan   I-303, II-286
Toscano, Rosario   II-111
Tsoy, Dana   I-150
Tuceryan, Mihran   II-179
Turini, Giuseppe   I-344, I-376

Ubilluz, Jonathan I.   II-393

Valencia, Francisco F.   II-146
Varela-Aldás, José   I-175
Vasev, Pavel   I-60
Viglialoro, Rosanna Maria   I-344
Villani, Nicholas J.   I-10
Villarroel, Mayra L.   II-239

Walczak, Krzysztof   I-158, I-220
Wendorff, Nils   II-71
Wessel, Daniel   II-84
Wideström, Josef   I-186
Witte, Thomas   II-43

Yugcha, Edison P.   II-393

Zambrano, Víctor D.   II-146
Zapata, M.   II-310
Zhang, WeiWei   I-3
Zhang, YanXiang   I-3, II-13, II-277
Zhang, Yanxiang   II-3

Printed in the United States
By Bookmasters